Alternatives to Antibiotics against Pathogens in Poultry

Awad A. Shehata •
Guillermo Tellez-Isaias •
Wolfgang Eisenreich
Editors

Alternatives to Antibiotics against Pathogens in Poultry

Springer

Editors
Awad A. Shehata
Structural Membrane Biochemistry,
Bavarian NMR Center
Technical University of Munich (TUM)
Garching, Bayern, Germany

Guillermo Tellez-Isaias
Division of Agriculture, Department of
Poultry Science
University of Arkansas
Fayetteville, AR, USA

Wolfgang Eisenreich
Structural Membrane Biochemistry,
Bavarian NMR Center
Technical University of Munich (TUM)
Garching, Bayern, Germany

ISBN 978-3-031-70479-6   ISBN 978-3-031-70480-2   (eBook)
https://doi.org/10.1007/978-3-031-70480-2

Preface

Regardless of the source or type of stress (biological, environmental, nutritional, physical, chemical, or psychological), it can lead to inflammation and further harmful reactions. In poultry farming, it is important to pay attention to chronic stress caused by various factors, called secret killers, such as heat stress, dysbiosis, mycotoxins, endotoxins, and an oxidized diet that can reduce the performance of the animal and increase its susceptibility to infections. Secret killers can disrupt gut-tight junctions and increase the production of reactive oxygen species (ROS), which leads to increased intestinal permeability, endotoxemia, and systemic inflammation. These pathologies also increase the need for antimicrobials in poultry due to the invasion of opportunistic bacteria in the intestinal epithelium.

Historically, growth-promoting antibiotics were used to control chronic stress in poultry. However, antibiotic growth promotors (AGPs) can also increase the prevalence of drug-resistant bacteria. Thus, the development of antibiotic resistance in bacteria, which is common in both animals and humans, will be a continuous public health hazard. Based on the "Precautionary Principle" and experiences made in some European countries, the EU completely banned the use of growth-promoting antibiotics in the feed of food-producing animals by January 2006.

Field observations in Europe showed that the poultry industry faced several problems after the AGP ban. The impact of the ban has been seen in the performances (reduced body weight and low feed conversion rate) as well as in the rearing husbandry (wet litter and high ammonia level), animal welfare problems (increased footpad dermatitis), and general health issues on the birds (enteric disorders due to dysbacteriosis and high clostridial infections). However, several investigations have indicated that competitive exclusion, prebiotics, probiotics, enzymes, and acids can impact poultry's incidence and severity of clostridial infections. Combinations of different approaches are necessary to enhance the performance and health status of the birds, such as using highly digestible feed ingredients, using a special pre-starter diet, improving climate conditions in the poultry house, and keeping litter quality in optimal condition.

The first paradigm of using alternatives to antimicrobials focusing only on the bioactive substances that possess antimicrobial activities is the main cause of missing opportunities. Thus, antimicrobial activity is not bioactive substances' sole mechanism of action. Additionally, the oversimplification and lack of science in which the industry can recommend bioactive substances for treating acute infections results in consumer dissatisfaction. Therefore, nowadays, the second generation of alternatives to antimicrobials focuses more on the host-mediated effects, not on the pathogens only. This paradigm is interested in using antimicrobials for several beneficial effects, such as modulation of host immunity and modulation of and restoring intestinal microbiota, with well-balanced physiology and metabolism, which subsequently increase the resistance to pathogens and finally improve animal performance. This new paradigm provides a clear opportunity for next-generation phytotherapy.

The book *Alternatives to Antibiotics Against Pathogens in Poultry* addresses poultry production's main challenges and the potential use of alternatives to antimicrobials to ameliorate antibiotics. The book is organized into 15 chapters. In the first chapter, we give an overview of the gut health of poultry. The second chapter highlights the intestinal microbiota as a forgotten organ and its role in pathogen colonization, immune responses, and inflammatory diseases. The third chapter discusses the main secret killers affecting poultry, including heat stress, endotoxins, mycotoxins and rancid feed, and dysbiosis. Then, we present the potential alternatives to antimicrobials or biotics, including probiotics (Chap. 4), prebiotics (Chap. 5), postbiotics (Chap. 6), enzymes (Chap. 7), phytogenic substances (Chap. 8), essential oils (Chap. 9), bacteriophages (Chap. 10), and peptides (Chap. 11). Chapter 12 discusses the recent advances in antimycotoxins, while Chap. 13 highlights the potential strategies to control coccidiosis in biofarms. Finally, in Chap. 14, we give an overview of nanotechnology and its potential applications in poultry.

This book was designed as a handbook for undergraduate students and a resource for researchers, practical poultry specialists, and nutritionists. Additionally, it may be instrumental as a guide for production and health problems in poultry. At the end of each chapter, we provide the reader with selected literature covering the topic. Comprehensive citation of the references was minimized, and the presentation of literature data was based on the interpretation or correlation of research findings.

This book can serve as a valuable guide to the knowledge of alternatives to antimicrobials. We hope that readers will find this book useful and interesting to read.

Garching, Bayern, Germany   Awad A. Shehata
Fayetteville, AR, USA   Guillermo Tellez-Isaias
Garching, Bayern, Germany   Wolfgang Eisenreich

Contents

About the Editors

Awad A. Shehata gained his Bachelor in Veterinary Medicine from the Faculty of Veterinary Medicine, Alexandria University, Egypt, and "Approbation als Tierarzt" Ludwig –Maximilian University, Munich, Germany. In 2005, he obtained a master's degree in avian diseases at the Faculty of Veterinary Medicine, Sadat City University, Egypt. In 2011, he completed his Ph.D. (*Dr. med.* vet.) in "Virology" at the Faculty of Medicine, Leipzig University, Germany. In 2015, he obtained his *Dr. habilitatus* (*Dr. med. vet. habil.*) in "Bacteriology and Poultry diseases" at Leipzig University, Germany. In 2022, he was honored with a Professor of "Avian Diseases" title from Sadat City University, Egypt. Besides his academic work at several universities, Dr. Shehata has six years of industrial experience in vaccine production and the development of alternatives to antimicrobials. He is a certified quality manager, auditor, project manager, European Business Competence License level EBCL, "Good Manufacturing Practice" (GMP), FELASA-B and -C, and phytotherapist. Dr. Shehata's research interests lie mainly in developing alternatives or complementary to antimicrobials (developing and evaluating recombinant vaccines, live attenuated bacterial vaccines, inactivated vaccines, probiotics, and prebiotics) and diagnostic strategies and studying avian pathogens' epidemiology and molecular epidemiology. Dr. Shehata has taught Avian diseases and Microbiology courses in English and German at several universities worldwide. He is currently a project leader at the Structural Membrane Biochemistry, Technical University of Munich, Garching, Germany, to study the analysis and mechanism of actions of natural products, metabolic pathways, and fluxes in various organisms by ^{13}C-labeling experiments and profiling.

Guillermo Tellez-Isaias received his MS and DVM in Veterinary Sciences from the National Autonomous University of Mexico (UNAM) and his Ph.D. from Texas A&M University, USA. He worked as a professor at UNAM for 16 years, 8 as head of the Avian Medicine Department, College of Veterinary Medicine. Dr. Tellez-Isaias was president of the National Poultry Science Association of Mexico and a member of the Mexican Veterinary Academy and the Mexican National Research System. Currently, he is a research professor at the Center of Excellence in Poultry Science, University of Arkansas, USA. His research is focused on poultry gastrointestinal models to evaluate the beneficial effects of functional foods to enhance intestinal health and disease resistance.

Wolfgang Eisenreich gained his diploma in Chemistry (Dipl. Chem.) from the Technical University of Munich, Germany (TUM), in 1987. In 1989, he worked as a guest scientist at Procter & Gamble, Brussels, Belgium. In 1990, he completed his Dr. rerum naturalium (Dr. rer. nat) at the Faculty of Chemistry, TUM. In 2006, he gained the venia legendi, and in 2014, he was appointed as apl. Professor for Biochemistry at TUM. For more than 20 years, Eisenreich has headed an independent research group studying natural products, metabolic pathways, and fluxes in various organisms, including plants. Eisenreich and his co-workers became welcome partners on a national and international level in sharing their analytical expertise and techniques to determine the structures of bioactive compounds, including triterpenes, and to study their formation and function by ^{13}C-labeling experiments and profiling.

An Overview of the Gut Health in Chickens

1

Awad A. Shehata, Guillermo Tellez-Isaias, Wolfgang Eisenreich, and Shereen Basiouni

Contents

A. A. Shehata · W. Eisenreich
Structural Membrane Biochemistry, Bavarian NMR Center, Technical University of Munich (TUM), Garching, Bayern, Germany

G. Tellez-Isaias
Division of Agriculture, Department of Poultry Science, University of Arkansas, Fayetteville, AR, USA

S. Basiouni (✉)
Cilia Cell Biology, Institute of Molecular Physiology, Johannes-Gutenberg University, Mainz, Germany
e-mail: sbasioun@uni-mainz.de

Abstract

Maintaining gut health is essential to reduce disease incidence and improve production performance. This chapter focuses on how to enhance production by maintaining intestinal health in chickens. Several components, such as goblet cells, mucin, tight junctions, enterocytes, and Paneth cells, play a significant role in maintaining gut health. They work together to maximize the utilization and acquisition of dietary nutrients, ultimately leading to improved performance. Moreover, this chapter provides insights into the various aspects which is a crucial factor affecting their performance in the field. Moreover, the potential markers to assess gut health will be discussed.

Keywords

GIT · Goblet cells · Mucins · Tight junctions · Paneth cells

1   Introduction: A Journey Through the Gut in Chickens

The digestive system is essential in digesting and absorbing nutrients in domestic chickens (*Gallus gallus*). Compared to mammals, poultry have a shorter intestinal tract, which means they have less digestive capacity and a shorter retention time for their food (Pan and Yu 2014). The alimentary tract of chickens includes the crop, gizzard, proventriculus, small intestine, and large intestine.

Chickens get their food with their beaks. Unlike mammals, birds have an oral cavity that allows food particles from the beak to pass through to the toothless pharynx and esophagus. The tongue is keratinized, has no stratum granulosum, and contains Langerhans cells that produce the antimicrobial peptides ß-defensin and gallinacin-3. It also plays a crucial role in local immunity (Pan and Yu 2014; Scanes and Pierzchala-Koziec 2014). Following swallowing, feed is partially fermented in the crop by lactobacilli. Feed is then transferred to the glandular stomach (proventriculus), where digestion begins with gastric juices, HCL, and enzymes (Alshamy et al. 2018). In the gizzard, mechanical digestion takes place, where food is ground into smaller particles with the aid of insoluble fibers.

The alimentary tract contains several antimicrobial peptides, including beta-defensin-gallicin-6, which has been documented in the esophagus, crop, and proventriculus. (van Dijk et al. 2007). Other antimicrobial peptides like Cathelicidins-1, liver-expressed antimicrobial peptides, gallinacin 11, and ß-defensin also play a pivotal role in innate immunity. A vital component of mechanical digestion is the spasmolytic polypeptide found on the cellular surface of proventriculus, where feedstuffs are ground into smaller particles. Insoluble fibers promote the healing of epithelial cells and restore gut health (Tabata and Yasugi 1998).

Nutrient breakdown, comprising proteins, lipids, and carbohydrates, primarily occurs in the small intestine with the support of bile and pancreatic enzymes. Epithelial villi, small finger-like projections, are instrumental in promoting the processes of digestion and nutrient uptake (Scanes and Pierzchala-Koziec 2014).

Intestinal epithelium serves two main functions: (i) Selectively filtering luminal contents, including nutrients, water, and electrolytes, to enter the circulation. (ii) Selective barrier to preclude the penetration of the luminal toxins, commensal or opportunistic microorganisms into the circulation (Groschwitz and Hogan 2009). However, this homeostatic barrier can be injured due to any stressors, leading to disease emergence and reduced animal performance. Indeed, understanding the mechanisms that govern is critical for maintaining the gut health. This chapter provides an overview of the main components of the intestinal barriers in chickens.

2 Intestinal Epithelium as a Selective Barrier

Four intestinal barriers are functional in the alimentary tract (Hooper 2009; Coleman and Haller 2018) (Fig. 1.1):

- Microbiological barrier, made of microbiota (will be discussed in the next chapter)

Physical barrier, made of the epithelial cell layer and the tight junctions. The crypt forms a pocket, while the villus protrudes as a finger-like structure. It comprises several epithelial cell populations such as goblet, enterocytes, enteroendocrine, and Paneth cells (Zhang et al. 2019a, b).

Chemical barrier, made of outer and inner mucus. Although gut microbiota colonizes the outer layer, the inner layer is sterile and contains antimicrobial peptides, IgA, and other bioactive molecules.

The immunological barrier comprises several immunological cells, including B and T cells, dendritic, and phagocytic cells. While chickens lack lymph nodes, they

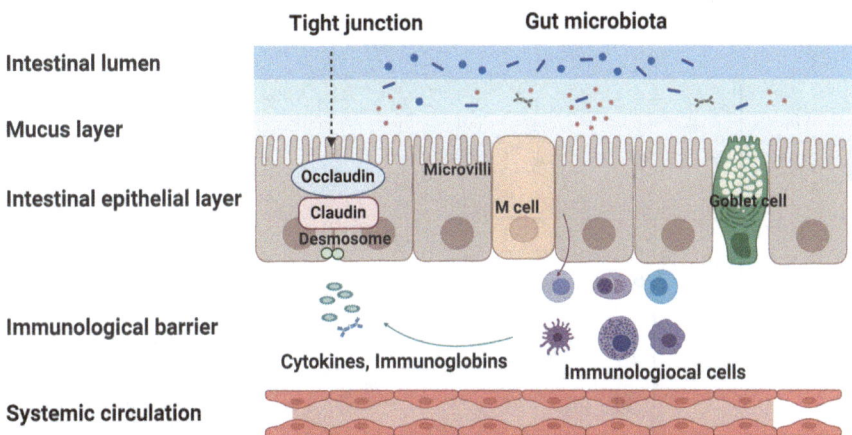

Fig. 1.1 Alimentary tract barriers; intestinal microbiota, mucus, epithelial cells, and immunological barriers. The Figure was designed by BioRender

possess numerous lymphoid aggregates. The notion of "gut-associated lymphoid tissue" (GALT) denotes the various lymphoid structures of the gut. The bursa of Fabricius, Peyer's patches, lymphocyte follicles, Meckel's diverticulum, and cecal tonsils are the main lymphoid organs of chickens. Indeed, the gut immune systems of baby chicks are not as developed as those of adult chickens; they reach complete development 1–1.5 months after hatching. Preventing infection can vary significantly between young and old chickens. In this context, chickens' guts are active organs with several immune compartments. It experiences profound changes that are impacted by diet, microbiome, and physiology prior- and post-hatching. Therefore, the methods employed to prevent infection may vary significantly according to the age of the birds.

3   Cellular Elements of the Gut Epithelium

3.1   Enterocytes

About 80% of intestinal epithelial cells are enterocytes. They are polarized and form a lining on the villi (Fig. 1.2). Their physiological functions are digestion and absorption, creating a barrier that can distinguish between digestible nutrients and antigens (De Santa Barbara et al. 2003). The enterocyte apical membrane has microvilli that enhance the surface area. In chickens, the development of villi begins in the second half of incubation. Prior to hatching, the tight junctions (TJs) are not well developed and permit the uptake of macromolecules. The mature enterocytes are also present from the day of the hatch. Once hatched, the TJs become tight, which prevents bacterial invasion but also enables nutrient uptake selectively (Karcher and Applegate 2008). The enzymatic activity increases as the crypt depth decreases. This is why the villous height crypt depth ratio is used as a biomarker of intestinal health (Fig. 1.2). Aminopeptidases, members of the membrane-bound

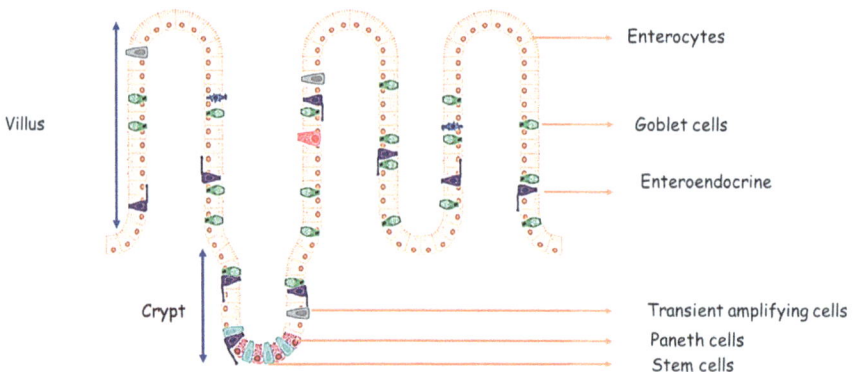

Fig. 1.2 The main cellular components of the gut epithelium. The figure was designed with BioRender

metallopeptidase family, released by enterocytes ensure the final nutrient processing for absorption. Notably, aminopeptidase Ey, present in the contents of the gut, could potentially use as a biomarker for assessment of gut health (De Meyer et al. 2019).

Apoptosis may be triggered by stressors, including endotoxins, pathogens, heat stress, and mycotoxins. Due to increased cell loss, villi shorten, and crypts elongate. Several feed additives can improve the villus length and villus-to-crypt ratio. Supplementing the diet with a zinc-amino acid complex instead of inorganic zinc sulfate into the broiler meal increased villus length and villus: crypt ratio. This implies that even inorganic zinc may cause oxidative stress within the intestinal tract (De Grande et al. 2020). Enterocytes are interconnected through the TJs. However, gut leakage could result from insufficient energy to uphold the TJs. Highly metabolically active small intestine villus epithelium cells use glutamine as a source of energy (Labow and Souba 2000). Of note, under intestinal stress, such as necrotic enteritis, glutamine supplementation has been shown to maintain healthy mucosa in chickens (Xue et al. 2018).

3.2 Enteroendocrine Cells

Only 1% of enteroendocrine cells comprise intestinal cells. These cells regulate animal feeding habits, enzyme activity, and hormone release. Several types of enteroendocrine cells are known and named alphabetically based on their primary secretory products. These cell types include gastric inhibitory peptide-secreting K cells, cholecystokinin-secreting I cells, secretin-secreting S cells, and somatostatin-secreting D cells (Latorre et al. 2016).

3.3 Paneth Cells

The first documented evidence of Paneth cells dates back to 1888. Although the location of these cells may vary in different species, in humans, they are found in the crypts of the intestinal epithelium. These cells are crucial for innate immunity as they produce antimicrobial peptides, regulate gut microbiota, and provide support to intestinal stem cells (Günther et al. 2011; Cui et al. 2023). Indeed, Paneth cells were first identified in chickens in 1974 (Humphrey and Turk 1974); however, several later studies failed to identify these cells, making their presence in chickens a debate. Recently, this debate was resolved with documentation of lysozyme c, g, and g2 in intestinal crypts via qPCR and in situ hybridization (Wang et al. 2016), followed by further confirmation using transmission electron microscopy (TEM) and phloxine tartrazine staining (Bar Shira and Friedman 2018). Chicken Paneth cells contain lysozyme, β-defensin, and angiogenin-4 (ANG4), which belong to the ribonuclease A superfamily (Losada-Medina et al. 2020). It was proposed that ANG4 may be involved in innate immunity and gut microbiota regulation (Losada-Medina et al. 2020). Several recent studies indicate that dietary supplementations can modulate the number and/or diameter of Paneth cells as follows: (i) Probiotics (*Lactobacillus salivarius* and *Lactobacillus agilis*) promote intestinal Paneth and stem cell growth (Hong et al. 2021). (ii) Methionine deficiency affects Paneth cell proliferation and upregulates lysozyme expression (Wang et al. 2022), highlighting the diet's impact on Paneth cell functions. (iii) Black corn soluble extract, empire apple (juice, pomace, and pulp), quinoa soluble fiber, quercetin, and saffron flower water extract significantly increased the number and diameter of Paneth cells duodenum (Agrizzi Verediano et al. 2022).

3.4   Goblet Cells

The immature goblet cells are large and pyramidal and contain mucin granules. They are located at the crypt's base. However, mature goblet cells migrate toward the tips of the villi, undergo apoptosis, or are damaged and shed into the lumen. During maturation, the goblet cells become cup-like in shape and accumulate more mucin granules at the apical portion (Collier et al. 2008; Birchenough et al. 2015). The density and activity of goblet cells increase in freshly hatched chicks during their first week to adapt to environmental and dietary demands. (Proszkowiec-Weglarz et al. 2020). When chickens reach 21 days of age, their goblet cells mature and produce mucin that regenerates the external layer of mucus (Duangnumsawang et al. 2021). Alongside mucins, goblet cell secretion contains IgA, avidin, and lysozyme, which are important for immunity (Bar Shira and Friedman 2018).

4   Intestinal Mucus Barrier

The intestinal mucus acts as a barrier that protects the epithelium from damaging substances such as digestive enzymes, gut microbes, and forces generated during digestion. The mucus serves as the first level of defense against pathogenic microorganisms (Fig. 1.2). It also helps colonize commensal bacteria, aids digestion, and facilitates nutrient transport to reach the underlying cells of epithelial cells. Mucous is mostly composed of water (90–95%), electrolytes, lipids (1–2%), proteins, and mucins (1–5%) (Johansson and Hansson 2016; Bansil and Turner 2018).

4.1   Structure and Function of Mucin

Mucins are glycosylated proteins with a "mucin domain." The core contains a tandem proline, threonine, and serine (PTS) repeat. Proline keeps mucin unfolded in the Golgi apparatus for glycosylation, which is crucial to the O-glycosylation process (Johansson et al. 2011a). The O-glycosylation forms a glycan coat that safeguards it from endogenous protease degradation (Johansson et al. 2011b; Etienne-Mesmin et al. 2019). Additionally, it can bind and dissolve in water and form a gel. Mucin is made up of a variety of monosaccharides, including mannose,

	Neutral mucin	Acidic mucin
Terminal O-glycan	• GluNAc, GalNAc, galactose, fucose and mannose	• Sialic acid and sulfated group
Charge	• Neutral	• Negative
Degradation	• Bacterial enzymes (proteases and glycosidases) and proteolytic host enzymes	• Resist bacterial enzymes
Dominant	• More dominant after hatch	• Dominant in late embryonic stage

Fig. 1.3 Neutral vs. acidic mucins based upon O-glycan

galactose, fucose, sialic acid, and gluNAc. Figure 1.3 provides a summary of the neutral and acidic mucin compositions. Neutral mucin is more dominant after hatch while acidic mucin is dominant in the late embryonic stage (Osho et al. 2016; Duangnumsawang et al. 2021).

Specific gut sites contain distinct forms of mucin, each of which has a particular purpose in sustaining intestinal balance and health. Although MUC2, gel-mucins are also categorized as transmembrane and secretory, where the former can be detached mechanically (Fig. 1.4) (Johansson et al. 2013). However, secretory mucins momentarily bind to goblet cells before the enzymatic cleavage (Birchenough et al. 2015). The crypt has a higher concentration of secretory or gel-forming mucins than the villi. Indeed, microbiota influences the gut lumen's secretory mucin cleavage and release (Schütte et al. 2014), highlighting the link between both microbiota and mucins. Gel-forming mucin is the fundamental element of intestinal mucus (Duangnumsawang et al. 2021). However, only small amounts of it were exclusively found in trace concentrations in the crops (Jiang et al. 2013). Though being predominant in the proventriculus, MUC5ac is also distributed throughout the gut (Smirnov et al. 2004). The principal variations between the transmembrane and secretory mucins are illustrated in Fig. 1.4

4.2 Alterations of Mucin

Multiple regulatory mechanisms within the body collaborate to preserve intestinal equilibrium. The regulation mechanisms of the body play multiple roles in preserving intestinal homeostasis. Mucus is regularly cleared and replaced throughout the day; however, the rate and composition vary based on their location. For instance, nasal mucus is cleared every 10–20 min through mucociliary clearance (Krishnamoorthy and Mitra 1998), while intestinal mucus is renewed every 50–270 min (Illum 2003). However, some factors can hamper these mechanisms, causing a dysfunctional gastrointestinal barrier. Altered feeding habits and diet can impact the gut flora and nutrient digestibility. Furthermore, the feed can control the formation of mucin and goblet cells.

Transmembrane mucin	Secretory mucin
Character	• Protein has extracellular membrane and cytoplasmic parts, adhered to the surface of the enterocytes	• Before being secreted, the mucin polymers combine and form secretory granules
Forming gel-like structure	• Not gel-forming and act as pillars supporting the secretory type	• Gel-forming type and create a layer on the gut epithelium
Mucin types	• MUC4/MUC13 in small intestine	• MUC2/MUC5ac in small intestine and MUC2/MUC6 in large intestine

Fig. 1.4 Transmembrane vs. secretory mucins

Starvation or feed restriction. Acetylcholine causes the release of mucus. Acetylcholinesterase activity may be elevated by starvation, which would decrease mucus output (Smirnov et al. 2004). The decrease of mucus released in a specific area could be explained by a change in the composition of mucus, namely a decrease in water retention leading to a higher density (Smirnov et al. 2004). It has been demonstrated that delaying the feeding of baby chicks might disrupt the mucus layer. This could be attributed to the underdeveloped gut barriers that are more vulnerable to harmful substances and pathogens.

The following practical considerations should be taken into account in rearing chickens: (i) When baby chicks are starved, their goblet cells proliferate more slowly, they have more goblet cells overall, and there are fewer enterocytes per villus (Geyra et al. 2001; Uni et al. 2003). (ii) When a feed restriction lasts 24–36 h, fewer goblet cells migrate from crypts to villi (Liu et al. 2020), disrupting mucins. (iii) When a feed restriction lasts 72 h, a reduction of intestinal MUC2 leads to decreased mucus production (Liu et al. 2020; Proszkowiec-Weglarz et al. 2020). (iv) Studies have been done on how fasting affects the mucous in broiler chicks that are 4 weeks old. After 72 h of starvation, the quantity of mucus released per intestinal cells diminished. But there was a rise in mucin concentrations (Smirnov et al. 2004; Thompson and Applegate 2006). Collectively, to reinforce the mucosal barriers and to prevent damage from toxins and infections, the hatchlings must be fed as soon as possible.

Proteins. Dietary protein modulates the mucin by two mechanisms: (i) Synthesis and structure of mucin requires threonine, proline, and serine (Visscher et al. 2018a). (ii) Providing broilers with a low-protein diet leads to a decrease in mucin release. Mucin in the excreta was significantly decreased by 50% and 20%, respectively, when the dietary intake of threonine was lowered by 60% and 30% of the amount needed for broilers (8.2 g/kg) diet (Visscher et al. 2018b). (iii) Specific dietary proteins and amino acids may influence how the mucus layer regenerates (Ravindran et al. 2008) and could be considered to maintain gut health.

Carbohydrates. It is also found that indigestible carbohydrates impact mucin release, TJs, and microbiota (Montagne et al. 2003). Insoluble fibers (cellulose, rice hull) and soluble fibers (carboxymethyl cellulose) enhanced mucin secretion (Rahmatnejad and Saki 2016; Murai et al. 2018). Dietary fibers may influence mucins via the following pathways: (i) Crude fibers affect the physicochemical characteristics and one protective property of mucus (Horn et al. 2009). (ii) Non-starch polysaccharides, which are indigestible for poultry, ferment in the colon to create SCFAs such as acetate, propionate, and butyrate. The latter is a vital source of energy for the metabolic processes of the cell as well as for the development and differentiation of goblet cells, which increase MUC2 production (Burger-van Paassen et al. 2009; Wrzosek et al. 2013).

High-fat diets. It is unclear how fat supplementation affects mucus in poultry. It may, however, indirectly impact mucus release and TJs. When compared to a meal high in soybean oil (4.0%), palm kernel fatty acid (4.3%) can increase the content of SCFA as well as lower the pH in the ileum (Józefiak et al. 2014). This boosts mucus release and goblet cell numbers. Additional investigations are still needed to investigate the impacts of fats on mucin dynamics.

Probiotics. Intestinal microbiota can modulate mucin release by activating various signaling cascades and secretory elements. Some probiotics, such as Lactobacilli and Bifidobacterium, have proven mucus-modulating action (Laval et al. 2015). *Faecalibacterium prausnitzii* exerts a complementary action with Bacteroides as acetate consumer and butyrate producer to balance the mucus by modifying goblet cell differentiation, mucin gene expression, and glycosylation (Wrzosek et al. 2013).

5   Tight Junction

In chickens, QPepT1 mRNA is expressed in the small intestine. The oligopeptides are transported from crypts to villi by this mRNA (Zhang and Wong 2017). However, glucose transport is handled by SGLT1 mRNA. The chicken's small intestine contains crypts as well as the villus, where glucose is ingested (Zhang et al. 2019a, b).

The epithelial monolayer is a protective barrier of enterocytes connected by TJs formed by integral and peripheral membranes. TJs comprise transmembrane proteins that link to the actin cytoskeleton via cytoplasmic proteins (Aijaz et al. 2006). The TJ proteins are distributed among enterocytes and regulate the intestinal barrier to prevent harmful substances from entering the bloodstream. Various intracellular signaling pathways regulate TJs, such as myosin light chain (MLC) kinase, mitogen-activated protein kinases, protein kinase C, and the Rho family of small GTPases (Scott et al. 2002). Indeed, the bacterial lipopolysaccharide (LPS) can disrupt gut epithelium's tight junctions, allowing harmful substances, including bacteria, into the bloodstream.

Under normal physiological conditions, TJs hinder the uptake of potentially harmful substances. Nonetheless, silent killers, such as stress, endotoxins, and mycotoxins, can harm the TJs. Pathogens can also trigger stress reactions and pro-inflammatory cytokines in the gut and immune cells, causing phosphorylation of the MLC. This can ultimately disrupt the TJs (Turner 2009; Awad et al. 2017;

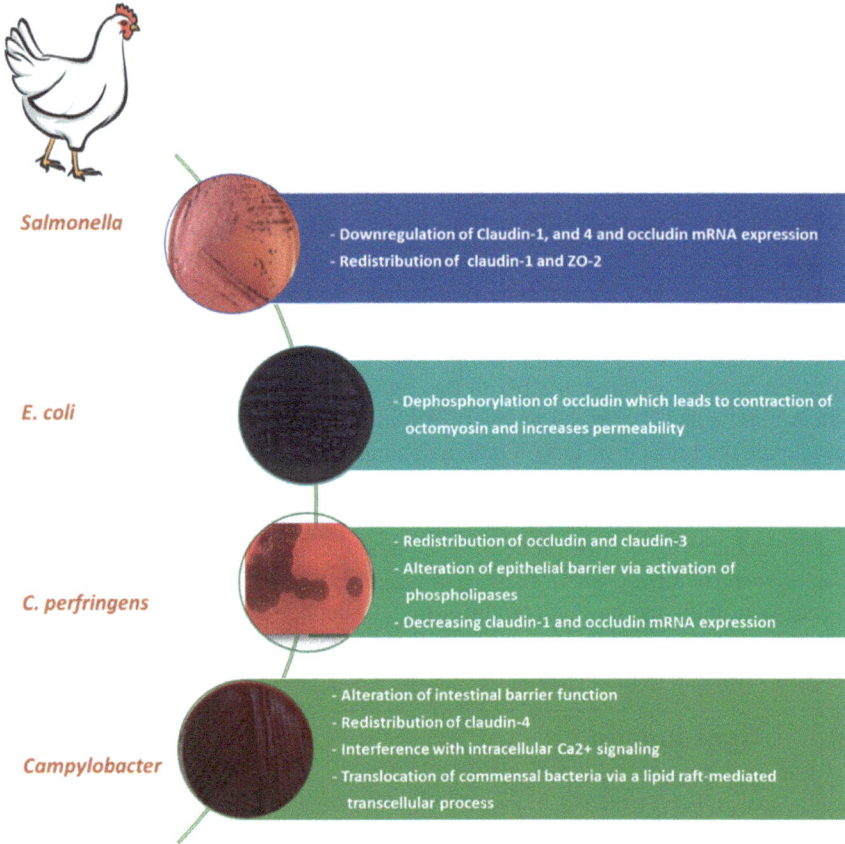

Fig. 1.5 Disruption of the TJs by avian pathogens. © Hafez M Hafez (morphology of bacteria)

Tellez-Isaias et al. 2023), Fig. 1.5 The impacts of chronic stress, including mycotoxins, heat stress, endotoxins, and dysbiosis in TJ disruption, will be discussed in detail in Chap. 3.

6   Biomarkers for Assessment of Intestinal Health

Several indirect biomarkers can be employed to assess the intestinal health of chickens (Shehata et al. 2022) (Fig. 1.6).

6.1   Intestinal Permeability and Electrocyte Function

Citrulline. Citrulline is a non-essential amino acid produced by enterocytes as a byproduct of glutamine and its derivatives metabolism (Wu et al. 1994; Barzał et al.

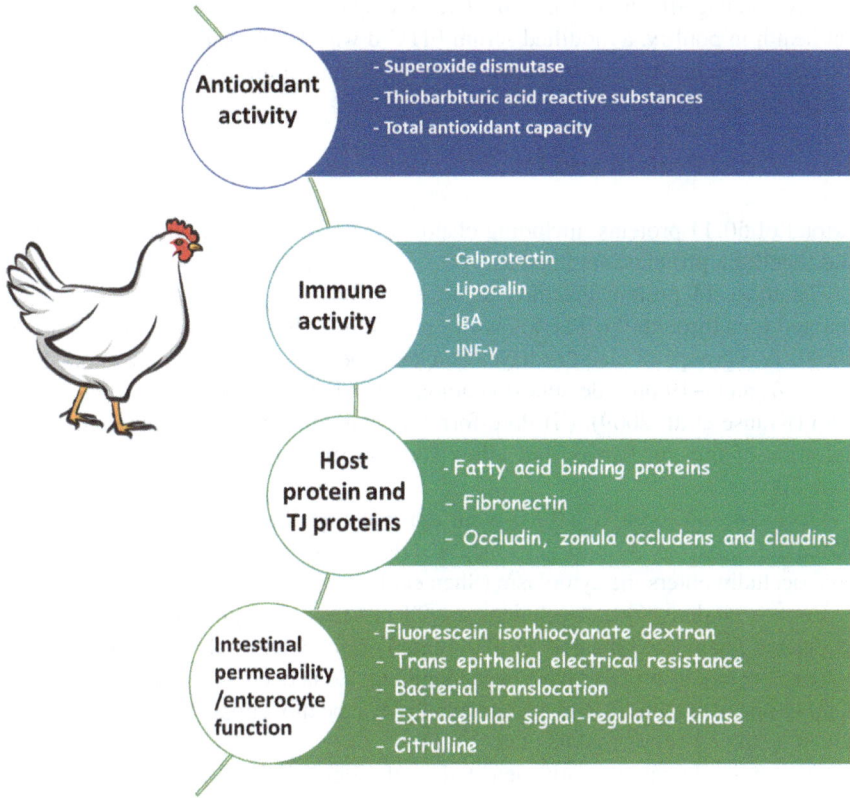

Fig. 1.6 Biomarkers for assessment of intestinal health

2014). Monitoring intestinal function can be assessed based on plasma citrulline, by which decreased levels of citrulline indicate damage to enterocytes. Serum citrulline may be implemented as a potential marker to assess intestinal health in poultry (Baxter et al. 2019).

Extracellular signal-regulated kinase (ERK). ERK belongs to the signaling cascade of mitogen-activated protein kinase (MAPK). It plays a role in intestinal health and tissue repair (Shaul and Seger 2007). Upregulation of it in serum Erk indicates intestinal disturbance (Iizuka 2011).

Transepithelial electrical resistance (TEER). TEER is an electrical resistance test that confirms the integrity and permeability of an acellular monolayer (Zucco et al. 2005). The flux of non-electrolyte tracers (permeability coefficient) reveals the paracellular water flow and the tight junction pore size (Zucco et al. 2005). The TEER test is frequently employed since it is non-invasive and may also be implemented to monitor cells at various phases of growth and differentiation.

Fluorescein isothiocyanate dextran (FITC-d). FITC-d can also be used as an intestinal permeability marker (Ruff et al. 2020, 2021; Vuong et al. 2021). FITC-d is too big (4–6 kDa) to pass through the intestinal wall unless the epithelial barrier has been damaged (Kuttappan et al. 2015; Vuong et al. 2021). In order to determine gut health in poultry, a modified serum FITC-d was developed to detect FITC-d in the circulation in case of a damaged intestine (Vuong et al. 2021).

6.2 Host Protein and TJs

A total of 50 TJ proteins, including claudins, occludin, Zona occludens-1 (ZO-1), and tricellulin proteins, are known.

Claudins. TJ protein claudins regulate the flow of positively and negatively charged ions through the TJs, known as paracellular flow (Groschwitz and Hogan 2009). Two groups of claudins have been described: (i) Pore-sealing: claudin-1, -3, -4, -5, -7, and −19 provide adhesion among cells and strict solutes through epithelium (Krause et al. 2009). (ii) Pore-forming: claudin-2 and -15, which reduce the tightness of the gut barrier and allow sodium ions to pass through (Van Itallie et al. 2008).

Occludin. Occludin is a TJ protein that restains TJs between epithelial cells. When the TJ barrier is no longer functional due to oxidative damage and inflammation, occludin enters the cytoplasm (Shen et al. 2008).

Zona occludens. Zona occludens are TJ proteins belonging to the membrane-associated guanylate kinase (MAGUK) family (Hunziker et al. 2009). Three types of Zona occludens proteins have been identified: Zona occludens 1, Zona occludens 2, and Zona occludens 3. Zona occludens 1 plays the most important role in forming TJs in epithelial cells (Tsukita et al. 2001).

Tricellulin. It was recently detected in the epithelial cells of several organs, including the kidney, intestine, and stomach (Schlüter et al. 2007). Although its exact function in TJs is unknown, it is proposed to play a role in actomyosin mobilization (Cho et al. 2022).

Fatty acid-binding proteins (FABPs). FABP is an energy-producing intracellular protein that is cytoplasmic and present in organs with high metabolic rates, such as the heart, muscle, gut, liver, brain, and heart (Glatz and van der Vusse 1996; Glatz and Storch 2001). By binding long-chain fatty acids and their metabolites, which can have lethal effects on cells, FABP plays a protective role. Subsequently, they activate the Peroxisome Proliferator-Activated Receptors (PPARs), which convey fatty acids to peroxisomes for degradation (Wolfrum et al. 2001). ELISA can be used for the detection of I-FABP concentrations; it showed high sensitivity and specificity. I-FABP is a biomarker for the early detection of gut damage (Cahyaningsih et al. 2018).

6.3 Biomarkers Associated with Immune Activity

These markers include acute phase proteins, calprotectin, and lipocalin. IgA and INF-gamma. Secretory immunoglobulin A (IgA) is an immunoglobulin that plays an important role in local immunity and is intimately associated with intestinal homeostasis.

6.4 Antioxidant Activities Markers

These markers include superoxide dismutase, thiobarbituric acid reactive substances, and total antioxidant capacity. More details are shown in Chap. 3.

7 Restoring of the Tight Junctions

Restoring TJs may be accomplished with a variety of natural products, Fig. 1.7 For more details, please see the respective chapters. Briefly, several strategies can be implemented for restoring the TJs, such as: (i) Dietary fiber positively affects intestinal health via its metabolic byproduct, butyrate (Molnár et al. 2015). (ii) Prebiotics, probiotics, and synbiotics can enhance the gut barrier (Awad et al. 2013). (iii) Glutamate supplementation can maintain the integrity of the intestinal barrier and be involved in protein synthesis (Li et al. 2003). (iv) Anti-inflammatory flavonoids

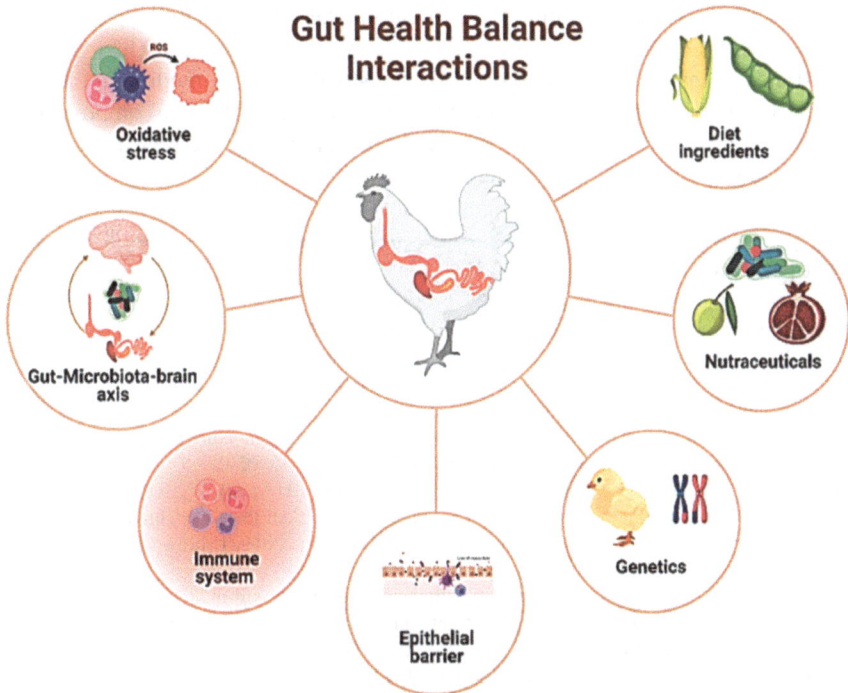

Fig. 1.7 Gut health balance in chickens. Determinants factors of gut health (Tellez-Isaias et al. 2023)

Table 1.1 Selected phytogenic substances that can restore tight junctions (Basiouni et al. 2022)

Phytogenic substance	Tight junction and biological effects
Carvacrol/Thymol	Upregulation of occludin, ZO-1, and claudin-1
Capsicum	Anti-inflammatory
Isoflavone	Antioxidative
Berberine	Increasing SCFA
	Acting as an antioxidant and anti-inflammatory compound
Resveratrol	Reducing the permeability of LPS
	Anti-inflammatory and antioxidant effects
	Increase butyrate and SCFA production
Fibers (i.e., inulin, pectin, and cellulose)	Increase SCFA such as acetate, propionate, and butyrate

and phytogenic substances could help lessen intestinal inflammation (Basiouni et al. 2022). (v) The intestinal mucosal barrier can be enhanced by dietary supplements that contain host defense peptides (HDPs) (Robinson et al. 2015) (Table 1.1).

8   Conclusions

The intestinal cells work together to maintain a healthy balance, protect the body from harmful diseases, and improve nutrient uptake. This activity helps to reduce the likelihood of digestive issues and increase overall productivity. There are various strategies that can be used to optimize intestinal health by modifying the environment in which it operates. By utilizing innovative research tools, a comprehensive understanding of the mechanisms underlying gut health can be obtained, paving the way for future scientific breakthroughs.

References

Agrizzi Verediano T, Stampini Duarte Martino H, Kolba N, Fu Y, Cristina Dias Paes M, Tako E (2022) Black corn (Zea mays L.) soluble extract showed anti-inflammatory effects and improved the intestinal barrier integrity in vivo (Gallus gallus). Food Res Int 157:111227. https://doi.org/10.1016/j.foodres.2022.111227

Aijaz S, Balda MS, Matter K (2006) Tight junctions: molecular architecture and function. Int Rev Cytol 248:261–298. https://doi.org/10.1016/S0074-7696(06)48005-0

Alshamy Z, Richardson KC, Hünigen H, Hafez HM, Plendl J, Al Masri S (2018) Comparison of the gastrointestinal tract of a dual-purpose to a broiler chicken line: a qualitative and quantitative macroscopic and microscopic study. PLoS One 13:e0204921. https://doi.org/10.1371/journal.pone.0204921

Awad W, Hess C, Hess M (2017) Enteric pathogens and their toxin-induced disruption of the intestinal barrier through alteration of tight junctions in chickens. Toxins 9:60. https://doi.org/10.3390/toxins9020060

Awad WA, Ghareeb K, Paßlack N, Zentek J (2013) Dietary inulin alters the intestinal absorptive and barrier function of piglet intestine after weaning. Res Vet Sci 95:249–254. https://doi.org/10.1016/j.rvsc.2013.02.009

Bansil R, Turner BS (2018) The biology of mucus: composition, synthesis and organization. Adv Drug Deliv Rev 124:3–15. https://doi.org/10.1016/j.addr.2017.09.023

Bar Shira E, Friedman A (2018) Innate immune functions of avian intestinal epithelial cells: response to bacterial stimuli and localization of responding cells in the developing avian digestive tract. PLoS One 13:e0200393. https://doi.org/10.1371/journal.pone.0200393

Barzał JA, Szczylik C, Rzepecki P, Jaworska M, Anuszewska E (2014) Plasma citrulline level as a biomarker for cancer therapy-induced small bowel mucosal damage. Acta Biochim Pol 61:615–631

Basiouni S, Tellez-Isaias G, Latorre DG, Graham DB, Petrone-Garcia MW, El-Sweedi H, Yalçın S, Wahab A, Visscher C, May-Simera LM, Huber C, Eisenreich W, Shehata AA (2022) Anti-inflammatory and antioxidative phytogenic substances against secret killers in poultry: current status and prospects. Vet Res

Baxter MFA, Latorre JD, Dridi S, Merino-Guzman R, Hernandez-Velasco X, Hargis BM, Tellez-Isaias G (2019) Identification of serum biomarkers for intestinal integrity in a broiler chicken malabsorption model. Front Vet Sci 6:144. https://doi.org/10.3389/fvets.2019.00144

Birchenough GMH, Johansson ME, Gustafsson JK, Bergström JH, Hansson GC (2015) New developments in goblet cell mucus secretion and function. Mucosal Immunol 8:712–719. https://doi.org/10.1038/mi.2015.32

Burger-van Paassen N, Vincent A, Puiman PJ, van der Sluis M, Bouma J, Boehm G, van Goudoever JB, van Seuningen I, Renes IB (2009) The regulation of intestinal mucin MUC2 expression by short-chain fatty acids: implications for epithelial protection. Biochem J 420:211–219. https://doi.org/10.1042/BJ20082222

Cahyaningsih U, Satyaningtijas AS, Tarigan R, Nugraha AB (2018) Chicken I-FABP as biomarker of chicken intestinal lesion caused by coccidiosis. IOP Conf Ser: Earth Environ Sci 196:012032. https://doi.org/10.1088/1755-1315/196/1/012032

Cho Y, Haraguchi D, Shigetomi K, Matsuzawa K, Uchida S, Ikenouchi J (2022) Tricellulin secures the epithelial barrier at tricellular junctions by interacting with actomyosin. J Cell Biol 221:e202009037. https://doi.org/10.1083/jcb.202009037

Coleman OI, Haller D (2018) Bacterial signaling at the intestinal epithelial interface in inflammation and cancer. Front Immunol 8:1927. https://doi.org/10.3389/fimmu.2017.01927

Collier CT, Hofacre CL, Payne AM, Anderson DB, Kaiser P, Mackie RI, Gaskins HR (2008) Coccidia-induced mucogenesis promotes the onset of necrotic enteritis by supporting Clostridium perfringens growth. Vet Immunol Immunopathol 122:104–115. https://doi.org/10.1016/j.vetimm.2007.10.014

Cui C, Li L, Wu L, Wang X, Zheng Y, Wang F, Wei H, Peng J (2023) Paneth cells in farm animals: current status and future direction. J Animal Sci Biotechnol 14:118. https://doi.org/10.1186/s40104-023-00905-5

De Grande A, Leleu S, Delezie E, Rapp C, De Smet S, Goossens E, Haesebrouck F, Van Immerseel F, Ducatelle R (2020) Dietary zinc source impacts intestinal morphology and oxidative stress in young broilers. Poult Sci 99:441–453. https://doi.org/10.3382/ps/pez525

De Meyer F, Eeckhaut V, Ducatelle R, Dhaenens M, Daled S, Dedeurwaerder A, De Gussem M, Haesebrouck F, Deforce D, Van Immerseel F (2019) Host intestinal biomarker identification in a gut leakage model in broilers. Vet Res 50:46. https://doi.org/10.1186/s13567-019-0663-x

De Santa Barbara P, Van Den Brink GR, Roberts DJ (2003) Development and differentiation of the intestinal epithelium. Cell Mol Life Sci 60:1322–1332. https://doi.org/10.1007/s00018-003-2289-3

Duangnumsawang Y, Zentek J, Goodarzi Boroojeni F (2021) Development and functional properties of intestinal mucus layer in poultry. Front Immunol 12:745849. https://doi.org/10.3389/fimmu.2021.745849

Etienne-Mesmin L, Chassaing B, Desvaux M, De Paepe K, Gresse R, Sauvaitre T, Forano E, de Wiele TV, Schüller S, Juge N, Blanquet-Diot S (2019) Experimental models to study intestinal microbes–mucus interactions in health and disease. FEMS Microbiol Rev 43:457–489. https://doi.org/10.1093/femsre/fuz013

Geyra A, Uni Z, Sklan D (2001) The effect of fasting at different ages on growth and tissue dynamics in the small intestine of the young chick. Br J Nutr 86:53–61. https://doi.org/10.1079/BJN2001368

Glatz JF, Storch J (2001) Unravelling the significance of cellular fatty acid-binding proteins. Curr Opin Lipidol 12:267–274. https://doi.org/10.1097/00041433-200106000-00005

Glatz JF, van der Vusse GJ (1996) Cellular fatty acid-binding proteins: their function and physiological significance. Prog Lipid Res 35:243–282. https://doi.org/10.1016/s0163-7827(96)00006-9

Groschwitz KR, Hogan SP (2009) Intestinal barrier function: molecular regulation and disease pathogenesis. J Allergy Clin Immunol 124:3–20; quiz 21–22. https://doi.org/10.1016/j.jaci.2009.05.038

Günther C, Martini E, Wittkopf N, Amann K, Weigmann B, Neumann H, Waldner MJ, Hedrick SM, Tenzer S, Neurath MF, Becker C (2011) Caspase-8 regulates TNF-α-induced epithelial necroptosis and terminal ileitis. Nature 477:335–339. https://doi.org/10.1038/nature10400

Hong Y, Zhou Z, Yu L, Jiang K, Xia J, Mi Y, Zhang C, Li J (2021) Lactobacillus salivarius and lactobacillus agilis feeding regulates intestinal stem cells activity by modulating crypt niche in hens. Appl Microbiol Biotechnol 105:8823–8835. https://doi.org/10.1007/s00253-021-11606-2

Hooper LV (2009) Do symbiotic bacteria subvert host immunity? Nat Rev Microbiol 7:367–374. https://doi.org/10.1038/nrmicro2114

Horn NL, Donkin SS, Applegate TJ, Adeola O (2009) Intestinal mucin dynamics: response of broiler chicks and white Pekin ducklings to dietary threonine. Poult Sci 88:1906–1914. https://doi.org/10.3382/ps.2009-00009

Humphrey CD, Turk DE (1974) The ultrastructure of normal chick intestinal epithelium. Poult Sci 53:990–1000. https://doi.org/10.3382/ps.0530990

Hunziker W, Kiener TK, Xu J (2009) Vertebrate animal models unravel physiological roles for zonula occludens tight junction adaptor proteins. Ann N Y Acad Sci 1165:28–33. https://doi.org/10.1111/j.1749-6632.2009.04033.x

Iizuka M (2011) Wound healing of intestinal epithelial cells. WJG 17:2161. https://doi.org/10.3748/wjg.v17.i17.2161

Illum L (2003) Nasal drug delivery–possibilities, problems and solutions. J Control Release 87:187–198. https://doi.org/10.1016/s0168-3659(02)00363-2

Jiang Z, Applegate TJ, Lossie AC (2013) Cloning, annotation and developmental expression of the chicken intestinal MUC2 gene. PLoS One 8:e53781. https://doi.org/10.1371/journal.pone.0053781

Johansson MEV, Ambort D, Pelaseyed T, Schütte A, Gustafsson JK, Ermund A, Subramani DB, Holmén-Larsson JM, Thomsson KA, Bergström JH, Van Der Post S, Rodriguez-Piñeiro AM, Sjövall H, Bäckström M, Hansson GC (2011a) Composition and functional role of the mucus layers in the intestine. Cell Mol Life Sci 68:3635–3641. https://doi.org/10.1007/s00018-011-0822-3

Johansson MEV, Hansson GC (2016) Immunological aspects of intestinal mucus and mucins. Nat Rev Immunol 16:639–649. https://doi.org/10.1038/nri.2016.88

Johansson MEV, Larsson JMH, Hansson GC (2011b) The two mucus layers of colon are organized by the MUC2 mucin, whereas the outer layer is a legislator of host–microbial interactions. Proc Natl Acad Sci USA 108:4659–4665. https://doi.org/10.1073/pnas.1006451107

Johansson MEV, Sjövall H, Hansson GC (2013) The gastrointestinal mucus system in health and disease. Nat Rev Gastroenterol Hepatol 10:352–361. https://doi.org/10.1038/nrgastro.2013.35

Józefiak D, Kierończyk B, Rawski M, Hejdysz M, Rutkowski A, Engberg RM, Højberg O (2014) Clostridium perfringens challenge and dietary fat type affect broiler chicken performance and fermentation in the gastrointestinal tract. Animal 8:912–922. https://doi.org/10.1017/S1751731114000536

Karcher DM, Applegate T (2008) Survey of enterocyte morphology and tight junction formation in the small intestine of avian embryos. Poult Sci 87:339–350. https://doi.org/10.3382/ps.2007-00342

Krause G, Winkler L, Piehl C, Blasig I, Piontek J, Müller SL (2009) Structure and function of extracellular claudin domains. Ann N Y Acad Sci 1165:34–43. https://doi.org/10.1111/j.1749-6632.2009.04057.x

Krishnamoorthy R, Mitra AK (1998) Prodrugs for nasal drug delivery. Adv Drug Deliv Rev 29:135–146. https://doi.org/10.1016/S0169-409X(97)00065-3

Kuttappan VA, Berghman LR, Vicuña EA, Latorre JD, Menconi A, Wolchok JD, Wolfenden AD, Faulkner OB, Tellez GI, Hargis BM, Bielke LR (2015) Poultry enteric inflammation model with dextran sodium sulfate mediated chemical induction and feed restriction in broilers. Poult Sci 94:1220–1226. https://doi.org/10.3382/ps/pev114

Labow BI, Souba WW (2000) Glutamine. World J Surg 24:1503–1513. https://doi.org/10.1007/s002680010269

Latorre R, Sternini C, De Giorgio R, Greenwood-Van Meerveld B (2016) Enteroendocrine cells: a review of their role in brain-gut communication. Neurogastroenterol Motil 28:620–630. https://doi.org/10.1111/nmo.12754

Laval L, Martin R, Natividad J, Chain F, Miquel S, De Maredsous CD, Capronnier S, Sokol H, Verdu E, Van Hylckama Vlieg J, Bermúdez-Humarán L, Smokvina T, Langella P (2015) Lactobacillus rhamnosus CNCM I-3690 and the commensal bacterium Faecalibacterium prausnitzii A2-165 exhibit similar protective effects to induced barrier hyper-permeability in mice. Gut Microbes 6:1–9. https://doi.org/10.4161/19490976.2014.990784

Li N, DeMarco VG, West CM, Neu J (2003) Glutamine supports recovery from loss of transepithelial resistance and increase of permeability induced by media change in Caco-2 cells. J Nutr Biochem 14:401–408. https://doi.org/10.1016/s0955-2863(03)00071-8

Liu K, Jia M, Wong EA (2020) Delayed access to feed affects broiler small intestinal morphology and goblet cell ontogeny. Poult Sci 99:5275–5285. https://doi.org/10.1016/j.psj.2020.07.040

Losada-Medina D, Yitbarek A, Nazeer N, Uribe-Diaz S, Ahmed M, Rodriguez-Lecompte JC (2020) Identification, tissue characterization, and innate immune role of Angiogenin-4 expression in young broiler chickens. Poult Sci 99:2992–3000. https://doi.org/10.1016/j.psj.2020.03.022

Molnár A, Hess C, Pál L, Wágner L, Awad WA, Husvéth F, Hess M, Dublecz K (2015) Composition of diet modifies colonization dynamics of campylobacter jejuni in broiler chickens. J Appl Microbiol 118:245–254. https://doi.org/10.1111/jam.12679

Montagne L, Pluske JR, Hampson DJ (2003) A review of interactions between dietary fibre and the intestinal mucosa, and their consequences on digestive health in young non-ruminant animals. Anim Feed Sci Technol 108:95–117. https://doi.org/10.1016/S0377-8401(03)00163-9

Murai A, Kitahara K, Terada H, Ueno A, Ohmori Y, Kobayashi M, Horio F (2018) Ingestion of paddy rice increases intestinal mucin secretion and goblet cell number and prevents dextran sodium sulfate-induced intestinal barrier defect in chickens. Poult Sci 97:3577–3586. https://doi.org/10.3382/ps/pey202

Osho SO, Wang T, Horn NL, Adeola O (2016) 0439 comparison of intestinal goblet cell staining methods in Turkey poults. J Anim Sci 94:211–211. https://doi.org/10.2527/jam2016-0439

Pan D, Yu Z (2014) Intestinal microbiome of poultry and its interaction with host and diet. Gut Microbes 5:108–119. https://doi.org/10.4161/gmic.26945

Proszkowiec-Weglarz M, Schreier LL, Kahl S, Miska KB, Russell B, Elsasser TH (2020) Effect of delayed feeding post-hatch on expression of tight junction–and gut barrier–related genes in the small intestine of broiler chickens during neonatal development. Poult Sci 99:4714–4729. https://doi.org/10.1016/j.psj.2020.06.023

Rahmatnejad E, Saki AA (2016) Effect of dietary fibres on small intestine histomorphology and lipid metabolism in young broiler chickens. J Anim Physiol Nutr 100:665–672. https://doi.org/10.1111/jpn.12422

Ravindran V, Morel PCH, Rutherfurd SM, Thomas DV (2008) Endogenous flow of amino acids in the avian ileum as influenced by increasing dietary peptide concentrations. Br J Nutr 101:822–828. https://doi.org/10.1017/S0007114508039974

Robinson K, Deng Z, Hou Y, Zhang G (2015) Regulation of the intestinal barrier function by host defense peptides. Front Vet Sci 2. https://doi.org/10.3389/fvets.2015.00057

Ruff J, Barros TL, Tellez G, Blankenship J, Lester H, Graham BD, Selby CAM, Vuong CN, Dridi S, Greene ES, Hernandez-Velasco X, Hargis BM, Tellez-Isaias G (2020) Research note: evaluation of a heat stress model to induce gastrointestinal leakage in broiler chickens. Poult Sci 99:1687–1692. https://doi.org/10.1016/j.psj.2019.10.075

Ruff J, Tellez G, Forga AJ, Señas-Cuesta R, Vuong CN, Greene ES, Hernandez-Velasco X, Uribe ÁJ, Martínez BC, Angel-Isaza JA, Dridi S, Maynard CJ, Owens CM, Hargis BM, Tellez-Isaias

G (2021) Evaluation of three formulations of essential oils in broiler chickens under cyclic heat stress. Animals 11:1084. https://doi.org/10.3390/ani11041084

Scanes CG, Pierzchala-Koziec K (2014) Biology of the gastrointestinal tract in poultry. Avian Biol Res 7:193–222. https://doi.org/10.3184/175815514X14162292284822

Schlüter H, Moll I, Wolburg H, Franke WW (2007) The different structures containing tight junction proteins in epidermal and other stratified epithelial cells, including squamous cell metaplasia. Eur J Cell Biol 86:645–655. https://doi.org/10.1016/j.ejcb.2007.01.001

Schütte A, Ermund A, Becker-Pauly C, Johansson MEV, Rodriguez-Pineiro AM, Bäckhed F, Müller S, Lottaz D, Bond JS, Hansson GC (2014) Microbial-induced meprin β cleavage in MUC2 mucin and a functional CFTR channel are required to release anchored small intestinal mucus. Proc Natl Acad Sci USA 111:12396–12401. https://doi.org/10.1073/pnas.1407597111

Scott KG-E, Meddings JB, Kirk DR, Lees-Miller SP, Buret AG (2002) Intestinal infection with Giardia spp. reduces epithelial barrier function in a myosin light chain kinase-dependent fashion. Gastroenterology 123:1179–1190. https://doi.org/10.1053/gast.2002.36002

Shaul YD, Seger R (2007) The MEK/ERK cascade: from signaling specificity to diverse functions. Biochim Biophys Acta 1773:1213–1226. https://doi.org/10.1016/j.bbamcr.2006.10.005

Shehata AA, Attia Y, Khafaga AF, Farooq MZ, El-Seedi HR, Eisenreich W, Tellez-Isaias G (2022) Restoring healthy gut microbiome in poultry using alternative feed additives with particular attention to phytogenic substances: challenges and prospects. Ger J Vet Res 2:32–42. https://doi.org/10.51585/gjvr.2022.3.0047

Shen L, Weber CR, Turner JR (2008) The tight junction protein complex undergoes rapid and continuous molecular remodeling at steady state. J Cell Biol 181:683–695. https://doi.org/10.1083/jcb.200711165

Smirnov A, Sklan D, Uni Z (2004) Mucin dynamics in the chick small intestine are altered by starvation. J Nutr 134:736–742. https://doi.org/10.1093/jn/134.4.736

Tabata H, Yasugi S (1998) Tissue interaction regulates expression of a spasmolytic polypeptide gene in chicken stomach epithelium. Develop Growth Differ 40:519–526. https://doi.org/10.1046/j.1440-169x.1998.t01-3-00006.x

Tellez-Isaias G, Eisenreich W, Petrone-Garcia VM, Hernandez-Velasco X, Castellanos-Huerta C-H, Tellez G Jr, Latorre JD, Bottje WG, Senas-Cuesta R, Coles ME, Hargis BM, El-Ashram S, Graham BD, Shehata AA (2023) Effects of chronic stress and intestinal inflammation on commercial poultry health and performance: a review. Ger J Vet Res 3:38–57. https://doi.org/10.51585/gjvr.2023.1.0051/

Thompson KL, Applegate TJ (2006) Feed withdrawal alters small-intestinal morphology and mucus of broilers. Poult Sci 85:1535–1540. https://doi.org/10.1093/ps/85.9.1535

Tsukita S, Furuse M, Itoh M (2001) Multifunctional strands in tight junctions. Nat Rev Mol Cell Biol 2:285–293. https://doi.org/10.1038/35067088

Turner JR (2009) Intestinal mucosal barrier function in health and disease. Nat Rev Immunol 9:799–809. https://doi.org/10.1038/nri2653

Uni Z, Smirnov A, Sklan D (2003) Pre- and posthatch development of goblet cells in the broiler small intestine: effect of delayed access to feed. Poult Sci 82:320–327. https://doi.org/10.1093/ps/82.2.320

van Dijk A, Veldhuizen EJA, Kalkhove SIC, Tjeerdsma-van Bokhoven JLM, Romijn RA, Haagsman HP (2007) The beta-defensin gallinacin-6 is expressed in the chicken digestive tract and has antimicrobial activity against food-borne pathogens. Antimicrob Agents Chemother 51:912–922. https://doi.org/10.1128/AAC.00568-06

Van Itallie CM, Holmes J, Bridges A, Gookin JL, Coccaro MR, Proctor W, Colegio OR, Anderson JM (2008) The density of small tight junction pores varies among cell types and is increased by expression of claudin-2. J Cell Sci 121:298–305. https://doi.org/10.1242/jcs.021485

Visscher C, Klingenberg L, Hankel J, Brehm R, Langeheine M, Helmbrecht A (2018a) Feed choice led to higher protein intake in broiler chickens experimentally infected with campylobacter jejuni. Front Nutr 5:79. https://doi.org/10.3389/fnut.2018.00079

Visscher C, Klingenberg L, Hankel J, Brehm R, Langeheine M, Helmbrecht A (2018b) Influence of a specific amino acid pattern in the diet on the course of an experimental campylobacter jejuni infection in broilers. Poult Sci 97:4020–4030. https://doi.org/10.3382/ps/pey276

Vuong CN, Mullenix GJ, Kidd MT, Bottje WG, Hargis BM, Tellez-Isaias G (2021) Research note: modified serum fluorescein isothiocyanate dextran (FITC-d) assay procedure to determine intestinal permeability in poultry fed diets high in natural or synthetic pigments. Poult Sci 100:101138. https://doi.org/10.1016/j.psj.2021.101138

Wang L, Li J, Li J, Li RX, Lv CF, Li S, Mi YL, Zhang CQ (2016) Identification of the Paneth cells in chicken small intestine. Poult Sci 95:1631–1635. https://doi.org/10.3382/ps/pew079

Wang Y, Hou Q, Wu Y, Xu Y, Liu Y, Chen J, Xu L, Guo Y, Gao S, Yuan J (2022) Methionine deficiency and its hydroxy analogue influence chicken intestinal 3-dimensional organoid development. Anim Nutr 8:38–51. https://doi.org/10.1016/j.aninu.2021.06.001

Wolfrum C, Borrmann CM, Borchers T, Spener F (2001) Fatty acids and hypolipidemic drugs regulate peroxisome proliferator-activated receptors alpha–and gamma-mediated gene expression via liver fatty acid binding protein: a signaling path to the nucleus. Proc Natl Acad Sci USA 98:2323–2328. https://doi.org/10.1073/pnas.051619898

Wrzosek L, Miquel S, Noordine M-L, Bouet S, Chevalier-Curt MJ, Robert V, Philippe C, Bridonneau C, Cherbuy C, Robbe-Masselot C, Langella P, Thomas M (2013) Bacteroides thetaiotaomicron and Faecalibacterium prausnitziiinfluence the production of mucus glycans and the development of goblet cells in the colonic epithelium of a gnotobiotic model rodent. BMC Biol 11:61. https://doi.org/10.1186/1741-7007-11-61

Wu G, Knabe DA, Flynn NE (1994) Synthesis of citrulline from glutamine in pig enterocytes. Biochem J 299:115–121. https://doi.org/10.1042/bj2990115

Xue GD, Barekatain R, Wu SB, Choct M, Swick RA (2018) Dietary L-glutamine supplementation improves growth performance, gut morphology, and serum biochemical indices of broiler chickens during necrotic enteritis challenge. Poult Sci 97:1334–1341. https://doi.org/10.3382/ps/pex444

Zhang H, Li D, Liu L, Xu L, Zhu M, He X, Liu Y (2019b) Cellular composition and differentiation signaling in chicken small intestinal epithelium. Animals 9:870. https://doi.org/10.3390/ani9110870

Zhang H, Li H, Kidrick J, Wong EA (2019a) Localization of cells expressing SGLT1 mRNA in the yolk sac and small intestine of broilers. Poult Sci 98:984–990. https://doi.org/10.3382/ps/pey343

Zhang H, Wong EA (2017) Spatial transcriptional profile of PepT1 mRNA in the yolk sac and small intestine in broiler chickens. Poult Sci 96:2871–2876. https://doi.org/10.3382/ps/pex056

Zucco F, Batto A-F, Bises G, Chambaz J, Chiusolo A, Consalvo R, Cross H, Dal Negro G, de Angelis I, Fabre G, Guillou F, Hoffman S, Laplanche L, Morel E, Pinçon-Raymond M, Prieto P, Turco L, Ranaldi G, Rousset M, Sambuy Y, Scarino ML, Torreilles F, Stammati A (2005) An inter-laboratory study to evaluate the effects of medium composition on the differentiation and barrier function of Caco-2 cell lines. Altern Lab Anim 33:603–618. https://doi.org/10.1177/026119290503300618

Intestinal Microbiota: A Hidden Metabolic and Immune Organ

2

Awad A. Shehata, Shereen Basiouni, Guillermo Tellez-Isaias, and Wolfgang Eisenreich

Contents

A. A. Shehata (✉) · W. Eisenreich
Structural Membrane Biochemistry, Bavarian NMR Center, Technical University of Munich (TUM), Garching, Bayern, Germany
e-mail: Awad.shehata@tum.de

S. Basiouni
Cilia Cell Biology, Institute of Molecular Physiology, Johannes-Gutenberg University, Mainz, Germany

G. Tellez-Isaias
Division of Agriculture, Department of Poultry Science, University of Arkansas, Fayetteville, AR, USA

21

Abstract

The gut microbiota, a hidden metabolic organ, is the immune system's largest and most complex part. It is vital to maintaining the host organisms' health. The gut microbial community comprises hundreds of trillions of bacteria, viruses, fungi, archaea, and bacteriophages. Generally, the predominant bacterial phyla in chickens include Firmicutes, Proteobacteria, Actinobacteria, and Bacteroidetes. However, each gut compartment may be a distinct ecosystem for specific microbiota. Additionally, various factors such as animal species, breed, age, nutrition, environment, stocking density, stress, and medication can all influence the delicate composition of gut microbiota. Maintaining animal health and enhancing animal performance requires a focus on microbiota, which provides a mutually beneficial and dynamic interaction. This review focuses on the dynamics and functions of the gut microbiota in chickens.

Keywords
Intestinal microbiota · synbiosis · Dysbiosis · Functions · Microbiota

1   Introduction

The majority of the biodiversity on our planet is made up of microorganisms. They are critical in controlling chemical equilibrium and cycles within their particular ecosystems. Prokaryotic bacteria, which have existed for billions of years, have had a significant influence on the evolution of eukaryotic life forms. Furthermore, early eukaryotes' nature has significantly changed due to symbiotic connections with microbes. More bacteria are encountered by the mammalian gut than by any other tissue in the body, making it a complicated immune system component. Numerous trillions of bacteria, viruses, fungi, archaea, and bacteriophages live and co-function in a healthy gut (Sartor 2008; Guinane and Cotter 2013).

The term "microbiome" describes the entire population of microorganisms present, their genes, and their ecological role (Rychlik 2020), while "gastrointestinal microbiota" describes the elements comprising the intestinal ecosystem community. Because of the profound influences of gut microbiota on the host's physiology, nutrition, immune system, and metabolism, the gut microbiota is sometimes referred to as the "hidden metabolic organ" (Guinane and Cotter 2013). Indeed, the interaction between microbiota and the host constantly evolves and is mutually beneficial. Additionally, microbiota is a major determinant of health and disease and critical in maintaining mucosal immune function, epithelial barrier integrity, motility, and nutrient absorption (Gieryńska et al. 2022).

In a healthy state, the microbiota and the host have a symbiotic relationship that determines the well-being of the intestines. However, a disruption in this balance causes an imbalanced host–microbe relationship known as "dysbiosis" (Zoetendal et al. 2008). Various factors, including antinutritional substances, heavy metals, toxic substances, bacterial toxins, xenobiotics, and antibiotics, can cause dysbiosis.

These factors can cause local inflammation, systemic infection, or even intoxication (Krüger et al. 2014; Schrödl et al. 2014; Ackermann et al. 2015).

Indeed, there is a great interest in better understanding the interconnection between the gut microbiota and the host, particularly after the ban of antibiotic growth promotors in several countries, to explore bioactive substances that can restore gut health and subsequently reduce the need for antibiotics. This review highlights the intestinal microbiota dynamics in poultry and their role in maintaining gut health.

2   Intestinal Microbial in Chickens: A Distinct Ecosystem for each Compartment of the Avian Digestive System

Over 100 trillion (>1020) microorganisms, including bacteria, fungi, viruses, and protozoa, are found in the chicken gut. The majority of these microbes are only detectable by high-throughput sequencing methods and cannot be cultivated in a laboratory (Ballou et al. 2016). The most prevalent phyla in chickens are Proteobacteria, Firmicutes, Actinobacteria, and Bacteroidetes (Clavijo and Flórez 2018).

Interestingly, the health and overall metabolism can be predicted by analyzing the ratio of *Firmicutes* to *Bacteroidetes*. Although certain *Bacteroidetes* spp. ferment carbohydrates to create propionate, *Firmicutes* contribute to the digestion of polysaccharides and the production of butyrate. Of note, each gut compartment has a distinct ecosystem for microbiota (van der Wielen et al. 2002; Rychlik 2020). This section will, therefore, provide insights into the microbiota of different gut compartments.

2.1   Crop Microbiota

The crop is an exclusive organ to birds and plays a role in the storage and fermentation of the feed. The starch hydrolysis starts in the crop, and the microbial fermentation of sugar into organic acids lowers the pH and restricts the other bacteria (Feye et al. 2020). The crop is often dominated by *Lactobacillus* spp., such as *Lactobacillus reuteri*, *Lactobacillus salivarius*, and *Lactobacillus crispatus*, although coliforms, streptococci, and bifidobacteria were also identified. The bacterial colonization started immediately before or within 4 h post-hatching and can reach 108–109 CFU/g (Lev and Briggs 1956; Abbas Hilmi et al. 2007; Rehman et al. 2007).

Lactobacilli bind to the non-secretory stratified squamous epithelium of the crop to form biofilms. However, several factors could impact colonization, including the *Lactobacillus* strains, organic acids, prebiotics, medication, and nutrient content. Some lactobacillus strains can prevent the colonization of *Escherichia coli* by competitive exclusion (Fuller 1977), highlighting that providing chickens with stable and dominant lactobacilli in the crop is essential for gut health and a balanced crop microbiota. It is important to note that lactobacilli are fastidious bacteria. Lack of feed, such as during withdrawal before slaughter or molting procedures, decreased lactobacillus and increased the crop's susceptibility to colonization by *Salmonella* and *Campylobacter* spp. (Classen et al. 2016).

2.2 Gizzard and Proventriculus Microbiota

Enterococci and *Lactobacillus* spp. are the main dominant bacteriome in the gizzard, with approximately 107–108 bacteria/gram. Nonetheless, the type of nutrition determines the kind and densities of bacteriomes. Fermentation and formation of short-chain fatty acids (SCFA) are restricted in the proventriculus and gizzard due to their high pH levels (reviewed by Fathima et al. 2022).

2.3 Microbiota of the Intestine

The growth of bacteria in the small intestine is restricted by various factors, including chemical inhibitors like bile acids, competition with the host for nutrients, rapid intestinal transit rate, continuous turnover of epithelial cells, and host immune responses (Apajalahti et al. 2004). The predominant phylum in the broiler small intestine is the Firmicutes, largely represented by *Lactobacillus, Arthromitus, Clostridium, Streptococcus*, and *Ruminococcus* spp. (Munyaka et al. 2016; Wang et al. 2016), Fig. 2.1.

In the small intestine, the type and densities of microbiota are not uniform throughout. The highest bacterial densities are present in the ileum and the lowest in the duodenum (Shang et al. 2018). Streptococci, Bacteroides, Clostridia, *Escherichia coli*, Ruminococci, Enterococci, and Lactobacilli are numerous in the ileum (Yadav

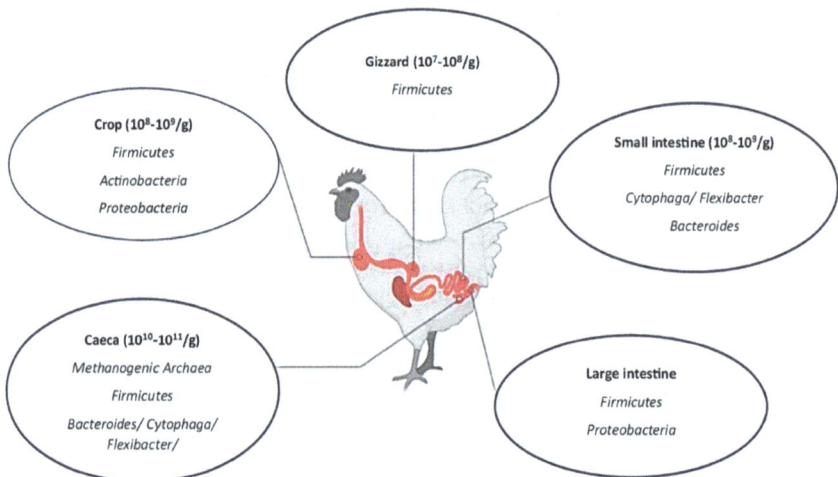

Fig. 2.1 Gut microbiota in chickens. The figure is modified after (Shehata et al. 2022)

and Jha 2019). However, Lactobacilli and Bifidobacteria are more dominant in the duodenum (Waite and Taylor 2015). In the jejunum, Lactobacilli, Enterococci, and Clostridiaceae are more prevalent (Waite and Taylor 2015). Converting bile acids into secondary bile acids and digesting and absorbing lipids are critical functions of the jejunum (Calefi et al. 2014). The jejunum cannot operate without microbiota because lipase and bile salt hydrolase are found in *Lactobacillus*, *Enterococcus*, and *Staphylococcus* spp. (Borsoi et al. 2015).

The bacterial families Clostridiaceae, Bacteroidacea, Lactobacillus, and Proteobacteria are commonly found in high numbers, 1010 and 1011 bacteria/g digesta, within the ceca (Borda-Molina et al. 2018). Due to the anti-peristalsis of the ceca that limits the movement of the digest, the ceca is evacuated twice daily (Svihus 2014), allowing the fermentation of fibers such as beta-glucans, starch, and non-starch polysaccharides (Clench and Mathias 1995).

A number of aspects, such as the anaerobic condition, the prolonged transit duration, and partially digested metabolites that enter the ceca, promote the fermentation process and SCFA generation in the ceca (Rehman et al. 2007).

3   Factors Affecting Gut Microbiota in Chickens

A number of variables can influence the composition of the gut microbiota. These factors could be classified into the host (species, breed, age), management and nutritional (feed, flock density, temperature, relative humidity), and health interventions (antibiotics, probiotics, prebiotics, feed additives) factors (Xu et al. 2007). Figure 2.2 illustrates the factors that modulate intestinal microbiota.

Age of birds. Intestinal microbiota in birds are affected by age. They have been detected in 1-day-old chicks (Ballou et al. 2016). At an age of 3 days, *Campylobacter coli*, *L. delbrueckii*, and *C. perfringens* were the most prevalent gut bacteria. *L. acidophilus*, *Enterococcus*, and *Streptococcus* were prominent from 7 to 21 days of age (Lu et al. 2003). However, there is no consistent data since several other factors play an important role in the dynamics of microbiota.

Environment. The litter is surely a source of bacteria for the birds and, consequently, a possible source of pathogenic bacteria. Inadequate hygiene practices lead to a rise in enteric infections and the related issues of moist litter (Dawkins et al. 2004).

Management. Heat stress and overcrowding—both frequently observed in intensive poultry farming—substantially affect the microbiota of chickens. The host genotype seems essential for preserving a wholesome gut environment. Though the exact cause of this phenomenon is unknown, host-mediated mechanisms that affect the intestinal environment and its occupants are considered to be involved. Van der Wielen et al. (van der Wielen et al. 2002) conducted a thorough investigation on chickens and found that each bird had a different microbial makeup.

Health status. An important factor in modifying gut microbiota is the birds' health. Opportunistic pathogens can infect tissues when the natural defenses and immunity are compromised or come under stress from an illness challenge. An

Managemental and nutritional factors
- Stocking density
- Temperature
- Photoperiod
- Ventilation
- Litter type and humidity
- Feed

Factors affecting
gut microbiota
composition

Host factors
- Age
- Breed
- Species
- Sex

Health interventions
- Antibiotic growth promoters
- Vaccination
- Probiotics/Prebiotics
- Phytogenic substances

Gut development
- Immune system
- Gut morphology
- Microbiota acquisition

Gut microbiota
- Functional output (production of SCFA, added metabolic potentials, competitive exclusion of pathogens
- Composition (Richness species, population structure and complexity, and synbiosis)

Gut Health
- Nutrient assimilation
- Intestinal barrier integrity
- Inflammatory balance
- Susceptibility to enteric pathogens

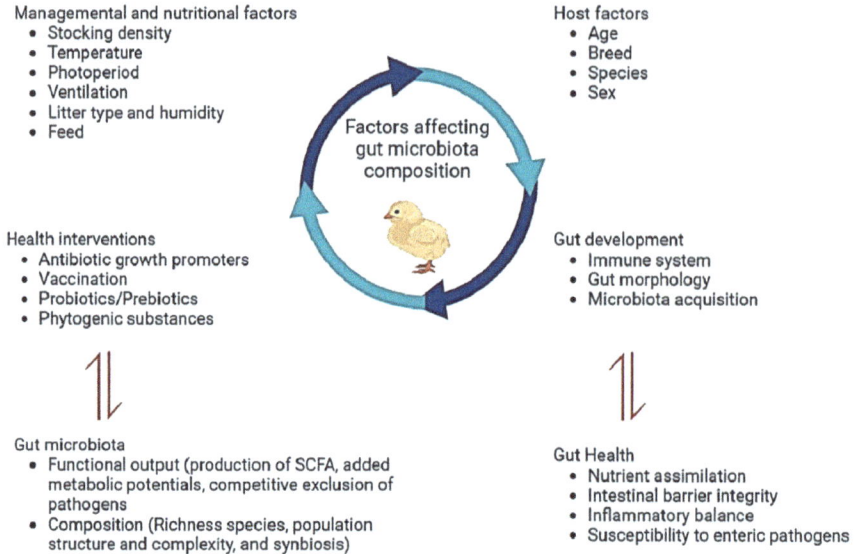

Fig. 2.2 Factors affecting the Gut microbiota in chickens composition, modified after (Shehata et al. 2022)

animal's immunocompetence may be weakened by exposure to stressors, including heat, handling during transit, crowding, or relocation to a new home.

Stress. The release of chemicals, including cortisol, adrenaline, and noradrenaline, increases in reaction to stress (Mayer 2000). These hormones weaken the immune system, which may create a window of opportunity for infection.

Diet. Food components provide the majority of the energy needed by intestinal bacteria. Therefore, the diet has a significant impact on bacterial populations. Diets based mostly on maize were shown to raise populations of lactobacilli and coliforms as compared to diets with wheat and barley (Mathlouthi et al. 2002). Additionally, the ileal microbiota was impacted by dietary fat sources, including the lactobacilli population and *Clostridium perfringens* (Knarreborg et al. 2002).

Antibiotics. Antibiotics can modulate the gut microbiota. Subtherapeutic antibiotics were used for poultry to delay the onset of endemic illnesses and to enhance growth (Gustafson and Bowen 1997). Antibiotic growth promotors (AGPs) help to control poultry's chronic stress and to improve performance. However, due to the ban on AGPs, there is now a need for "natural" alternatives.

Probiotics and prebiotics. Common supplements are used to modify the gut flora. Numerous studies have demonstrated the health advantages of giving probiotics, prebiotics, or a combination of the two (synbiotics) to broilers. See Chap. 5 for more details.

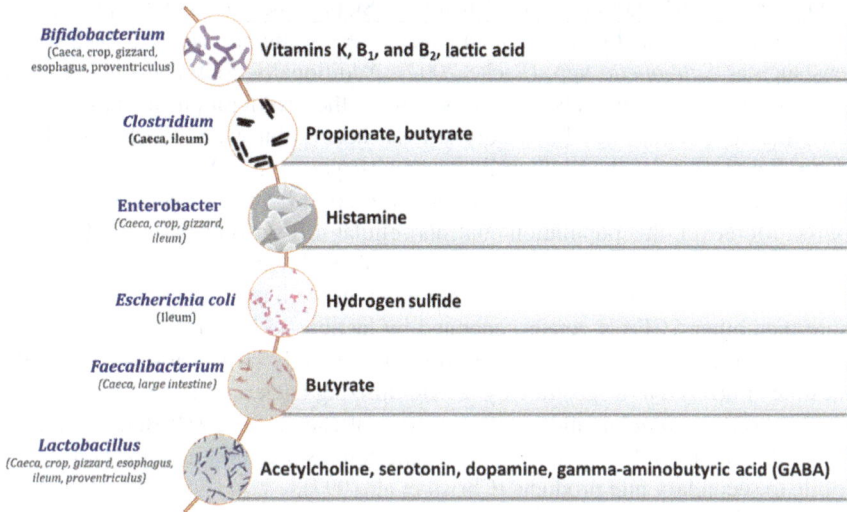

Fig. 2.3 The main gut microbiota in broiler chickens and their metabolites

4   Functions of Intestinal Microbiota

4.1   Metabolic Function

A significant metabolic function of microbiota is fermenting non-digestible nutrients, including polysaccharides and oligosaccharides that escape digestion, unabsorbed sugars, and alcohol, as well as mucus and shed epithelial cells, leading to the production of energy and nutrients such as vitamins, amino acids, and SCFA, (Beldowska et al. 2023). The main metabolic functions of some selected intestinal microbiota are summarized in Fig. 2.3.

4.1.1   Microbial Fermentation

Unabsorbed feed flows into the hindgut and partially into the caeca and may become a substrate for microbial fermentation. In contrast to other mammals, fermentation occurs primarily in the caeca of chickens rather than in the colon (Adil and Magray 2012). Thus, the colon has a short retention period for digesta, estimated 4–56 min, with an average of around 30 min (Van Krimpen et al. 2011). However, the ceca are likely to have lengthy digesta retention due to infrequent emptying (Svihus et al. 2013).

Additionally, the intestinal microbiota is critical in converting uric acid to ammonia, which the bird absorbs to produce amino acids like glutamine (Vispo and Karasov 1997). Additionally, intestinal microbiota have a role in synthesizing proteins and amino acids by incorporating nitrogen from the diet into their cellular protein (Metges 2000).

Microbial fermentations of feed result in SCFA, such as acetate, propionate, butyrate, succinate, and lactate, which have bacteriostatic properties against pathogens such as *Salmonella* spp. (Ricke 2003). Additionally, SCFA also serves as a source of energy for animals and can stimulate the proliferation of gut epithelial cells, thereby increasing the absorption surface (Dibner and Richards 2005). Clostridial clusters of the phylum Firmicutes (Pryde et al. 2002), *Roseburia* spp., and *Eubacterium rectale* produce butyrate, preferred by enterocytes as an energy source. Moreover, this population controls cellular differentiation and proliferation inside the intestinal mucosa (Fukunaga et al. 2003). Lactic acid is the most potent of the major SCFA generated by gut microbiota; hence it tends to lower residual pH more than other SCFA. Lactate generated by lactic acid bacteria is rarely seen in high concentrations in the hindgut because it is generally swiftly absorbed from the intestine or utilized as a substrate by lactate-utilizing bacteria such as *Eubacterium*, *Anaerostipes*, *Veillonella*, and *Megasphaera* (Harmsen et al. 2002; Belenguer et al. 2007). Production of SCFA helps to lower the colon's pH, inhibiting the conversion of bile to secondary bile products (Christl et al. 1997).

Protein fermentation is assumed to be hazardous due to the production of biogenic amines, phenols, indoles, cresol, sulfur-containing compounds, and ammonia (Apajalahti et al. 2004; Qaisrani et al. 2014). High protein intake could be a risk factor for necrotic enteritis caused by *C. perfringens* (Wilkie et al. 2005), since *C. perfringens* lacks the genes that produce arginine, aromatic amino acids, branched-chain amino acids, glutamic acid, histidine, lysine, methionine, serine, and threonine (Shimizu et al. 2002). In addition, some protein fermentation metabolites harm gut epithelial cells (Gilbert et al. 2018). This is crucial because the use of less digestible protein sources for poultry is expected to increase due to the food-feed dispute.

Of note, gut microbiota compete with the host for energy and protein in the proximal gut, produce toxic metabolites, and catabolize bile acids in both the proximal and distal gut, which might lead to decreased growth and fat digestibility in birds (Gaskins et al. 2002).

4.2   Protective Functions

The intestinal microbiota has a crucial function as an immune organ: (i) Prevention of the colonization of harmful pathogens and other non-native microbes through competitive exclusion (van der Waaij et al. 1971). Non-pathogenic bacteria attach to the brush border of gut cells, obstructing pathogens from attaching and entering the cells. (ii) Suppression of the growth of pathogens by secreting organic acids and bacteriocins, which stimulate the immune system while competing for nutrition and attachment points on the mucosal wall (Kelly 2001). (iii) Inhibition of the growth of harmful pathogens. *Lactobacillus* spp. inhibits *Salmonella*, *Shigella*, *Clostridium*, and *Listeria* by producing the bacteriocin reuterin (Naidu et al. 1999). (iv) Development of the intestinal host's defenses, including the mucus layer, epithelial layer, and lamina propria (Kelly 2001; Gaskins et al. 2002; Snel et al. 2002), which subsequently act as a barrier. (v) Prevention of inflammation and maintaining immune homeostasis (Lei et al. 2015).

4.3 Microbiota–Gut–Brain Axis

The interconnection between the gut and brain is well studied in mammals; however, the gut–brain axis in chickens lacks sufficient information. In 2021, Aruwa et al. presented a chicken-specific model of the gut–brain axis (Wickramasuriya et al. 2022), illustrated in Fig. 2.4. In brief, when intestinal stress and inflammatory conditions occur, various intestinal cells, including epithelial cells, enteric muscles, and immune cells, transmit signals to the brain via the vagus nerve. This process stimulates the secretion of cytokines from the leucocytes into blood circulation, which triggers the activation of the central nervous system (CNS) (Kabiersch et al. 1988). Additionally, beneficial gut microbes produce neurotransmitters that stimulate the CNS. These intestinal stimuli activate the hypothalamic–pituitary–adrenal (HPA) axis and increase the levels of corticosterone in the blood (Villageliu and Lyte 2017). Subsequently, this helps to control inflammation by modulating the

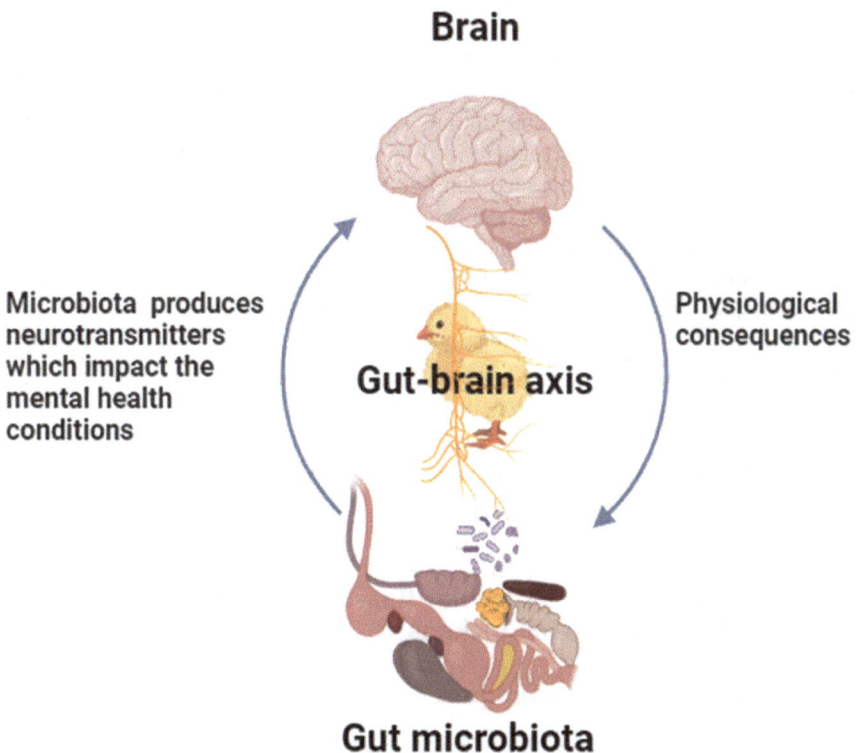

Brain

Microbiota produces neurotransmitters which impact the mental health conditions

Gut-brain axis

Physiological consequences

Gut microbiota

Fig. 2.4 Proposed gut–brain axis model in chickens. The figure was generated by BioRender

translocation of white blood cells to the site of inflammation in the gut (Calefi et al. 2014). Elevated cortisone may cause unhealthy behavioral circumstances in chickens, including reduced food intake, lethargy, and drowsiness (Calefi et al. 2019).

Intestinal infections in chickens activate the serotonergic pathway in the midbrain by increasing the levels of 5-hydroxytryptamine and 5-hydroxyindoleacetic acid (Borsoi et al. 2015). Recently, the gut–brain interconnection was studied in several avian pathogens like necrotic enteritis and *Eimeria*. It was found that *C. perfringens* infection activates the dopaminergic pathway, while *Eimeria* spp. and *C. perfringens* mixed infection cause increased tissue damage and activation of the HPA axis, leading to increased noradrenaline and norepinephrine levels in the CNS (Calefi et al. 2019).

5   Dysbacteriosis

The term "dysbacteriosis" was initially used in the middle of the 1990s to refer to intestinal dysbiosis in chickens that was suggested to be connected to several changes in animal husbandry and environment. Dysbacteriosis has increased in intensive chicken production since AGPs were prohibited. A general description of this pathological condition, which lacks specificity, is an overabundance of microbiota resulting in non-specific enteritis (McMullin 2004). The imbalance between intestinal microbiota leads to impaired meal absorption and worse litter quality since the afflicted birds' droppings have more moisture content. Bacteriosis has numerous non-specific clinical symptoms, making the diagnosis challenging. The primary sign is moist feces, characterized by foamy cecal droppings and a characteristic orange mucoid appearance. Undigested feed in the feces, indicative of a malfunctioning digestive system, can be observed. Several diseases usually accompany bacteriosis. It is suggested that inadequate management, dietary changes, and overcrowding are the main causes of the condition's onset, often occurring between 20 and 30 days of age.

Since there is currently no reliable test to diagnose dysbacteriosis, the health of the flock, variations in feed and water intake, and visual examination of the feces within a broiler house are used to make the diagnosis. To assess intestinal integrity, however, there are techniques to measure the water content in feces; this might allow identification of the intestinal integrity and enable early diagnosis and treatment before the illness worsens.

6   Culture- Versus Molecular-Based Techniques to Study Gut Microbiota in Chickens

6.1   Culture-Based Techniques

It is greatly accepted that microbiota is significantly underestimated when using culture-based approaches, which were once utilized to characterize these bacteria.

There are several challenges to the culture-based gut microbiota analysis: (i) Up to 90% of the gut microbiota are unculturable (Zoetendal et al. 2004). (ii) Some anaerobic bacterial species grow slowly for up to 2 weeks (Toivanen et al. 2001), which might significantly increase the time and labor involved. (iii) Fastidious bacteria need specific enriched media and might not be cultivated easily. (iv) Targeting a certain bacterium with selective media may give a prior prejudice and may provide erroneous findings (Simpson et al. 2000; Amit-Romach et al. 2004; Zoetendal et al. 2004; Bibiloni et al. 2005).

6.2   Molecular-Based Techniques

Fortunately, molecular techniques enable an increasingly precise picture of the intestinal microbiota without culture. To date, several molecular techniques have been developed, such as DNA microarrays, denaturing gradient gel electrophoresis (DGGE), quantitative PCR, temperature gradient gel electrophoresis, Dot/slot hybridization, sequencing of 16S rDNA genes, terminal restriction fragment length polymorphism (TRFLP), temperature gradient gel electrophoresis (TGGE), and next generation sequencing. Most molecular techniques depend on the 16S ribosome, and its encoding gene is amplified through PCR. Since all bacterial species have ribosomes that consistently function over extended periods, ribosomes are an attractive target for detecting nucleotide sequence variations that could be implemented to determine phylogeny or relatedness (Patel 2001). Because the 16S rRNA gene has conserved and variable regions, it is possible to create PCR primers targeting all or only a subset of bacterial DNA. The 16S rDNA gene, 1.5 kbp in length, comprises nine variable regions (V1-V9) with a range of heterogeneity (Neefs et al. 1993).

The varieties of intestinal microbiota have been better defined based on the tremendous progress in molecular biology. This allows us to identify the unculturable bacteria present in extremely low quantities in samples. However, the technology also has some drawbacks. Among these constraints are the following: (i) The physical, biological, and chemical processes used in molecular analysis can introduce unwanted bias and distort the nature of the reported ecological niche. (ii) Careful sample collection and management are required to ensure that the resulting community profiles accurately portray the sample location. (iii) Extraction of DNA from environmental materials could be a cause of inaccuracy (Wintzingerode et al. 1997). Environmental samples frequently contain additional biotic and abiotic substances that might reduce PCR effectiveness if left behind after DNA extraction. (iv) The PCR reaction itself can be a multifactorial source of bias. Quantitative conclusions can only be drawn from DNA amplification of mixtures of different templates if one assumes the following (Suzuki and Giovannoni 1996): All molecules are equally accessible to primer hybridization, i.e., primer and templates hybridize with equal efficiency. DNA polymerase extension efficiency is uniform for all templates. The template concentration is consistent, so substrate exhaustion is uniform. (v) Several factors can negate the above assumptions. Poor DNA template quality (Liesack et al. 1991), high or low DNA concentration (Chandler et al. 1997), poor choice of primers and target region of the 16S rRNA gene (Suzuki and Giovannoni 1996; Yu and Morrison 2004), and increasing replication cycles (Qiu et al. 2001) can all influence the amplification of the template DNA and result in the formation of chimeric and heteroduplex molecules. Fromin et al. (Fromin et al. 2002) proposed that the best way to ensure accuracy in the comparisons of microbial profiling is to follow the same standardized methodology for all samples strictly. Thus, it is deemed that any bias occurs uniformly across all sample populations.

7 Conclusion

Increasing awareness of gut health, including microbiota homeostasis, is essential for successful chicken production. Additional studies are necessary to understand better the gut microbiota's fundamental composition and effects on the host animal's health. Because microbiota perform certain metabolic processes specific to the location of the digestive tract in which they reside, a description of the members of the community across the gastrointestinal system is thus required. This is important because feed additives like prebiotics, depending on their composition, may be digested differently in different chicken gastrointestinal system compartments (Ricke 2018). The resident gut microbiota populations in GIT compartments may also change as the birds mature (Rychlik 2020).

Additionally, based on the feed composition, the metabolism can still change even when microbiota compositional profiles seem quite consistent (Dittoe et al. 2022). Knowledge about the microbiota, the dynamics in its metabolic networks, and the beneficial effects of dietary supplements for gut health would improve feed efficiency, growth, and general health. In response to the prohibition of AGPs, there is a need to develop efficient alternatives to antibiotics to maintain eubiotics and avoid dysbiosis in poultry. In the next book chapters, the authors address the state-of-the-art challenges and opportunities of diverse alternatives to antimicrobials, i.e., probiotics, prebiotics, postbiotics, organic acids, enzymes, organic acids, bacteriophages, and antimicrobial peptides that can be used in the modern poultry industry.

References

Abbas Hilmi HT, Surakka A, Apajalahti J, Saris PEJ (2007) Identification of the most abundant lactobacillus species in the crop of 1- and 5-week-old broiler chickens. Appl Environ Microbiol 73:7867–7873. https://doi.org/10.1128/AEM.01128-07
Ackermann W, Coenen M, Schrödl W, Shehata AA, Krüger M (2015) The influence of glyphosate on the microbiota and production of botulinum neurotoxin during ruminal fermentation. Curr Microbiol 70:374–382. https://doi.org/10.1007/s00284-014-0732-3
Adil S, Magray SN (2012) Impact and manipulation of gut microflora in poultry: a review. J Anim Vet Adv 11:873–877. https://doi.org/10.3923/javaa.2012.873.877
Amit-Romach E, Sklan D, Uni Z (2004) Microflora ecology of the chicken intestine using 16S ribosomal DNA primers. Poult Sci 83:1093–1098. https://doi.org/10.1093/ps/83.7.1093

Apajalahti J, Kettunen A, Graham H (2004) Characteristics of the gastrointestinal microbial communities, with special reference to the chicken. Worlds Poult Sci J 60:223–232. https://doi.org/10.1079/WPS20040017

Ballou AL, Ali RA, Mendoza MA, Ellis JC, Hassan HM, Croom WJ, Koci MD (2016) Development of the chick microbiome: how early exposure influences future microbial diversity. Front Vet Sci 3:2. https://doi.org/10.3389/fvets.2016.00002

Beldowska A, Barszcz M, Dunislawska A (2023) State of the art in research on the gut-liver and gut-brain axis in poultry. J Anim Sci Biotechnol 14:37. https://doi.org/10.1186/s40104-023-00853-0

Belenguer A, Duncan SH, Holtrop G, Anderson SE, Lobley GE, Flint HJ (2007) Impact of pH on lactate formation and utilization by human fecal microbial communities. Appl Environ Microbiol 73:6526–6533. https://doi.org/10.1128/AEM.00508-07

Bibiloni R, Simon MA, Albright C, Sartor B, Tannock GW (2005) Analysis of the large bowel microbiota of colitic mice using PCR/DGGE. Lett Appl Microbiol 41:45–51. https://doi.org/10.1111/j.1472-765X.2005.01720.x

Borda-Molina D, Seifert J, Camarinha-Silva A (2018) Current perspectives of the chicken gastrointestinal tract and its microbiome. Comput Struct Biotechnol J 16:131–139. https://doi.org/10.1016/j.csbj.2018.03.002

Borsoi A, Quinteiro-Filho WM, Calefi AS, Piantino Ferreira AJ, Astolfi-Ferreira CS, Florio JC, Palermo-Neto J (2015) Effects of cold stress and salmonella Heidelberg infection on bacterial load and immunity of chickens. Avian Pathol 44:490–497. https://doi.org/10.1080/03079457.2015.1086976

Calefi AS, da Fonseca JGS, de Nunes CAQ, Lima APN, Quinteiro-Filho WM, Flório JC, Zager A, Ferreira AJP, Palermo-Neto J (2019) Heat stress modulates brain monoamines and their metabolites production in broiler chickens co-infected with Clostridium perfringens type a and Eimeria spp. Vet Sci 6:4. https://doi.org/10.3390/vetsci6010004

Calefi AS, Honda BTB, Costola-de-Souza C, de Siqueira A, Namazu LB, Quinteiro-Filho WM, da Silva Fonseca JG, Aloia TPA, Piantino-Ferreira AJ, Palermo-Neto J (2014) Effects of long-term heat stress in an experimental model of avian necrotic enteritis. Poult Sci 93:1344–1353. https://doi.org/10.3382/ps.2013-03829

Chandler DP, Fredrickson JK, Brockman FJ (1997) Effect of PCR template concentration on the composition and distribution of total community 16S rDNA clone libraries. Mol Ecol 6:475–482. https://doi.org/10.1046/j.1365-294X.1997.00205.x

Christl SU, Bartram P, Paul A, Kelber E, Scheppach W, Kasper H (1997) Bile acid metabolism by colonic bacteria in continuous culture: effects of starch and pH. Ann Nutr Metab 41:45–51. https://doi.org/10.1159/000177977

Classen HL, Apajalahti J, Svihus B, Choct M (2016) The role of the crop in poultry production. Worlds Poult Sci J 72:459–472. https://doi.org/10.1017/S004393391600026X

Clavijo V, Flórez MJV (2018) The gastrointestinal microbiome and its association with the control of pathogens in broiler chicken production: a review. Poult Sci 97:1006–1021. https://doi.org/10.3382/ps/pex359

Clench MH, Mathias JR (1995) The avian cecum: a review. Wilson J Ornithol 107:93–121

Dawkins MS, Donnelly CA, Jones TA (2004) Chicken welfare is influenced more by housing conditions than by stocking density. Nature 427:342–344. https://doi.org/10.1038/nature02226

Dibner JJ, Richards JD (2005) Antibiotic growth promoters in agriculture: history and mode of action. Poult Sci 84:634–643. https://doi.org/10.1093/ps/84.4.634

Dittoe DK, Olson EG, Ricke SC (2022) Impact of the gastrointestinal microbiome and fermentation metabolites on broiler performance. Poult Sci 101:101786. https://doi.org/10.1016/j.psj.2022.101786

Fathima S, Shanmugasundaram R, Adams D, Selvaraj RK (2022) Gastrointestinal microbiota and their manipulation for improved growth and performance in chickens. Food Secur 11:1401. https://doi.org/10.3390/foods11101401

Feye KM, Baxter MFA, Tellez-Isaias G, Kogut MH, Ricke SC (2020) Influential factors on the composition of the conventionally raised broiler gastrointestinal microbiomes. Poult Sci 99:653–659. https://doi.org/10.1016/j.psj.2019.12.013

Fromin N, Hamelin J, Tarnawski S, Roesti D, Jourdain-Miserez K, Forestier N, Teyssier-Cuvelle S, Gillet F, Aragno M, Rossi P (2002) Statistical analysis of denaturing gel electrophoresis (DGE) fingerprinting patterns. Environ Microbiol 4:634–643. https://doi.org/10.1046/j.1462-2920.2002.00358.x

Fukunaga T, Sasaki M, Araki Y, Okamoto T, Yasuoka T, Tsujikawa T, Fujiyama Y, Bamba T (2003) Effects of the soluble fibre pectin on intestinal cell proliferation, fecal short chain fatty acid production and microbial population. Digestion 67:42–49. https://doi.org/10.1159/000069705

Fuller R (1977) The importance of lactobacilli in maintaining normal microbial balance in the crop. Br Poult Sci 18:85–94. https://doi.org/10.1080/00071667708416332

Gaskins HR, Collier CT, Anderson DB (2002) Antibiotics as growth promotants: mode of action. Anim Biotechnol 13:29–42. https://doi.org/10.1081/ABIO-120005768

Gieryńska M, Szulc-Dąbrowska L, Struzik J, Mielcarska MB, Gregorczyk-Zboroch KP (2022) Integrity of the intestinal barrier: the involvement of epithelial cells and microbiota—a mutual relationship. Animals 12:145. https://doi.org/10.3390/ani12020145

Gilbert MS, Ijssennagger N, Kies AK, Van Mil SWC (2018) Protein fermentation in the gut; implications for intestinal dysfunction in humans, pigs, and poultry. Am J Physiol-Gastrointest Liver Physiol 315:G159–G170. https://doi.org/10.1152/ajpgi.00319.2017

Guinane CM, Cotter PD (2013) Role of the gut microbiota in health and chronic gastrointestinal disease: understanding a hidden metabolic organ. Ther Adv Gastroenterol 6:295–308. https://doi.org/10.1177/1756283X13482996

Gustafson RH, Bowen RE (1997) Antibiotic use in animal agriculture. J Appl Microbiol 83:531–541. https://doi.org/10.1046/j.1365-2672.1997.00280.x

Harmsen HJM, Raangs GC, He T, Degener JE, Welling GW (2002) Extensive set of 16S rRNA-based probes for detection of bacteria in human feces. Appl Environ Microbiol 68:2982–2990. https://doi.org/10.1128/AEM.68.6.2982-2990.2002

Kabiersch A, del Rey A, Honegger CG, Besedovsky HO (1988) Interleukin-1 induces changes in norepinephrine metabolism in the rat brain. Brain Behav Immun 2:267–274. https://doi.org/10.1016/0889-1591(88)90028-1

Kelly D (2001) Bacteria KTPL. Regulation of gut function and immunity. In: Piva A, Knudsen KEB, Lindberg JE (eds) Gut environment of pigs. Nottingham University Press, Nottingham, UK, pp 113–131

Knarreborg A, Simon MA, Engberg RM, Jensen BB, Tannock GW (2002) Effects of dietary fat source and subtherapeutic levels of antibiotic on the bacterial community in the ileum of broiler chickens at various ages. Appl Environ Microbiol 68:5918–5924. https://doi.org/10.1128/AEM.68.12.5918-5924.2002

Krüger M, Neuhaus J, Herrenthey AG, Gökce MM, Schrödl W, Shehata AA (2014) Chronic botulism in a Saxony dairy farm: sources, predisposing factors, development of the disease and treatment possibilities. Anaerobe 28:220–225. https://doi.org/10.1016/j.anaerobe.2014.06.010

Lei YMK, Nair L, Alegre M-L (2015) The interplay between the intestinal microbiota and the immune system. Clin Res Hepatol Gastroenterol 39:9–19. https://doi.org/10.1016/j.clinre.2014.10.008

Lev M, Briggs CAE (1956) The gut flora of the chick. II. The establishment of the flora. J Appl Bacteriol 19:224–230. https://doi.org/10.1111/j.1365-2672.1956.tb00070.x

Liesack W, Weyland H, Stackebrandt E (1991) Potential risks of gene amplification by PCR as determined by 16S rDNA analysis of a mixed-culture of strict barophilic bacteria. Microb Ecol 21:191–198. https://doi.org/10.1007/BF02539153

Lu J, Idris U, Harmon B, Hofacre C, Maurer JJ, Lee MD (2003) Diversity and succession of the intestinal bacterial community of the maturing broiler chicken. Appl Environ Microbiol 69:6816–6824. https://doi.org/10.1128/AEM.69.11.6816-6824.2003

Mathlouthi N, Lallès JP, Lepercq P, Juste C, Larbier M (2002) Xylanase and beta-glucanase supplementation improve conjugated bile acid fraction in intestinal contents and increase villus size of small intestine wall in broiler chickens fed a rye-based diet. J Anim Sci 80:2773–2779. https://doi.org/10.2527/2002.80112773x

Mayer EA (2000) The neurobiology of stress and gastrointestinal disease. Gut 47:861–869. https://doi.org/10.1136/gut.47.6.861

McMullin P (2004) A pocket guide to poultry health and disease. Rev Ed edition (Jan 2004) ed. 5M Enterprises, pp. 101–102

Metges CC (2000) Contribution of microbial amino acids to amino acid homeostasis of the host. J Nutr 130:1857S–1864S. https://doi.org/10.1093/jn/130.7.1857S

Munyaka PM, Nandha NK, Kiarie E, Nyachoti CM, Khafipour E (2016) Impact of combined β-glucanase and xylanase enzymes on growth performance, nutrients utilization and gut microbiota in broiler chickens fed corn or wheat-based diets. Poult Sci 95:528–540. https://doi.org/10.3382/ps/pev333

Naidu AS, Bidlack WR, Clemens RA (1999) Probiotic spectra of lactic acid bacteria (LAB). Crit Rev Food Sci Nutr 39:13–126. https://doi.org/10.1080/10408699991279187

Neefs J-M, Van de Peer Y, De Rijk P, Chapelle S, De Wachter R (1993) Compilation of small ribosomal subunit RNA structures. Nucleic Acids Res 21:3025–3049. https://doi.org/10.1093/nar/21.13.3025

Patel JB (2001) 16S rRNA gene sequencing for bacterial pathogen identification in the clinical laboratory. Mol Diagn J Devoted Underst Hum Dis Clin Appl Mol Biol 6:313–321. https://doi.org/10.1054/modi.2001.29158

Pryde SE, Duncan SH, Hold GL, Stewart CS, Flint HJ (2002) The microbiology of butyrate formation in the human colon. FEMS Microbiol Lett 217:133–139. https://doi.org/10.1111/j.1574-6968.2002.tb11467.x

Qaisrani SN, Moquet PCA, Van Krimpen MM, Kwakkel RP, Verstegen MWA, Hendriks WH (2014) Protein source and dietary structure influence growth performance, gut morphology, and hindgut fermentation characteristics in broilers. Poult Sci 93:3053–3064. https://doi.org/10.3382/ps.2014-04091

Qiu X, Wu L, Huang H, McDonel PE, Palumbo AV, Tiedje JM, Zhou J (2001) Evaluation of PCR-generated chimeras, mutations, and heteroduplexes with 16S rRNA gene-based cloning. Appl Environ Microbiol 67:880–887. https://doi.org/10.1128/AEM.67.2.880-887.2001

Rehman HU, Vahjen W, Awad WA, Zentek J (2007) Indigenous bacteria and bacterial metabolic products in the gastrointestinal tract of broiler chickens. Arch Anim Nutr 61:319–335. https://doi.org/10.1080/17450390701556817

Ricke S (2003) Perspectives on the use of organic acids and short chain fatty acids as antimicrobials. Poult Sci 82:632–639. https://doi.org/10.1093/ps/82.4.632

Ricke SC (2018) Impact of prebiotics on poultry production and food safety. Yale J Biol Med 91:151–159

Rychlik I (2020) Composition and function of chicken gut microbiota. Animals 10:103. https://doi.org/10.3390/ani10010103

Sartor RB (2008) Microbial influences in inflammatory bowel diseases. Gastroenterology 134:577–594. https://doi.org/10.1053/j.gastro.2007.11.059

Schrödl W, Krüger S, Konstantinova-Müller T, Shehata AA, Rulff R, Krüger M (2014) Possible effects of glyphosate on Mucorales abundance in the rumen of dairy cows in Germany. Curr Microbiol 69:817–823. https://doi.org/10.1007/s00284-014-0656-y

Shang Y, Kumar S, Oakley B, Kim WK (2018) Chicken gut microbiota: importance and detection technology. Front Vet Sci 5:254. https://doi.org/10.3389/fvets.2018.00254

Shehata AA, Yalçın S, Latorre JD, Basiouni S, Attia YA, Abd El-Wahab A, Visscher C, El-Seedi HR, Huber C, Hafez HM, Eisenreich W, Tellez-Isaias G (2022) Probiotics, prebiotics, and phytogenic substances for optimizing gut health in poultry. Microorganisms 10:395. https://doi.org/10.3390/microorganisms10020395

Shimizu T, Ohtani K, Hirakawa H, Ohshima K, Yamashita A, Shiba T, Ogasawara N, Hattori M, Kuhara S, Hayashi H (2002) Complete genome sequence of Clostridium perfringens, an anaerobic flesh-eater. Proc Natl Acad Sci USA 99:996–1001. https://doi.org/10.1073/pnas.022493799

Simpson JM, McCracken VJ, Gaskins HR, Mackie RI (2000) Denaturing gradient gel electrophoresis analysis of 16S ribosomal DNA amplicons to monitor changes in fecal bacterial populations of weaning pigs after introduction of lactobacillus reuteri strain MM53. Appl Environ Microbiol 66:4705–4714. https://doi.org/10.1128/AEM.66.11.4705-4714.2000

Snel J, Harmsen HJM, van der Wielen PWJJ, Williams BA (2002) Dietary strategies to influence the gastrointestinal microflora of young animals, and its potential to improve intestinal health., in. In: Blok MC, Vahl HA, de Lange L, van de Braak AE, Hemke G, Hessing M (eds) Nutrition and health in the gastrointestinal tract. Wageningen, Wageningen Academic Publishers, pp 37–69

Suzuki MT, Giovannoni SJ (1996) Bias caused by template annealing in the amplification of mixtures of 16S rRNA genes by PCR. Appl Environ Microbiol 62:625–630. https://doi.org/10.1128/aem.62.2.625-630.1996

Svihus B (2014) Function of the digestive system. J Appl Poult Res 23:306–314. https://doi.org/10.3382/japr.2014-00937

Svihus B, Choct M, Classen HL (2013) Function and nutritional roles of the avian caeca: a review. Worlds Poult Sci J 69:249–264. https://doi.org/10.1017/S0043933913000287

Toivanen P, Vaahtovuo J, Eerola E (2001) Influence of major histocompatibility complex on bacterial composition of fecal flora. Infect Immun 69:2372–2377. https://doi.org/10.1128/IAI.69.4.2372-2377.2001

Wintzingerode V, Göbel UB, Stackebrandt E (1997) Determination of microbial diversity in environmental samples: pitfalls of PCR-based rRNA analysis. FEMS Microbiol Rev 21:213–229. https://doi.org/10.1111/j.1574-6976.1997.tb00351.x

van der Waaij D, Berghuis-de Vries JM, Lekkerkerk-v L (1971) Colonization resistance of the digestive tract in conventional and antibiotic-treated mice. J Hyg (Lond) 69:405–411. https://doi.org/10.1017/s0022172400021653

van der Wielen PWJJ, Keuzenkamp DA, Lipman LJA, van Knapen F, Biesterveld S (2002) Spatial and temporal variation of the intestinal bacterial community in commercially raised broiler chickens during growth. Microb Ecol 44:286–293. https://doi.org/10.1007/s00248-002-2015-y

Van Krimpen MM, Kwakkel RP, Van Der Peet-Schwering CMC, Den Hartog LA, Verstegen MWA (2011) Effects of dietary energy concentration, nonstarch polysaccharide concentration, and particle sizes of nonstarch polysaccharides on digesta mean retention time and gut development in laying hens. Br Poult Sci 52:730–741. https://doi.org/10.1080/00071668.2011.638620

Villageliu DN, Lyte M (2017) Microbial endocrinology: why the intersection of microbiology and neurobiology matters to poultry health. Poult Sci 96:2501–2508. https://doi.org/10.3382/ps/pex148

Vispo, C., Karasov, W.H., 1997. The interaction of avian gut microbes and their host: an elusive symbiosis., in: Gastrointestinal microbiology, Mackie R White B. Springer, New York, NY, pp. 116–155

Waite DW, Taylor MW (2015) Exploring the avian gut microbiota: current trends and future directions. Front Microbiol 6. https://doi.org/10.3389/fmicb.2015.00673

Wang L, Lilburn M, Yu Z (2016) Intestinal microbiota of broiler chickens as affected by litter management regimens. Front Microbiol 7. https://doi.org/10.3389/fmicb.2016.00593

Wickramasuriya SS, Park I, Lee K, Lee Y, Kim WH, Nam H, Lillehoj HS (2022) Role of physiology, immunity, microbiota, and infectious diseases in the gut health of poultry. Vaccine 10:172. https://doi.org/10.3390/vaccines10020172

Wilkie DC, Van Kessel AG, White LJ, Laarveld B, Drew MD (2005) Dietary amino acids affect intestinal Clostridium perfringens populations in broiler chickens. Can J Anim Sci 85:185–193. https://doi.org/10.4141/A04-070

Xu J, Mahowald MA, Ley RE, Lozupone CA, Hamady M, Martens EC, Henrissat B, Coutinho PM, Minx P, Latreille P, Cordum H, Van Brunt A, Kim K, Fulton RS, Fulton LA, Clifton SW, Wilson RK, Knight RD, Gordon JI (2007) Evolution of symbiotic bacteria in the distal human intestine. PLoS Biol 5:e156. https://doi.org/10.1371/journal.pbio.0050156

Yadav S, Jha R (2019) Strategies to modulate the intestinal microbiota and their effects on nutrient utilization, performance, and health of poultry. J Anim Sci Biotechnol 10:2. https://doi.org/10.1186/s40104-018-0310-9

Yu Z, Morrison M (2004) Comparisons of different hypervariable regions of RRS genes for use in fingerprinting of microbial communities by PCR-denaturing gradient gel electrophoresis. Appl Environ Microbiol 70:4800–4806. https://doi.org/10.1128/AEM.70.8.4800-4806.2004

Zoetendal EG, Collier CT, Koike S, Mackie RI, Gaskins HR (2004) Molecular ecological analysis of the gastrointestinal microbiota: a review. J Nutr 134:465–472. https://doi.org/10.1093/jn/134.2.465

Zoetendal EG, Rajilic-Stojanovic M, de Vos WM (2008) High-throughput diversity and functionality analysis of the gastrointestinal tract microbiota. Gut 57:1605–1615. https://doi.org/10.1136/gut.2007.133603

Secret Killers in Poultry as Drivers for Intestinal Inflammation and Oxidative Stress

3

Shereen Basiouni, Awad A. Shehata, Wolfgang Eisenreich, Guillermo Tellez-Isaias, and Helen L. May-Simera

Contents

S. Basiouni (✉)
Cilia Cell Biology, Institute of Molecular Physiology, Johannes-Gutenberg University, Mainz, Germany
e-mail: sbasiouni@uni-mainz.de

A. A. Shehata · W. Eisenreich · H. L. May-Simera
Structural Membrane Biochemistry, Bavarian NMR Center, Technical University of Munich (TUM), Garching, Bayern, Germany
e-mail: awad.shehata@tum.de; may-simera@uni-mainz.de

G. Tellez-Isaias
Division of Agriculture, Department of Poultry Science, University of Arkansas, Fayetteville, AR, USA

Abstract

Stress can trigger inflammation and provoke adverse reactions. In poultry, chronic stress may be caused by various causes, including dysbiosis. Heat stress, mycotoxins, and an oxidized diet can also reduce the animal's performance and heighten its vulnerability to infections. Heat stress can disrupt gut–tight junctions and generate free radicals, such as reactive oxygen species, which results in increased intestinal permeability, endotoxemia, and systemic inflammation. Historically, the ban on using antibiotics as growth promoters is a practical employed to manage pathological conditions. This situation has prompted the necessity to explore alternative approaches to antimicrobials for the preservation of gut health. This review highlights the key elements that might cause chronic stress in poultry. Additionally, we shed light on the potential negative impacts on animal health and the link between chronic stress and the emergence of antimicrobial resistance, which can threaten public health.

Keywords

Chronic stress · Heat stress · Endotoxins · Mycotoxins · Dysbiosis · Oxidized diet · Antimicrobial resistance · Oxidative stress biomarkers

1   Introduction

Animal studies have extensively demonstrated the pivotal roles played by the interactions between dietary components, intestinal microbiota, and immunity in various health conditions. These conditions encompass metabolic and gut disorders, diabetes, cancers, autoimmune diseases, malnutrition, obesity, myopathies, cardiovascular issues, muscle function, and even neurological diseases in humans (Gostner et al. 2013; Fasano 2020).

Interestingly, these metabolic diseases and neurological pathologies in humans, such as autism, schizophrenia, bipolar disorder, dementia, Alzheimer's disease, Parkinson's disease, epilepsy, and stroke, exhibit strong associations with mitochondrial problems (Cryan and Dinan 2012). Consequently, it is imperative to recognize chronic oxidative stress as "the silent killer" of humans and animals (Stecher 2015).

In the realm of animal farming, it is important to investigate the consequences of chronic stressors, including heat stress, dysbiosis, leaky gut, mycotoxins, endotoxins, and oxidized diet. These factors not only diminish the animals' performance but also stimulate their vulnerability to infections (Shehata et al. 2022b).

Specifically, in poultry, heat stress induces disruption of gut-tight junctions and the generation of reactive oxygen species (ROS). This can impact the permeability of the intestine, which subsequently causes systemic inflammation (Yu et al. 2006), triggers hepatic inflammation (Liu et al. 2022), reduction of calcium absorption, and impede the metabolism of vitamin D3 (cholecalciferol) to 1,25-dihydroxy vitamin D3 (1,25 (OH)2D3) (Petruk and Korver 2004) and contribute to growth retardation (Saracila et al. 2021). This cascade of effects also amplifies the necessity for substitutes to antimicrobials to control these pathological conditions and enhance animal performance.

2   Oxidative Stress

Free radicals are perpetually present in cells throughout normal oxygen consumption (Estévez 2015). These substances play roles as signaling molecules crucial for maintaining homeostasis. However, under extreme stress conditions, the ROS levels may escalate, leading to several pathological disorders, including lipid peroxidation. Cell membranes, DNA damage, and modulation of small GTPases are common results of high ROS levels (Ferro et al. 2012; Pizzino et al. 2017; Shehata et al. 2022a).

Consequently, these processes have a crucial role in the emergence of chronic stress manifestations. The intricate interplay between stress and cellular functions underscores the need to maintain a balance in ROS levels for overall cellular health and well-being. Reactive species are classified into two categories: (i) ROS consists of a group of free radicals and (ii) Reactive nitrogen species (RNS) (Costa et al. 2021), Fig. 3.1.

Reactive species are primarily produced by two sources: electron transport and the nicotinamide adenine dinucleotide phosphate oxidases (NADPH oxidase or NOX). The electron transport is found in the chain in the mitochondria, whereas NOX is present in the cell membrane. Indeed, seven transmembrane enzymes are found in the cell membrane, including dual oxidase 1 (DUOX1), DUOX2, and NOX1–NOX5 (Bedard and Krause 2007; Brandes et al. 2014).

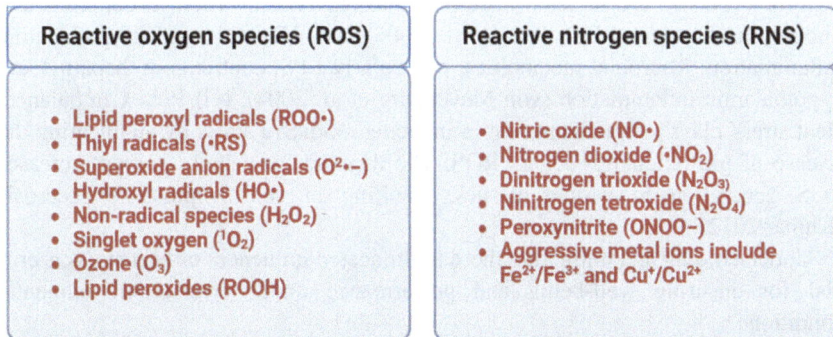

Reactive oxygen species (ROS)	Reactive nitrogen species (RNS)
• Lipid peroxyl radicals (ROO·) • Thiyl radicals (·RS) • Superoxide anion radicals ($O_2{}^{·-}$) • Hydroxyl radicals (HO·) • Non-radical species (H_2O_2) • Singlet oxygen ($^1O_2$) • Ozone (O_3) • Lipid peroxides (ROOH)	• Nitric oxide (NO·) • Nitrogen dioxide (·NO_2) • Dinitrogen trioxide (N_2O_3) • Ninitrogen tetroxide (N_2O_4) • Peroxynitrite (ONOO–) • Aggressive metal ions include Fe^{2+}/Fe^{3+} and Cu^+/Cu^{2+}

Fig. 3.1 Reactive species: ROS and RNS

3 Potential Causes of Chronic Stress in Poultry

3.1 Heat Stress

High temperature is one of the major stressors in chickens (Lara and Rostagno 2013). It has several negative impacts, particularly for poultry bred in tropical and subtropical regions, where temperatures are persistent, and Formularbeginn moderate climate zones encompassing central and eastern Europe. In these diverse geographical settings, these negative impacts are critical for successful and sustainable poultry production (Hirakawa et al. 2020). An animal experiences heat stress when its physiological heat-dissipation mechanisms—sweating, breathing, or panting—are unable to cope with ambient temperature above its thermoneutral zone (Mount 1978). Heat stress can cause several pathological conditions in poultry and negatively impacts performance, underscoring the need for meticulous management approaches to alleviate its impact across diverse climates. Implementing careful and targeted approaches is crucial to safeguard the well-being and productivity of poultry subjected to elevated temperatures. Chickens are particularly vulnerable to elevated ambient temperatures resulting from the absence of sweat glands and due to significant heat production—features that set them apart from mammals. This vulnerability makes them particularly prone to the challenges posed by elevated temperatures in their environment. To regulate their body temperature, chickens primarily employ panting to expel excess heat (Lasiewski 1969). Heat stress can have several impacts on chickens in more ways than just discomfort; it also negatively impacts several vital physiological processes, including TJs, mucus barrier, and gut immunity (Ortega and Szabó 2021), Fig. 3.2. These effects include (i) Reduction of mucin: Reducing goblet cell number and inhibiting mucin release, leading to thinning protective mucin layers and bacterial penetration (Yi et al. 2020). (ii) Disruption of TJs: Alteration of TJs proteins, including occludin, claudins, and zonula occludens-1, -2, and -3 (Dokladny et al. 2016; Yi et al. 2020). (iii) Impairment of the intestinal barriers: Impairment of TJs and intestinal barriers increases the permeability of the intestine (Wu et al. 2018). (iv) Systemic inflammation: Inflammatory cytokines, including interleukins (IL-1β and IL-6) and tumor necrosis factor-α (TNF-α), are secreted in response to bacterial invasion, endotoxins, and lipopolysaccharides (LPS) (Yu et al. 2006). (v) Hepatitis and Hypothalamic Inflammation: Microbial metabolites, particularly LPS, contribute to hepatitis and hypothalamic inflammation (von Meyenburg et al. 2004). (vi) Redox Imbalance: Heat stress plays a significant role in inducing oxidative stress by augmenting the release of pro-oxidants, such as ROS. The demand for cellular energy increases ROS generation in various tissues, resulting in cell apoptosis or necrosis (Kumar 2012).

Understanding and mitigating these multifaceted influences of heat stress is crucial for ensuring well-being and performance under diverse environmental conditions.

Heat-induced oxidative stress has profound implications for the intestinal environment, disrupting the intestine integrity and influencing various cellular

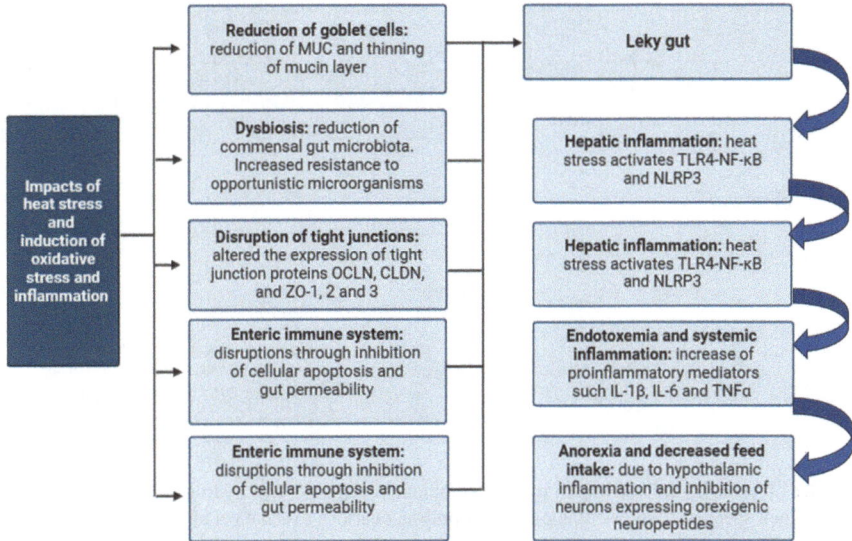

Fig. 3.2 Negative impacts of heat stress. *OCLN* occludin; *CLDN* claudins; *ZO* zonula occludens; *TLR4* toll-like receptor 4; *NF-κB* nuclear factor-kappa B; *IL* interleukin; *TNF-α* tumor necrosis factor-α; *SOD* superoxide dismutase 1; *GSH* glutathione. Adapted from Shehata et al. (2022b)

processes. ROS during heat stress contributes to increased intestinal permeability, a phenomenon that enables the migration and action of bacteria and their molecular patterns, such as LPS, from the gut. This disruption in the intestinal barrier function is commonly referred to as "leaky gut syndrome" (Quinteiro-Filho et al. 2010) (Fig. 3.2). The consequences of this increased permeability have far-reaching effects on gut health, systemic inflammation, and overall animal well-being. Efficient management strategies are essential to address the challenges heat stress poses and its impact on intestinal function in poultry production (Shehata et al. 2022b).

3.2 Dysbiosis

The success of poultry production heavily relies on the integrity and functionality of the intestine, crucial for optimizing nutrient absorption and facilitating growth, both of which directly impact overall animal performance. The gastrointestinal microbiota of poultry is predominantly composed of bacteria, fungi, and protozoa. The gut microbiota contains bacteria that also work together to produce ROS, which act as messengers in cellular communication. The intestinal epithelial cells are connected by tight junctions that act as a barrier to keep microorganisms from entering the host organism (Ulluwishewa et al. 2011).

Dysbiosis occurs when an imbalance of microbiota affects the relationship between the host and the microbe (Shehata et al. 2022b). This alteration can increase microbial metabolites (Fig. 3.3), contributing to oxidative damage and

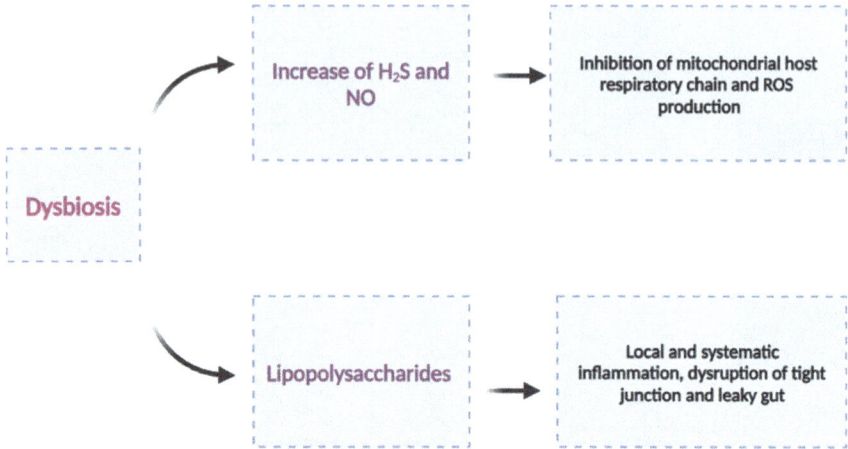

Fig. 3.3 Dysbiosis-derived inflammation and oxidative damage. *H₂S* hydrogen sulfide; *ROS* reactive oxygen species; *IL* interleukins; *LPS* lipopolysaccharides (Shehata et al. 2022b)

inflammatory conditions. It underscores the intricate interaction between microbiota and the host, emphasizing the critical role of maintaining a balanced and healthy gut environment for optimal poultry performance.

Specifically, in the context of the gut epithelium, several stressors enhance the generation of ROS, causing an imbalance in the redox system and inducing inflammation. These stressors include (Mishra and Jha 2019): (i) By converting nitrite and nitrate, the gut microbiota produces NO. Excessive production of NO brought on by dysbiosis produces ROS, which is linked to cellular damage (Oleskin and Shenderov 2016). This damage occurs, for example, due to inhibiting the host's mitochondrial respiratory chain (Tse 2017). (ii) Certain bacteria in the intestine, including *E. coli*, produce large amounts of H₂S by breaking down sulfur-containing peptides and amino acids in the intestine. When there is an imbalance in the gut bacteria (dysbiosis), H₂S increases, which can inhibit cytochrome oxidase. This, in turn, suppresses the respiratory chain of the host's mitochondria, causing the overproduction of cytokines (Saint-Georges-Chaumet and Edeas 2016). Dysbiosis can generate a new nutritional niche that promotes the replications of pathogenic bacteria, even though the cecal mucosa can detoxify H₂S (Winter et al. 2010; Tsolis and Bäumler 2020). (iii) Short-chain fatty acids (SCFAs): The tricarboxylic acid (TCA) cycle can be stimulated by SCFAs, especially butyrate. SCFAs also promote the expression of the signaling hormone GLP-1 and anti-inflammatory IL-10 cytokines, leading to decreased energy intake (Saint-Georges-Chaumet and Edeas 2016). (iv) LPS: When there is dysbiosis, LPS produced by Gram-negative bacteria increases. This increase leads to local and systemic inflammation by stimulating the gut epithelial cells and macrophages. As a result, tight junctions get damaged, which causes leaky gut syndrome. The complex interactions outlined here demonstrate how dysbiosis can trigger a chain of events, disturbing the redox balance, cellular function, and inflammatory reactions within the gastrointestinal epithelial cells (Gilani et al. 2021).

3.3   Mycotoxins-Derived Inflammation and Oxidative Damages

The most prevalent mycotoxins that influence poultry include polar mycotoxins (aflatoxin B1 and fumonisin B1) and non-polar mycotoxins (ochratoxin A, T-2 toxin, deoxynivalenol, and zearalenone). Mycotoxins can harm various aspects of the gut, such as digestion, absorption, the immune system, and the gut microbiota. Even at low levels, their constant presence can negatively affect intestinal balance and reduce performance.

Mycotoxins induce oxidative damage and inflammation by downregulation of intracellular antioxidants and upregulation of pro-inflammatory cytokines (Longobardi et al. 2020). These effects impact membrane integrity, potentially resulting in invasion of the intestinal epithelium by opportunistic microorganisms due to a leaky gut (Wang et al. 2017; Liu et al. 2022). Consequently, there is an immediate need for antioxidants and anti-inflammatory substances in poultry subjected to mycotoxins. The mechanism of mycotoxins-derived inflammation and oxidative damage is shown in Fig. 3.4 (da Silva et al. 2018).

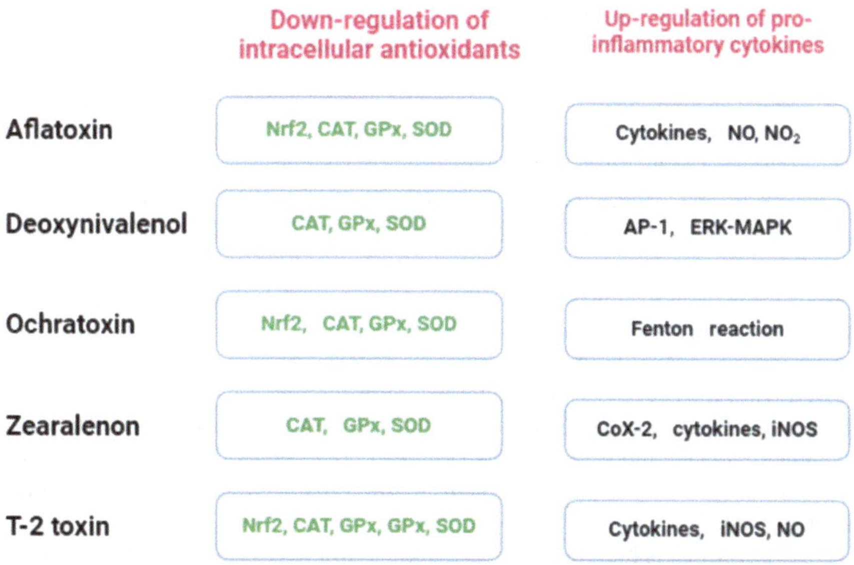

	Down-regulation of intracellular antioxidants	Up-regulation of pro-inflammatory cytokines
Aflatoxin	Nrf2, CAT, GPx, SOD	Cytokines, NO, NO$_2$
Deoxynivalenol	CAT, GPx, SOD	AP-1, ERK-MAPK
Ochratoxin	Nrf2, CAT, GPx, SOD	Fenton reaction
Zearalenon	CAT, GPx, SOD	CoX-2, cytokines, iNOS
T-2 toxin	Nrf2, CAT, GPx, GPx, SOD	Cytokines, iNOS, NO

Fig. 3.4 Mycotoxins-derived inflammation and oxidative damages. *Nrf2* erythroid 2-related factor 2, *CAT* catalase, *GPx* glutathione peroxidase, *SOD* superoxide dismutase. *NO* nitric oxide, *NO2* nitrogen dioxide, *AP-1* activator protein 1, *ERK-MAPK* extracellular signal-regulated kinase-mitogen-activated protein kinase, *CoX-2* cyclooygenase-2, *iNOS* inducible nitric oxide synthetase. Adapted from (Shehata et al. 2022b)

3.4   Diet

Diet and nutritional imbalance can cause intestinal inflammation and oxidative stress. Low-quality or contaminated feed could be a source of toxic compounds. In this part, we will highlight the role of rancid fats, non-starch polysaccharides (NSPs), and low protein quality in the induction of inflammation and oxidative stress in poultry.

3.4.1   Rancid Fats

Lipids are necessary in chicken feed to boost energy content and promote growth. However, the quality of oils and fats impacts avian health, particularly the gut. Oils rich in polyunsaturated fatty acids (PUFAs) are frequently incorporated into poultry diets as an energy source, enhancing pellet quality, palatability, and the absorption of fat-soluble vitamins (Chen et al. 2016). Unfortunately, PUFAs can be oxidized more quickly than saturated fats, which increases their susceptibility to rancidity. Reactive double bonds, which let molecular oxygen react with these bonds, are the source of this enhanced oxidation (Alagawany et al. 2019). Additionally, exposure to light, catalytic transition metal ions, and high temperatures during feed pelleting and storage can generate free radicals, which initiate lipid autoxidation (St. Angelo 1992).

Lipid oxidation produces reactive compounds with deleterious biological consequences and an unpleasant odor (Ben Hammouda et al. 2017). Toxic and physiologically harmful oxidation products can come from mild oxidation. Lipids decompose into peroxides, aldehydes, and polar chemicals through peroxidation. The body processes and absorbs these compounds in different ways. The temperature, length of the thermal processing stages, and oil content are some variables that affect the peroxidation result. Poultry may experience in vivo metabolic oxidative stress if fed peroxidized oils deficient in endogenous antioxidants (Boler et al. 2012).

Oxidative stress prevents ROS and free radicals from being efficiently changed into less reactive species, which causes the free radicals to attach to lipids, proteins, and DNA, causing tissue damage (Kalyanaraman 2013). Oxidized oils generate reactive aldehydes to build up in the stomach and are subsequently absorbed into the small intestine, concentrated, and broken down in the liver, according to studies (Kanazawa and Ashida 1998). Elevated thiobarbituric acid reactive substances (TBARS) in the blood plasma and tissues of broiler chicks fed oxidized oils suggested slower development rates and higher levels of lipid degradation. They also exhibited reduced antioxidant levels (Engberg et al. 1996). This emphasizes how important it is to control food ingredients to lower oxidative stress in chicken development properly.

Fish oil, which is high in PUFA, has been shown to promote growth and intestinal health (Chen et al. 2016). However, oxidation poses a risk because oxidized fish oil can harm birds due to the formation of peroxidation by-products such as hydroxyperoxide, aldehydes, ketones, dicarbonyls, furans, and hydrocarbons (Shibamoto 2006; Yangilar 2016).

It has been demonstrated that adding oxidized poultry fat to broiler feed negatively affects body weight, hematocrit, enterocyte life span, hepatocyte proliferation, and the efficiency of secretory IgA in the gut. Additionally, GALT (gut-associated lymphoid tissue) is negatively impacted by oxidized oil (Kubow 1990), causing oxidative stress in the jejunum and impacting cytokine expression and immune cells. Birds fed with oxidized oil experience peroxidation of their intestinal mucosa, which causes a decline in their antioxidative capacity and hinders the adequate removal of ROS. Furthermore, there is a reduction in tight junction proteins and an elevation in pro-inflammatory molecules, which leads to the compromise of the intestinal epithelial barrier integrity and triggers an inflammatory response.

Other studies showed that oxidized oil can cause oxidative stress, decrease enterocyte half-life, poor immunological response, inflammation, and epithelial barrier loss in chicken intestines. Even with a modest degree of oxidation, such as a 3.14-meqO$_2$/kg dietary peroxide value (POV) (Liang et al. 2015). For optimal bird performance and a healthy gut, it is important to consider the quality of lipid sources and high-fat foods carefully. Moreover, using the correct technique to evaluate the quality of oil or fat is crucial, as measuring only hydroperoxide molecules can lead to incorrect interpretations. Hydroperoxide molecules are intermediate compounds converted into end products during the process. Therefore, a thorough examination of lipid quality is necessary to minimize potential risks linked to oxidative stress in poultry (Shibamoto 2006).

3.4.2 Non-starch Polysaccharides

Non-starch polysaccharides (NSPs), a key component of plant cell walls, are essential in poultry diets, but their effects on intestinal health are complex. The NSPs include cellulose, non-cellulosic polysaccharides, and pectin polymers. They can also be categorized into soluble (arabinoxylans, glucans, fructans, pectins, and hemicelluloses) and insoluble (arabinoxylans, glucans, fructans, pectins, and hemicelluloses) (Bailey 1973). Poultry have a shorter retention time in the crop, limiting NSPs digestion (Bederska-Łojewska et al. 2017). NSPs can negatively impact gut health through several mechanisms: (i) Soluble NSPs form hydrocolloids in the intestine, reducing exposure to digestive secretions and enzymes, thus impeding the process of digestion and absorption (Bederska-Łojewska et al. 2017). This leads to increased digesta viscosity, decreased body weight gain, and suppression of the feed conversion ratio in broilers (Cardoso et al. 2014). Elevated digesta viscosity also prolongs gastric transit, reducing feed intake and overall transit time, fostering undesirable bacterial proliferation of, e.g., *E. coli* and *Clostridium perfringens*. (ii) High NSP diets may contribute to bacterial translocation, potentially causing systemic infection and inflammation due to a leaky gut (Latorre et al. 2015a; Tellez et al. 2015).

Cereals rich in NSPs have varying ratios of arabinoxylans, cellulose, and β-glucans (Choct 2015). Barley, wheat, rye, triticale, and oats are considered viscous cereals due to their soluble NSP content. The use of these cereals in poultry feed, particularly for broilers, requires careful consideration of the inflammatory effects of soluble fiber. Mitigating strategies include incorporating feed probiotics-producing enzymes like phytase, lipase, xylanase, and cellulases, adding insoluble fiber, and supplementing exogenous enzymes such as xylanase and β-glucanase to break down NSPs (Latorre et al. 2015b). These practices aim to alleviate the negative influences of higher NSP content in the diet and enhance overall intestinal health in poultry (Bederska-Łojewska et al. 2017).

3.4.3 Protein

In humans, high-protein food has been linked to inflammatory bowel illnesses such as ulcerative colitis and Crohn's disease (Neuman and Nanau 2012). Studies on broilers also showed that protein content and source substantially impact gut microbiota and shape. Increased indigestible protein, especially from rapeseed meal, lowers volatile fatty acid concentrations, increases protein fermentation products, and changes gut shape by lowering villus height and increasing crypt depths (Qaisrani et al. 2014). Furthermore, high-crude protein diets, particularly those containing animal ingredients, have been linked to conditions such as necrotic enteritis and increased colonization of pathogenic bacteria such as *C. perfringens* in the ileum and ceca (Drew et al. 2004; Dahiya et al. 2006).

Changing broiler diets' protein levels significantly impacts fecal bacterial counts. Aerobic mesophilic bacteria and *E. coli* counts are reduced when crude protein levels are reduced (Drew et al. 2004; Dahiya et al. 2006). Furthermore, the protein source in the diet affects *C. perfringens* colonization, with meat/bone meal, fish meal, feather meal, or potato protein fostering greater bacterial counts than corn gluten meal, soy, pea protein concentrates, or control diets (Wilkie et al. 2005). The presence of indigestible protein and unabsorbed amino acids in the lower digestive system, particularly in the ileum and ceca, promotes the growth of harmful microorganisms. Amino acids such as methionine and glycine have been shown to promote *C. perfringens* (Muhammed et al. 1975). The higher glycine content in animal ingredients relative to crude protein than in vegetal protein sources may be a factor in the increased growth of *C. perfringens* in diets containing animal-derived proteins (AminoDat 2001).

The link between high-protein diets and intestinal inflammation in humans extends to broilers, where diets high in crude protein, indigestible protein, and animal components might harm the GIT. Protein content and source should be taken into consideration to improve gut health. Readily digested vegetal foods and perhaps adding proteases to promote protein digestion in cost-effective methods should also be considered.

4 Biomarkers for Oxidative Stress and Inflammation in Poultry

The short half-life of ROS makes it challenging to assess oxidative stress using direct measurements of ROS and RNS (Poljsak et al. 2013). Several indirect biomarkers can be employed to assess the intestinal health of chickens (Shehata et al. 2022a).

Biomarkers for lipid peroxidation, including malondialdehyde (MDA), thiobarbituric acid reactive substances, isoprostanes, and 4-hydroxyalkenals, including 4-hydroxynonenal (Celi 2011). Isoprostanes are considered the most reliable markers for lipid peroxidation, as they represent distinctive end products resulting from the peroxidation of PUFAs (Montuschi et al. 2004).

Biomarkers indicating protein oxidation, such as carbonyl moieties in the side chains of amino acids, can be identified using ELISA, Western blot, or FPLC/HPLC (Dalle-Donne et al. 2003).

Biomarkers for DNA oxidation, including 8-hydroxy-20-deoxyguanosine, can be assessed by comet assays (Collins 2014).

5 The Contribution of Secret Killers in the Emergence of Antibiotic Resistance

Overall, secret killers are risk factors for several pathological disorders that may escalate the use of antibiotics in the chicken business, therefore endangering the public health of both humans and animals (Benrabia et al. 2020). Indeed, antibiotic usage is connected with the emergence of antimicrobial resistance (AMR). This makes sense because the gut microbiome is home to commensal and opportunistic microorganisms. Commensal bacteria frequently carry antibiotic-resistant genes, which might lead to the emergence of resistant clones of opportunistic diseases (Ellabaan et al. 2021; Forster et al. 2022). Antibiotic use eliminates susceptible bacteria, allowing resistant ones to persist and propagate through vertical inheritance or horizontal gene transfer, further expanding the reservoir of resistant genes (Lamberte and van Schaik 2022). Figure 3.5 shows the link between the secret killers in poultry and AMR acquisition.

6 Feed Additives Against Secret Killers in Poultry

The application of antibiotics in animals has undergone substantial changes in the last several decades because of three primary causes: (i) the development of antibiotic resistance, which can pose "One Health" problems (Zaman et al. 2017). (ii) Extended usage or misuse of antibiotics can leave behind antibiotic residues in animal-derived food, which can cause allergic responses in humans, a misbalance of microbiota in the gut, and the emergence of antibiotic resistance in human pathogens (Muaz et al. 2018). (iii) Organic poultry farming is becoming more popular,

Fig. 3.5 Procuring antibiotic resistance in poultry and its hidden killers, chronic stress. Removing susceptible strains from the gastrointestinal system following antibiotic administration will lead to the development of a resistant gene reservoir. ©adapted from Shehata et al. (2022a)

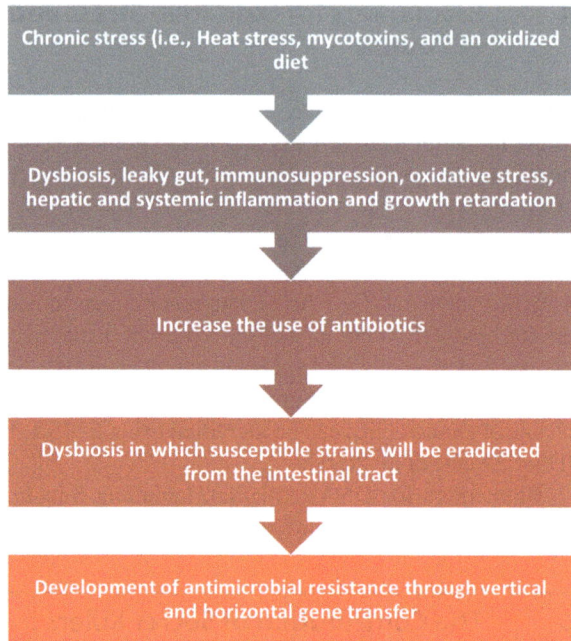

Chronic stress (i.e., Heat stress, mycotoxins, and an oxidized diet)

Dysbiosis, leaky gut, immunosuppression, oxidative stress, hepatic and systemic inflammation and growth retardation

Increase the use of antibiotics

Dysbiosis in which susceptible strains will be eradicated from the intestinal tract

Development of antimicrobial resistance through vertical and horizontal gene transfer

and production approaches that incorporate antibiotic-free management are part of this trend.

There is a trend for producing organic poultry farming that integrates antibiotic-free management into their production systems (Hafez and Shehata 2021). These developments have sped up research into substitute feed additives that can achieve some of the antibiotics' beneficial effects.

Due to the prohibition of sub-therapeutic antibiotics use in Europe and the USA, the incidence of enteric diseases in animals has significantly increased (Hao et al. 2014; Park et al. 2016), leading to a need for antibiotics at therapeutic doses for controlling and preventing diseases. This highlights the urgent need to explore antibiotic alternatives in poultry production (Fig. 3.6).

Candidates for alternative feed additives have undergone in-depth studies, have been created for real-world application, and are being actively marketed. While commercial uses are becoming more widespread, consistent benefits are not always guaranteed. For instance, the spotlight on reducing the development of foodborne pathogens in the GIT may not always result in improved broiler growth and efficiency or optimal layer hen egg production. This phenomenon is not surprising, considering that alternative feed additives might have more than one function, differ in structure, and have distinct bacterial targets and modes of action. Finding innovative methods to enhance growth and health practices has emerged as the main area of research. Dietary modification of feed additives, such as postbiotics, probiotics, prebiotics, and herbal items, might improve the growth and health of birds (Klemashevich et al. 2014, 2014; Arain et al. 2018; Nabi et al. 2020). As a result,

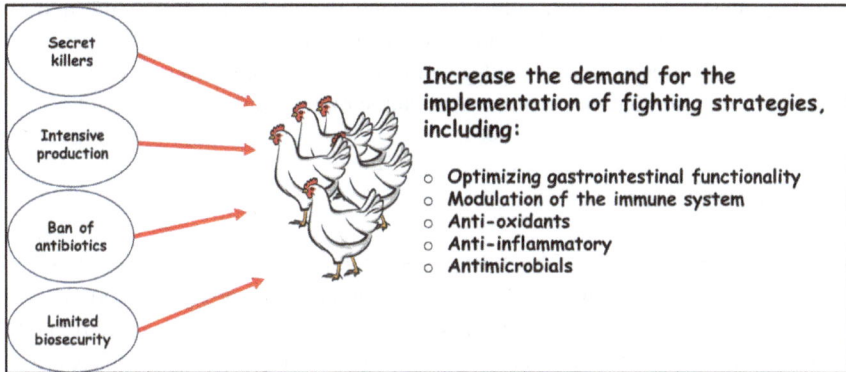

Fig. 3.6 Factors that necessitate the need to enhance gut health using ATA

these feed additives have been thoroughly studied as alternatives in recent years (Loh et al. 2014; Kareem et al. 2015).

Current research suggests that dietary approaches that counteract the debilitating effects of stress and chronic inflammation may occasionally be a useful substitute for antibiotics. Enhancing disease resistance in non-antibiotic-maintained animals has been proven advantageous to their health, welfare, and productivity. It is also a crucial strategy for improving the microbiological safety of animal products. Recent international legislation and growing consumer demands to ban growth-promoting antibiotics and restrict the therapeutic use of antimicrobials have prompted research and development of alternative feed additives, including probiotics, prebiotics, enzymes, short- and medium-chain fatty acids, and phytochemicals.

Thus, postbiotics essentially have all of the advantages of probiotics. Enhancing gut health is one of postbiotics' primary benefits for animal feeding. Other benefits include the prevention of the spread of infections, which allows the animal to consume the right nutrients, speeds up growth, and improves health.

Phytogenics, with their anti-inflammatory, antibacterial, antiviral, antifungal, immune-modulatory, and barrier integrity-enhancing properties, are essential in maintaining a healthy gut (Shehata et al. 2022a; Tellez-Isaias et al. 2022). Although natural alternatives to synthetic medications, such as antibiotic growth promoters (AGP), have been studied to enhance performance and gut health, there is no "magic bullet" for avoiding chronic stress-related diseases (Shehata et al. 2022a; Tellez-Isaias et al. 2022). The first paradigm of using alternatives to antimicrobials focusing only on the bioactive substances that possess antimicrobial activities is the main cause of missing opportunities. Thus, antimicrobial activity is not bioactive substances' sole mechanism of action. Additionally, the oversimplification and lack of science in which the industry can recommend bioactive substances for treating acute infections results in consumer dissatisfaction. However, nowadays, the second generation of alternatives to antimicrobials focuses on the host-mediated effects, not pathogens. This paradigm is interested in using antimicrobials for several beneficial outcomes, including modulation of host immunity modulation of and restoring intestinal microbiota, physiology, and metabolism, which subsequently increase the emergence of AMR and improve animal performance. This paradigm provides a clear opportunity for next-generation phytotherapy. The next book chapters will discuss alternatives to antimicrobials in more detail.

7 Conclusion

Altogether, stress is linked with a host of detrimental health outcomes, such as dysbiosis (an unbalanced intestinal bacterial population), systemic inflammation brought on by cytokines secretion, and "leaky gut syndrome," a condition that emerges when oxidative damage compromises the barrier function of the lining intestinal epithelial cells. ROS and RNS have a short half-life, which makes it challenging to evaluate oxidative stress directly. Nevertheless, various indirect biomarkers can be utilized to evaluate the health status of poultry intestines. Over the past few decades, the utilization of antibiotics in animals has undergone significant transformations due to the rise of antimicrobial resistance, and concerns about antibiotic residues in animal-derived food, and the increased demand for producing organic poultry. These challenges have sped up research into searching for alternative antimicrobials that can restore gut health. "Omics" methodologies may contribute to a better understanding of the mechanism of action of these alternatives.

References

Alagawany M, Elnesr SS, Farag MR, Abd El-Hack ME, Khafaga AF, Taha AE, Tiwari R, Yatoo M, Bhatt P, Khurana SK, Dhama K (2019) Omega-3 and Omega-6 fatty acids in poultry nutrition: effect on production performance and health. Animals 9:573. https://doi.org/10.3390/ani9080573

AminoDat TM (2001) In: Kennesaw GA (ed) Degussa feed additives. Degussa Corporation

Arain MA, Mei Z, Hassan FU, Saeed M, Alagawany M, Shar AH, Rajput IR (2018) Lycopene: a natural antioxidant for prevention of heat-induced oxidative stress in poultry. Worlds Poult Sci J 74:89–100. https://doi.org/10.1017/S0043933917001040

Bailey RW (1973) In: Butler GW (ed) Chemistry and biochemistry of herbage. Academic Press, London, UK, pp 157–211

Bedard K, Krause K-H (2007) The NOX Family of ROS-generating NADPH oxidases: physiology and pathophysiology. Physiol Rev 87:245–313. https://doi.org/10.1152/physrev.00044.2005

Bederska-Łojewska D, Świątkiewicz S, Arczewska-Włosek A, Schwarz T (2017) Rye non-starch polysaccharides: their impact on poultry intestinal physiology, nutrients digestibility and performance indices—a review. Ann Anim Sci 17:351–369. https://doi.org/10.1515/aoas-2016-0090

Ben Hammouda I, Zribi A, Ben Mansour A, Matthäus B, Bouaziz M (2017) Effect of deep-frying on 3-MCPD esters and glycidyl esters contents and quality control of refined olive pomace oil blended with refined palm oil. Eur Food Res Technol 243:1219–1227. https://doi.org/10.1007/s00217-016-2836-4

Benrabia I, Hamdi TM, Shehata AA, Neubauer H, Wareth G (2020) Methicillin-resistant Staphylococcus aureus (MRSA) in poultry species in Algeria: long-term study on prevalence and antimicrobial resistance. Vet Sci 7:54. https://doi.org/10.3390/vetsci7020054

Boler DD, Fernández-Dueñas DM, Kutzler LW, Zhao J, Harrell RJ, Campion DR, McKeith FK, Killefer J, Dilger AC (2012) Effects of oxidized corn oil and a synthetic antioxidant blend on performance, oxidative status of tissues, and fresh meat quality in finishing barrows. J Anim Sci 90:5159–5169. https://doi.org/10.2527/jas.2012-5266

Brandes RP, Weissmann N, Schröder K (2014) Nox family NADPH oxidases: molecular mechanisms of activation. Free Radic Biol Med 76:208–226. https://doi.org/10.1016/j.freeradbiomed.2014.07.046

Cardoso V, Ferreira AP, Costa M, Ponte PIP, Falcão L, Freire JP, Lordelo MM, Ferreira LMA, Fontes CMGA, Ribeiro T (2014) Temporal restriction of enzyme supplementation in barley-based diets has no effect in broiler performance. Anim Feed Sci Technol 198:186–195. https://doi.org/10.1016/j.anifeedsci.2014.09.007

Celi P (2011) Biomarkers of oxidative stress in ruminant medicine. Immunopharmacol Immunotoxicol 33:233–240. https://doi.org/10.3109/08923973.2010.514917

Chen J-R, Chen Y-L, Peng H-C, Lu Y-A, Chuang H-L, Chang H-Y, Wang H-Y, Su Y-J, Yang S-C (2016) Fish oil reduces hepatic injury by maintaining normal intestinal permeability and microbiota in chronic ethanol-fed rats. Gastroenterol Res Pract 2016:1–10. https://doi.org/10.1155/2016/4694726

Choct M (2015) Feed non-starch polysaccharides for monogastric animals: classification and function. Anim Prod Sci 55:1360. https://doi.org/10.1071/AN15276

Collins AR (2014) Measuring oxidative damage to DNA and its repair with the comet assay. Biochim Biophys Acta 1840:794–800. https://doi.org/10.1016/j.bbagen.2013.04.022

Costa M, Sezgin-Bayindir Z, Losada-Barreiro S, Paiva-Martins F, Saso L, Bravo-Díaz C (2021) Polyphenols as antioxidants for extending food shelf-life and in the prevention of health diseases: encapsulation and interfacial phenomena. Biomedicines 9:1909. https://doi.org/10.3390/biomedicines9121909

Cryan JF, Dinan TG (2012) Mind-altering microorganisms: the impact of the gut microbiota on brain and behaviour. Nat Rev Neurosci 13:701–712. https://doi.org/10.1038/nrn3346

da Silva EO, Bracarense APFL, Oswald IP (2018) Mycotoxins and oxidative stress: where are we? World Mycotoxin J 11:113–134. https://doi.org/10.3920/WMJ2017.2267

Dahiya JP, Wilkie DC, Van Kessel AG, Drew MD (2006) Potential strategies for controlling necrotic enteritis in broiler chickens in post-antibiotic era. Anim Feed Sci Technol 129:60–88. https://doi.org/10.1016/j.anifeedsci.2005.12.003

Dalle-Donne I, Rossi R, Giustarini D, Milzani A, Colombo R (2003) Protein carbonyl groups as biomarkers of oxidative stress. Clin Chim Acta 329:23–38. https://doi.org/10.1016/s0009-8981(03)00003-2

Dokladny K, Zuhl MN, Moseley PL (2016) Intestinal epithelial barrier function and tight junction proteins with heat and exercise. J Appl Physiol 120:692–701. https://doi.org/10.1152/japplphysiol.00536.2015

Drew MD, Syed NA, Goldade BG, Laarveld B, Van Kessel AG (2004) Effects of dietary protein source and level on intestinal populations of Clostridium perfringens in broiler chickens. Poult Sci 83:414–420. https://doi.org/10.1093/ps/83.3.414

Ellabaan MMH, Munck C, Porse A, Imamovic L, Sommer MOA (2021) Forecasting the dissemination of antibiotic resistance genes across bacterial genomes. Nat Commun 12:2435. https://doi.org/10.1038/s41467-021-22757-1

Engberg RM, Lauridsen C, Jensen SK, Jakobsen K (1996) Inclusion of oxidized vegetable oil in broiler diets. Its influence on nutrient balance and on the antioxidative status of broilers. Poult Sci 75:1003–1011. https://doi.org/10.3382/ps.0751003

Estévez M (2015) Oxidative damage to poultry: from farm to fork. Poult Sci 94:1368–1378. https://doi.org/10.3382/ps/pev094

Fasano A (2020) All disease begins in the (leaky) gut: role of zonulin-mediated gut permeability in the pathogenesis of some chronic inflammatory diseases. F1000Res 9:69. https://doi.org/10.12688/f1000research.20510.1

Ferro E, Goitre L, Retta SF, Trabalzini L (2012) The interplay between ROS and Ras GTPases: physiological and pathological implications. J Signal Trans 2012:1–9. https://doi.org/10.1155/2012/365769

Forster SC, Liu J, Kumar N, Gulliver EL, Gould JA, Escobar-Zepeda A, Mkandawire T, Pike LJ, Shao Y, Stares MD, Browne HP, Neville BA, Lawley TD (2022) Strain-level characterization of broad host range mobile genetic elements transferring antibiotic resistance from the human microbiome. Nat Commun 13:1445. https://doi.org/10.1038/s41467-022-29096-9

Gilani S, Chrystal PV, Barekatain R (2021) Current experimental models, assessment and dietary modulations of intestinal permeability in broiler chickens. Anim Nutr 7:801–811. https://doi.org/10.1016/j.aninu.2021.03.001

Gostner JM, Becker K, Fuchs D, Sucher R (2013) Redox regulation of the immune response. Redox Rep 18:88–94. https://doi.org/10.1179/1351000213Y.0000000044

Hafez HM, Shehata AA (2021) Turkey production and health: current challenges. Ger J Vet Res 1:3–14. https://doi.org/10.51585/gjvr.2021.0002

Hao H, Cheng G, Iqbal Z, Ai X, Hussain HI, Huang L, Dai M, Wang Y, Liu Z, Yuan Z (2014) Benefits and risks of antimicrobial use in food-producing animals. Front Microbiol 5:288. https://doi.org/10.3389/fmicb.2014.00288

Hirakawa R, Nurjanah S, Furukawa K, Murai A, Kikusato M, Nochi T, Toyomizu M (2020) Heat stress causes immune abnormalities via massive damage to effect proliferation and differentiation of lymphocytes in broiler chickens. Front Vet Sci 7:46. https://doi.org/10.3389/fvets.2020.00046

Kalyanaraman B (2013) Teaching the basics of redox biology to medical and graduate students: oxidants, antioxidants and disease mechanisms. Redox Biol 1:244–257. https://doi.org/10.1016/j.redox.2013.01.014

Kanazawa K, Ashida H (1998) Dietary hydroperoxides of linoleic acid decompose to aldehydes in stomach before being absorbed into the body. Biochim Biophys Acta 1393:349–361. https://doi.org/10.1016/s0005-2760(98)00089-7

Kareem KY, Loh TC, Foo HL, Asmara SA, Akit H, Abdulla NR, Ooi MF (2015) Carcass, meat and bone quality of broiler chickens fed with postbiotic and prebiotic combinations. Int J Probiotics Prebiotics:23–30

Klemashevich C, Wu C, Howsmon D, Alaniz RC, Lee K, Jayaraman A (2014) Rational identification of diet-derived postbiotics for improving intestinal microbiota function. Curr Opin Biotechnol 26:85–90. https://doi.org/10.1016/j.copbio.2013.10.006

Kubow S (1990) Toxicity of dietary lipid peroxidation products. Trends Food Sci Technol 1:67–71. https://doi.org/10.1016/0924-2244(90)90049-5

Kumar B (2012) Stress and its impact on farm animals. Front Biosci E4:1759–1767. https://doi.org/10.2741/e496

Lamberte LE, van Schaik W (2022) Antibiotic resistance in the commensal human gut microbiota. Curr Opin Microbiol 68:102150. https://doi.org/10.1016/j.mib.2022.102150

Lara L, Rostagno M (2013) Impact of heat stress on poultry production. Animals 3:356–369. https://doi.org/10.3390/ani3020356

Lasiewski R (1969) Physiological responses to heat stress in the poorwill. Am J Physiol-Legacy Content 217:1504–1509. https://doi.org/10.1152/ajplegacy.1969.217.5.1504

Latorre JD, Hernandez-Velasco X, Bielke LR, Vicente JL, Wolfenden R, Menconi A, Hargis BM, Tellez G (2015a) Evaluation of a bacillus direct-fed microbial candidate on digesta viscosity, bacterial translocation, microbiota composition and bone mineralisation in broiler chickens fed on a rye-based diet. Br Poult Sci 56:723–732. https://doi.org/10.1080/00071668.2015.1101053

Latorre JD, Hernandez-Velasco X, Kuttappan VA, Wolfenden RE, Vicente JL, Wolfenden AD, Bielke LR, Prado-Rebolledo OF, Morales E, Hargis BM, Tellez G (2015b) Selection of bacillus spp for cellulase and xylanase production as direct-fed microbials to reduce digesta viscosity and Clostridium perfringens proliferation using an in vitro digestive model in different poultry diets. Front Vet Sci 2. https://doi.org/10.3389/fvets.2015.00025

Liang F, Jiang S, Mo Y, Zhou G, Yang L (2015) Consumption of oxidized soybean oil increased intestinal oxidative stress and affected intestinal immune variables in yellow-feathered broilers. Asian Australas J Anim Sci 28:1194–1201. https://doi.org/10.5713/ajas.14.0924

Liu Y-L, Ding K-N, Shen X-L, Liu H-X, Zhang Y-A, Liu Y-Q, He Y-M, Tang L-P (2022) Chronic heat stress promotes liver inflammation in broilers via enhancing NF-κB and NLRP3 signaling pathway. BMC Vet Res 18:289. https://doi.org/10.1186/s12917-022-03388-0

Loh TC, Choe DW, Foo HL, Sazili AQ, Bejo MH (2014) Effects of feeding different postbiotic metabolite combinations produced by lactobacillus plantarum strains on egg quality and production performance, faecal parameters and plasma cholesterol in laying hens. BMC Vet Res 10:149. https://doi.org/10.1186/1746-6148-10-149

Longobardi C, Andretta E, Romano V, Lauritano C, Avantaggiato G, Schiavone A, Jarriyawattanachaikul W, Florio S, Ciarcia R, Damiano S (2020) Effects of some new antioxidants on apoptosis and ROS production in AFB1 treated chickens. Med Sci Forum 2:12. https://doi.org/10.3390/CAHD2020-08640

Mishra B, Jha R (2019) Oxidative stress in the poultry gut: potential challenges and interventions. Front Vet Sci 6:60. https://doi.org/10.3389/fvets.2019.00060

Montuschi P, Barnes PJ, Roberts LJ (2004) Isoprostanes: markers and mediators of oxidative stress. FASEB J 18:1791–1800. https://doi.org/10.1096/fj.04-2330rev

Mount LE (1978) Heat transfer between animal and environment. Proc Nutr Soc 37:21–27

Muaz K, Riaz M, Akhtar S, Park S, Ismail A (2018) Antibiotic residues in chicken meat: global prevalence, threats, and decontamination strategies: a review. J Food Prot 81:619–627. https://doi.org/10.4315/0362-028X.JFP-17-086

Muhammed SI, Morrison SM, Boyd WL (1975) Nutritional requirements for growth and sporulation of Clostridium perfringens. J Appl Bacteriol 38:245–253. https://doi.org/10.1111/j.1365-2672.1975.tb00529.x

Nabi F, Arain MA, Hassan F, Umar M, Rajput N, Alagawany M, Syed SF, Soomro J, Somroo F, Liu J (2020) Nutraceutical role of selenium nanoparticles in poultry nutrition: a review. Worlds Poult Sci J 76:459–471. https://doi.org/10.1080/00439339.2020.1789535

Neuman MG, Nanau RM (2012) Inflammatory bowel disease: role of diet, microbiota, life style. Transl Res 160:29–44. https://doi.org/10.1016/j.trsl.2011.09.001

Oleskin AV, Shenderov BA (2016) Neuromodulatory effects and targets of the SCFAs and gasotransmitters produced by the human symbiotic microbiota. Microb Ecol Health Dis 27. https://doi.org/10.3402/mehd.v27.30971

Ortega ADSV, Szabó C (2021) Adverse effects of heat stress on the intestinal integrity and function of pigs and the mitigation capacity of dietary antioxidants: a review. Animals 11:1135. https://doi.org/10.3390/ani11041135

Park YH, Hamidon F, Rajangan C, Soh KP, Gan CY, Lim TS, Abdullah WNW, Liong MT (2016) Application of probiotics for the production of safe and high-quality poultry meat. Korean J Food Sci Anim Resour 36:567–576. https://doi.org/10.5851/kosfa.2016.36.5.567

Petruk A, Korver DR (2004) Broiler breeder egg production and quality are affected by timing of increased dietary ca relative to photostimulation. Can J Anim Sci 84:411–420. https://doi.org/10.4141/A03-030

Pizzino G, Irrera N, Cucinotta M, Pallio G, Mannino F, Arcoraci V, Squadrito F, Altavilla D, Bitto A (2017) Oxidative stress: harms and benefits for human health. Oxidative Med Cell Longev 2017:8416763. https://doi.org/10.1155/2017/8416763

Poljsak B, Šuput D, Milisav I (2013) Achieving the balance between ROS and antioxidants: when to use the synthetic antioxidants. Oxidative Med Cell Longev 2013:956792. https://doi.org/10.1155/2013/956792

Qaisrani SN, Moquet PCA, Van Krimpen MM, Kwakkel RP, Verstegen MWA, Hendriks WH (2014) Protein source and dietary structure influence growth performance, gut morphology, and hindgut fermentation characteristics in broilers. Poult Sci 93:3053–3064. https://doi.org/10.3382/ps.2014-04091

Quinteiro-Filho WM, Ribeiro A, Ferraz-de-Paula V, Pinheiro ML, Sakai M, Sá LRM, Ferreira AJP, Palermo-Neto J (2010) Heat stress impairs performance parameters, induces intestinal injury, and decreases macrophage activity in broiler chickens. Poult Sci 89:1905–1914. https://doi.org/10.3382/ps.2010-00812

Saint-Georges-Chaumet Y, Edeas M (2016) Microbiota–mitochondria inter-talk: consequence for microbiota–host interaction. Pathog Dis 74:ftv096. https://doi.org/10.1093/femspd/ftv096

Saracila M, Panaite TD, Papuc CP, Criste RD (2021) Heat stress in broiler chickens and the effect of dietary polyphenols, with special reference to willow (Salix spp) bark supplements—a review. Antioxidants 10:686. https://doi.org/10.3390/antiox10050686

Shehata AA, Attia Y, Khafaga AF, Farooq MZ, El-Seedi HR, Eisenreich W, Tellez-Isaias G (2022a) Restoring healthy gut microbiome in poultry using alternative feed additives with particular attention to phytogenic substances: challenges and prospects. Ger J Vet Res 2:32–42. https://doi.org/10.51585/gjvr.2022.3.0047

Shehata AA, Yalçın S, Latorre JD, Basiouni S, Attia YA, Abd El-Wahab A, Visscher C, El-Seedi HR, Huber C, Hafez HM, Eisenreich W, Tellez-Isaias G (2022b) Probiotics, prebiotics, and phytogenic substances for optimizing gut health in poultry. Microorganisms 10:395. https://doi.org/10.3390/microorganisms10020395

Shibamoto T (2006) Analytical methods for trace levels of reactive carbonyl compounds formed in lipid peroxidation systems. J Pharm Biomed Anal 41:12–25. https://doi.org/10.1016/j.jpba.2006.01.047

St. Angelo AJ (1992) Lipid oxidation in food. American Chemical Society, Washington, DC, USA

Stecher B (2015) The roles of inflammation, nutrient availability and the commensal microbiota in enteric pathogen infection. Microbiol Spectr 3. https://doi.org/10.1128/microbiolspec.MBP-0008-2014

Tellez G, Latorre JD, Kuttappan VA, Hargis BM, Hernandez-Velasco X (2015) Rye affects bacterial translocation, intestinal viscosity, microbiota composition and bone mineralization in Turkey poults. PLoS One 10:e0122390. https://doi.org/10.1371/journal.pone.0122390

Tellez-Isaias G, Eisenreich W, Shehata AA (2022) Nutraceuticals to mitigate the secret killers in animals. Vet Sci 9:435. https://doi.org/10.3390/vetsci9080435

Tse JKY (2017) Gut microbiota, nitric oxide, and microglia as prerequisites for neurodegenerative disorders. ACS Chem Neurosci 8:1438–1447. https://doi.org/10.1021/acschemneuro.7b00176

Tsolis RM, Bäumler AJ (2020) Gastrointestinal host-pathogen interaction in the age of microbiome research. Curr Opin Microbiol 53:78–89. https://doi.org/10.1016/j.mib.2020.03.002

Ulluwishewa D, Anderson RC, McNabb WC, Moughan PJ, Wells JM, Roy NC (2011) Regulation of tight junction permeability by intestinal bacteria and dietary components1,2. J Nutr 141:769–776. https://doi.org/10.3945/jn.110.135657

von Meyenburg C, Hrupka BH, Arsenijevic D, Schwartz GJ, Landmann R, Langhans W (2004) Role for CD14, TLR2, and TLR4 in bacterial product-induced anorexia. Am J Physiol Regul Integr Comp Physiol 287:R298–R305. https://doi.org/10.1152/ajpregu.00659.2003

Wang W, Xu Z, Yu C, Xu X (2017) Effects of aflatoxin B1 on mitochondrial respiration, ROS generation and apoptosis in broiler cardiomyocytes. Anim Sci J 88:1561–1568. https://doi.org/10.1111/asj.12796

Wilkie DC, Van Kessel AG, White LJ, Laarveld B, Drew MD (2005) Dietary amino acids affect intestinal Clostridium perfringens populations in broiler chickens. Can J Anim Sci 85:185–193. https://doi.org/10.4141/A04-070

Winter SE, Thiennimitr P, Winter MG, Butler BP, Huseby DL, Crawford RW, Russell JM, Bevins CL, Adams LG, Tsolis RM, Roth JR, Bäumler AJ (2010) Gut inflammation provides a respiratory electron acceptor for salmonella. Nature 467:426–429. https://doi.org/10.1038/nature09415

Wu QJ, Liu N, Wu XH, Wang GY, Lin L (2018) Glutamine alleviates heat stress-induced impairment of intestinal morphology, intestinal inflammatory response, and barrier integrity in broilers. Poult Sci 97:2675–2683. https://doi.org/10.3382/ps/pey123

Yangilar F (2016) Effect of the fish oil fortified chitosan edible film on microbiological, chemical composition and sensory properties of Göbek Kashar cheese during ripening time. Korean J Food Sci Anim Resour 36:377–388. https://doi.org/10.5851/kosfa.2016.36.3.377

Yi H, Xiong Y, Wu Q, Wang M, Liu S, Jiang Z, Wang L (2020) Effects of dietary supplementation with l-arginine on the intestinal barrier function in finishing pigs with heat stress. J Anim Physiol Anim Nutr 104:1134–1143. https://doi.org/10.1111/jpn.13277

Yu Q, Tang C, Xun S, Yajima T, Takeda K, Yoshikai Y (2006) MyD88-dependent signaling for IL-15 production plays an important role in maintenance of CD8αα TCRαβ and TCRγδ intestinal intraepithelial lymphocytes. J Immunol 176:6180–6185. https://doi.org/10.4049/jimmunol.176.10.6180

Zaman SB, Hussain MA, Nye R, Mehta V, Mamun KT, Hossain N (2017) A review on antibiotic resistance: alarm bells are ringing. Cureus 9:e1403. https://doi.org/10.7759/cureus.1403

Probiotics as Alternative to Antibiotics in Poultry: Challenges and Prospects

4

Awad A. Shehata, Shereen Basiouni, Guillermo Tellez-Isaias, Wolfgang Eisenreich, and Hafez M. Hafez

Contents

A. A. Shehata (✉) · W. Eisenreich
Structural Membrane Biochemistry, Bavarian NMR Center, Technical University of Munich (TUM), Garching, Bayern, Germany
e-mail: Awad.shehata@tum.de

S. Basiouni
Institute of Molecular Physiology, Mainz, Germany

G. Tellez-Isaias
Division of Agriculture, Department of Poultry Science, University of Arkansas, Fayetteville, AR, USA

H. M. Hafez
Institute of Poultry Diseases, Faculty of Veterinary Medicine, Free University of Berlin, Berlin, Germany

© The Author(s), under exclusive license to Springer Nature Switzerland AG 2024   59
A. A. Shehata et al. (eds.), *Alternatives to Antibiotics against Pathogens in Poultry*, https://doi.org/10.1007/978-3-031-70480-2_4

Abstract

Probiotics are "live microbial feed additives that benefit the host by enhancing gut microbial balance" (eubiosis). Probiotics can be bacteria, fungi, or yeasts, and their use in poultry has progressively expanded over the years due to the rising demand for antibiotic alternatives. Although several microorganisms have been assessed to be used as probiotics, certain species can fulfill the assessment criteria such as (i) Resistance to the digestive system conditions (acid and bile), (ii) Ability to colonize the gastrointestinal tract (GIT) and triggering pathogen displacement (competitive exclusion), (iii) Inhibition of the growth of potential pathogens by producing antimicrobial metabolites, and (iv) Modulation of the immune system. Indeed, probiotics increase nutrient utilization and performance and reduce foodborne pathogens. Several factors affect the extent, magnitude, and duration of probiotic efficacy, including bacterial species, GIT ecological niche occupancy, metabolic products, rearing conditions, the dose, the frequency of daily administration, timing of administration (before, during, and after a meal), method of delivery (fermented feed, orally via drinking water, in the feed), the viability of the probiotic bacteria, and the duration of administration (short term or longer term). In the present review, we shed light on the mechanism of action of probiotics and their use in poultry. Additionally, we highlight the main challenges and possibilities for overcoming breakdowns of probiotics application.

Keywords

Probiotics · Performance · Immune functions · Gut function · Poultry

1   Introduction and History of Probiotics

The word probiotic comes from the Latin word "pro" and the Greek word "bios," both of which mean "for life." The theory of probiotics was introduced in the early twentieth century (Di Gioia and Biavati 2018). Fuller defined probiotics as " live microbial feed additives that benefit the host by enhancing gut microbial balance" (Fuller 1992; McFarland 2015). Of note, the terms "probiotics" and "direct-fed microbials (DFMs) are interchangeably used for beneficial microbes, but in practice, they are not synonyms. DFMs are recognized by the USA Food and Drug Administration as "products that are purported to contain live (viable) microorganisms" (Brashears et al. 2005). While the term "DFM" may be applied more broadly to any live microbe given to animals, probiotics might be viewed as a subset of DFM. For practical purposes, this review uses the terms probiotics and DFMs interchangeably.

Over a century ago, Eli Metchnikoff, also known as the "father of innate immunity," introduced a revolutionary idea that consuming live bacteria could enhance one's health by modifying the microbiota in the gut (Gordon 2008; Kaufmann 2008). He earned the Nobel Prize in Physiology or Medicine in 1908, which he shared with Paul Ehrlich. Probiotics enhance gut microflora by secreting beneficial enzymes, organic acids, vitamins, and non-toxic antibacterial substances upon ingestion, affecting local and systemic immune function (Tarabees et al. 2023). They are defined as "live strains of strictly selected microorganisms which, when administered in adequate amounts, confer a health benefit on the host (Tsuda and Miyamoto 2010)."

Several probiotics, including spore-forming (*Bacillus* spp) and lactic acid-producing bacteria (LAB), are used in poultry: *Lactobacillus* spp., *Streptococcus thermophilus*, *Enterococcus faecalis*, and *Bifidobacterium* spp. are the most frequent LABs in probiotic formulations. Commercially, probiotics have been marketed containing single or multimicrobial assemblies. Moreover, yeasts improve animal performance and feed quality by promoting the development of lactic acid bacteria, and their cell wall components competitively exclude harmful bacteria (Onifade and Babatunde 1996). In addition, *Saccharomyces cerevisiae* provides protein, trace minerals, vitamin B complex, and enzymes, including phytase and cellulase (Paryad and Mahmoudi 2008). It also improves the growth performance, immunological response, disease resistance, mineral retention, bone mineralization, and feed utilization of broilers (Elghandour et al. 2020; Attia et al. 2023). This chapter discusses the most common probiotics in poultry, the mechanism of action, and some challenges and prospects.

2 Selection of Probiotic Candidates

Due to the presence of probiotic strains that might pose negative side effects, probiotics can be unreliable, and their therapeutic importance may not be verified. The Food and Agricultural Organization (FAO) and World Health Organization (WHO) developed a guideline for assessing commercial probiotics. The steps of commercial probiotic development are illustrated in Fig. 4.1 (Perumalla et al. 2023). Ensuring that the probiotic strain is listed in the "Generally Recognized as Safe" (GRAS) list is a determinant criterion for selecting probiotic candidates.

Fig. 4.1 General steps for the development of probiotics

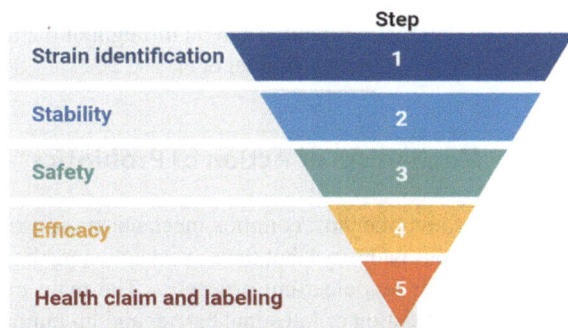

	Step
Strain identification	1
Stability	2
Safety	3
Efficacy	4
Health claim and labeling	5

Developing feed-stable commercial probiotics (resistant to the heat palletization process) is critically necessary. Bacterial spore formers, such as *Bacillus* spp., are stable—but not all—and can bear the gastric acids and enzymes. Additionally, *Bacillus* spores are heat tolerant in extremely hot conditions (Vreeland et al. 2000). Consequently, because they can tolerate extreme environments and extended storage times, particular *Bacillus* spp. have been employed as animal probiotics (Hong et al. 2005).

The main characteristics of a probiotic include: (i) Resistance to the digestive system conditions (acid and bile), (ii) Ability to colonize the gastrointestinal tract (GIT) and triggering pathogen invasion (competitive exclusion), (iii) Inhibition of the growth of potential pathogens via producing antimicrobial metabolites, and (iv) Modulation of the immune system (Harimurti and Hadisaputro 2015).

Certain probiotic strains can colonize the gut and have the ability to survive stomach acidity, intestinal enzymes, and bile (Fontana et al. 2013; Smith 2014). It is suggested that spores can germinate and survive throughout the gut. It can also adhere to feed particles, which helps to protect them during transit. Re-sporulation is the easiest method of transit for bacteria to survive throughout the animal's body. Germination and proliferation depend on the animal's diet, which requires plentiful nutrients to flourish (Hong et al. 2005). Certain probiotic species can colonize the digestive tract, including several strains of *Lactobacillus*, *Enterococcus*, and *Bifidobacterium* (Cummings and Macfarlane 1997; Guillot 1998). However, *Bacillus* spores are known as transients since they cannot colonize the stomachs of axenic and gnotobiotic animals.

Based on the colonization activity, there are two general approaches to using probiotics in poultry: (i) Probiotic strains can be continuously supplemented in chicken feed throughout their life, including in ovo administration for chicks (de Oliveira et al. 2014). However, high levels of probiotics must be administered continuously due to the loss of probiotics in feed and the stomach. (ii) Use of probiotic strain that can colonize the host. Indeed, a single treatment of newly hatched chicks with adult cecal contents can change host cecal protein expression (Volf et al. 2016). The main advantage of using strains capable of colonization is that a single treatment may result in a continuous effect of the probiotic in the chicken gut (Poudel et al. 2022), reducing the costs. Indeed, these strains might be supplemented with lower doses than non-colonizable strains, as they can replicate and colonize the gut. Moreover, these strains tend to spread throughout the flock, including chicks whose initial colonization failed (Poudel et al. 2022).

3   Mechanism of Action of Probiotics

Figure 4.2 shows the most common mechanisms of probiotics. Probiotics may work through various pathways, but their exact action method is probably a concoction of these and other less clear-cut processes. The main mechanisms of probiotics are through (i) regulation of intestinal barrier and immunomodulation and (ii) competitive exclusion. In this part, we will shed light on these mechanisms.

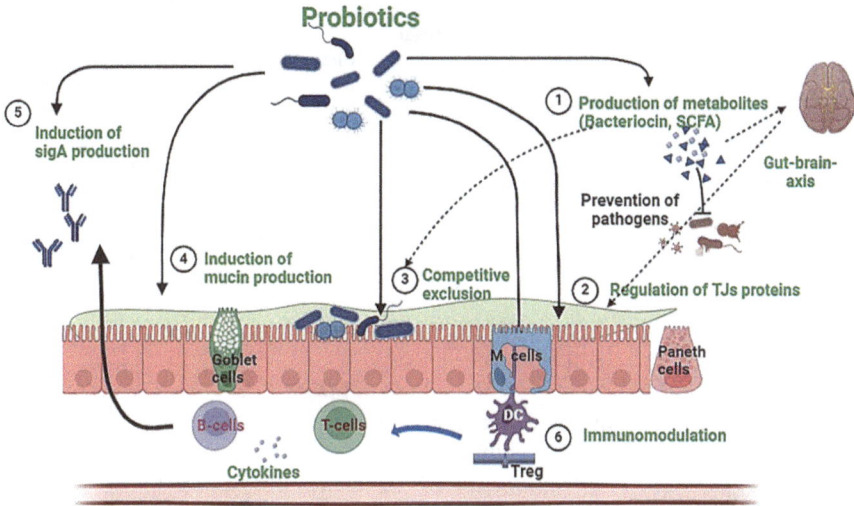

Fig. 4.2 Mechanism of action of probiotics, TJs proteins = Tight junctions proteins. The figure is modified after Latif et al. (2023) using BioRender. ©Permission obtained from Dr. Imran Mahmoud Khan, Dr. Habil. Tuba Esatbeyoglu and João Miguel Ferreira da Rocha

4 Regulation of Intestinal Barrier and Immunomodulation

Probiotics inhibit apoptosis and alleviate inflammation, resulting in homeostasis and regulation of the intestinal barrier (Liu et al. 2020) through two main mechanisms:

4.1 Microbial-Associated Molecular Patterns (MAMPs)

Microbial-associated molecular patterns (MAMPs), including lipopolysaccharide, capsular polysaccharide (CPS), pili, flagella, and surface layer proteins (SLPs) of probiotics, bind with the pattern recognition receptors (PRRs), which aid in the regulation of signaling pathways (Nagpal et al. 2012; Soltani et al. 2021).

The surface layer proteins (SLPs) of certain probiotic strains, such as *Lactobacillus acidophilus*, can bind to the dendritic cell-specific intercellular adhesion molecule, grabbing nonintegrin (DC-SIGN) and increasing the extracellular signal-regulated kinase (ERK) phosphorylation. This modulates the expression of nuclear factor kappa B (NF-κB) and, in turn, reduces cell apoptosis. Probiotic-treated birds demonstrated modulation of NF-κB (Higgins et al. 2011).

SLPs can also trigger the phagocytic capacity of macrophages and DCs (Higgins et al. 2011; Liu et al. 2020) and enhance NK cells and CD8+ T cells' cytotoxic activity (Aziz and Bonavida 2016; Yousefi et al. 2019; Mao et al. 2020).

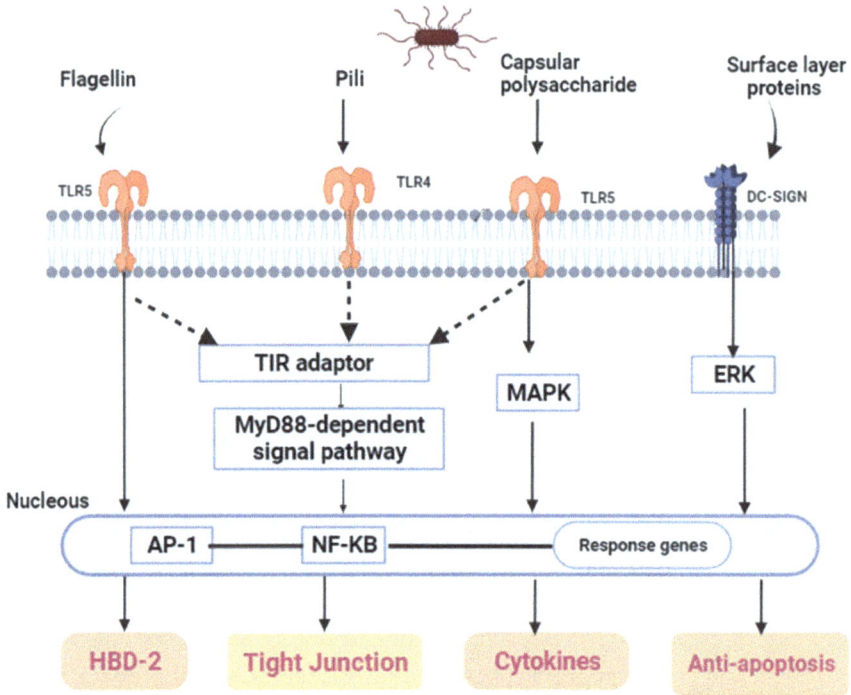

Fig. 4.3 Regulation of intestinal barrier through the microbial-associated molecular patterns. *TLR* toll-like receptors, *ERK* extracellular signal-regulated kinase, *MAPK* mitogen-activated protein kinase, NF-κB nuclear factor kappa B, *AP-1* activating protein-1, *HBD-2* beta-defensin 2. The figure was modified after (Liu et al. 2020) using BioRender. Permission was obtained from Prof. Dr. Qixiao Zhai

The interaction between probiotics enteric epithelial cells triggers the maturation of DCs and the subsequent induction of Tregs, the key mediators in maintaining gut homeostasis, stimulating the production of IgA by mature B cells (Rousseaux et al. 2023).

Bacillus subtilis PB6 also modulates the immune system in chickens by increasing IFN-γ and IL-6 (Abed et al. 2023) (Fig. 4.3).

4.2 Probiotics Metabolites

Several investigations have demonstrated the unique immunomodulatory characteristics of the probiotic's metabolites. Certain probiotic strains can produce metabolites, such as organic acids, and were demonstrated to reduce *Salmonella* spp., *E. coli*, and *Pseudomonas* spp. populations by pH decreases, cell membrane permeabilization, and potentiation of other antibacterial agents (Alakomi et al. 2000; Neal-McKinney et al. 2012). In addition, probiotics produce antimicrobial compounds, including hydrogen peroxide, carbon dioxide, diacetyl, acetaldehyde, reuterin, and bacteriocins, which have been demonstrated to reduce pathogens.

Short-Chain Fatty Acids
Short-chain fatty acids (SCFAs), produced by bacterial fermentation of dietary fiber (DF), are crucial indicators for tracking the health of poultry's intestines. They can strengthen poultry's immune system, control pathogens, improve barrier function, and alleviate gastrointestinal inflammation (Liu et al. 2021). However, excessive use of SCFAs will lead to enteritis in poultry production (Liu et al. 2021). Numerous studies have also demonstrated the positive effect of probiotics, such as *Lactobacillus* spp. (Meimandipour et al. 2010; Chang et al. 2019a) and *Bacillus* spp. (Aljumaah et al. 2020; Kan et al. 2021), in promoting the production of SCFAs. Nevertheless, fewer investigations addressed the combination of probiotics and their effects on SCFA synthesis in poultry.

The immunomodulatory effect of SCFAs can be explained by activating intestinal epithelial cells through G-protein-coupled receptor 1 (GPCR-1). Thoda and Touraki summarized the immune-modulatory effect of SCFAs: (i) Differentiation of naive CD4+ T cells to Th1, Th17, and IL-10-producing Tregs. (ii) Activation of dendritic cells enhances the production of IL-10-producing Tregs. (iii) Activation of goblet cells to produce mucin. (iv) Polarization of macrophages, also resulting in stimulation of IL-10 production. (v) Secretion of IL-18 through the activation of the NLRP3 inflammasome. (vi) Chemotaxis of neutrophils at inflammation sites. (vii) Activation of DCs, which activate IgA secretion by plasma cells (Thoda and Touraki 2023).

Bacteriocins
Bacteriocins, bacterial peptides mostly generated by LAB, display various biological properties (Fig. 4.4): (i) Antibacterial effects via several mechanisms. Bacteriocins can hinder the synthesis of cell walls by attaching to the primary peptidoglycan transporter, interfere with crucial enzymatic processes required for cell wall synthesis, or attach to cell membranes, causing pore formation and altering cell membrane permeability, ultimately resulting in rapid cellular death. Interestingly, bacteriocins do not specifically target commensal microbiota; they only affect pathogens (Simons et al. 2020). (ii) Immunomodulatory effects on cytokines by regulating various signaling cascades such as Toll-like receptors (TLR), nuclear factor kappa B (NF-κB), and mitogen-activated protein kinase (MAPK). For instance, sublancin, an antimicrobial peptide derived from *Bacillus subtilis*, has been shown to activate the innate immune response upregulation of IL-1β, IL-6, TNF-α, and NO in RAW264.7 cells or mouse macrophages (Dicks et al. 2018). Bacteriocins can also stimulate the immune response against viruses (Dreyer et al. 2019). (iii) Bacteriocins exhibit anticarcinogenic effects (Cesa-Luna et al. 2021).

Class I: Lantibiotics (<5 kDa)	Class II: Non-lantibiotics (<10 kDa)	Class III: Large peptides (>30 kDa)

• Heat-stable lanthionine-containing peptides
 	
• Antimicrobial activity through induction of pores in the bacterial cell membrane by using peptidoglycan precursor
 	
• Subclasses
  - 1a (positively charged)
  - 1b (negatively charged)
  - 1c (sulfur-containing negatively charged)
	• Heat-stable non-lanthionine-containing peptides
	
	• Antimicrobial activity through induction permeabilization in the cell membrane
	
	• Subclasses
	  - IIa
	  - IIb
	  - IIc
	  - IId
		• Heat-labile peptides
		
		• Antimicrobial activity through endopeptidase
		
		• Subclasses
		  - IIIa (bacteriolysis)
		  - IIIb (non-lytic peptides)

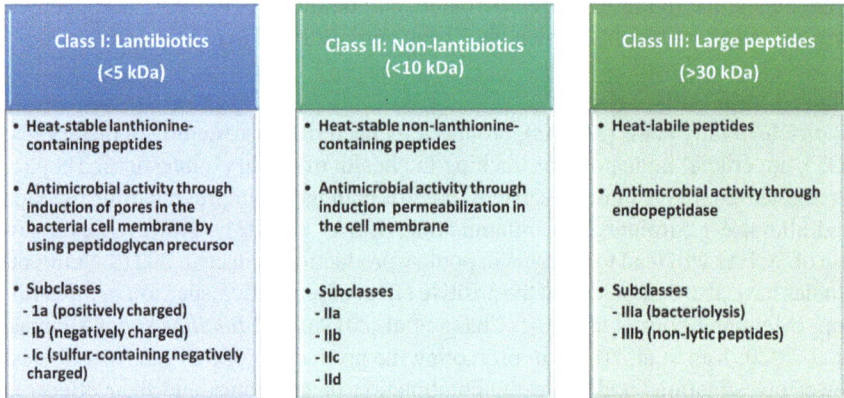

Fig. 4.4 Lactic acid bacteria bacteriocins, the figure is generated from Thoda and Touraki (2023). ©Permission was provided by Dr. Maria Touraki

5 Competitive Exclusion

The term competitive exclusion means administering probiotics (single or combinations of different probiotic strains) to animals' digestive tracts to prevent the colonization of pathogens. It was primarily introduced in 1973 to fight Salmonella in poultry farms (Nurmi and Rantala 1973), but it has subsequently been expanded to include other pathogens such as *E. coli*, *Clostridium*, and *Campylobacter*. Competitive exclusion acts against harmful germs through various processes, including competition for receptor sites and nutrients. The mechanism of competitive exclusion depends on the probiotic strain. Probiotics also exhibit antagonistic activity against pathogenic bacteria by modifying the pH of the gut and direct antimicrobial effect by secreting products that inhibit the growth of such bacteria (hydrogen peroxide, bacteriocins, and defensins) (Tiwari et al. 2012; Pan and Yu 2014).

6 Improvement of Digestion and Nutrient Utilization

Microbial communities utilize nutrients for their energy needs as soon as they reach the gut. The host may use the byproducts of microbial fermentation or exogenous enzyme production. The most common benefit is maintaining a healthy gut and intestinal barrier integrity, reflected by improving nutrient utilization and adequate nutrient absorption.

Probiotics have been shown to increase microbial enzymes and stimulate host enzymes, which in turn affects the host's enzyme activity (Wang et al. 2017). *Lactobacillus* spp. hydrolyzes amylopectin into maltose, maltotriose, and glucose, have been characterized (Jin et al. 2000). Additionally, *Bacillus coagulans* NJ0516 boosted the duodenum's amylase activity in broilers (Wang and Gu 2010). *B.*

licheniformis and *L. bulgaricus* have been shown to enhance the digestibility of amino acids, proteins, and starch (Cheng et al. 2017; Jha et al. 2020). Similarly, *Pediococcus acidilactici*, combined with butyric acid, increased amylase activity in chickens infected with Salmonella Typhimurium (Jazi et al. 2018). However, the process behind the effect of probiotics on enzymatic activity is still unclear.

The fermentation with certain *Lactobacillus* spp. reduces antinutritional chemicals such as tannins, phytate, and trypsin inhibitors, improving protein absorption (Manus et al. 2021; Walden et al. 2022). Additionally, several probiotics produce vitamins in the gut, including thiamin, B12, and folate (Hill 1997). The proteolytic activity facilitates the synthesis of bioactive soluble peptides by some *Lactobacillus* species, which enhance the nutritional value of proteins (Wang and Ji 2019; Manus et al. 2021). The use of probiotics may save 63 kcal/kg of feed (Harrington et al. 2016).

Probiotics improved the nutrient utilization in Cobb 500 broilers in which *Bacillus* spp. with enzymes improved amino acid utilization, except for arginine and serine (Jin et al. 2000). However, this effect might also be dose-dependent. High-dose probiotics (*Lactobacillus reuteri, Enterococcus faecium, Bifidobacterium animalis, Pediococcus acidilactici*, and *Lactobacillus salivarius*) at 1010 CFU/kg of feed decreased nutrient utilization in broilers compared to lower doses (108 CFU probiotic/kg of food), possibly due to the increased nutritional requirement of the probiotics (Mountzouris et al. 2010).

Probiotics also improve nutrient uptake by improving intestinal morphology and nutrient transport. Several probiotic species, including *Bacillus subtilis* (Samanya and Yamauchi 2002), *Enterococcus faecium* (Laudadio et al. 2012), *Lactobacillus acidophilus*, and *Bacillus licheniformis* can increase the villus height, which subsequently increases the surface area improving absorption (Cheng et al. 2021). Additionally, bacterial metabolites reduce gut inflammation in the mucosa (Beski and Al-Sardary 2014), which consequently improves the villus height and function (Adil et al. 2010).

7   Improvement of Growth Performance

Improved body weight growth is frequently linked to a high feed conversion ratio (FCR) and average daily intake (BWG). By preserving gut health, lowering the danger of pathogen infection, and boosting immunity, probiotics enhance growth performance (Tarabees et al. 2020, 2023; Jha et al. 2020; Sabry Abd Elraheam Elsayed et al. 2021). However, there is inconsistency regarding improving FCR using probiotics. Although several studies reported positive effects of probiotics supplementation on growth performance (Zaghari et al. 2020; Lokapirnasari et al. 2020; Deng et al. 2020), other studies could not prove this effect (Attia et al. 2023). These contradicting results might be related to the probiotic strain and hygienic measures (Attia et al. 2023), dose of probiotics, age, breed, species, the inoculation level of challenging pathogens, and external factors (Jha et al. 2020). Numerous investigations have been conducted to determine the optimal doses of probiotic inclusion and associated benefits. However, it is difficult to make definitive recommendations due to the varying dietary farming practices and stress levels observed in different settings. Further research is needed to determine the optimum dose for single- or multi-strain probiotics and to evaluate their effectiveness in birds with intestinal disorders and gut integrity issues (Jha et al. 2020).

8   Mitigation of Heat Stress

Probiotics are believed to alleviate problems caused by heat stress in poultry through several mechanisms: (i) Increasing the expression of genes responsible for transporting sugars, indicating an enhancement in sugar absorption. (ii) Restoring intestinal health and improving gut integrity (Deng et al. 2012), (iii) Reducing the production of reactive oxygen species in the ileum and the caecum and the serum concentration of malondialdehyde, a product of reactions occurring during oxidative stress (Neupane et al. 2019). Overall, probiotics can also help mitigate the negative effects of heat stress and improve the performance parameters of laying hens.

9   Improvement of Meat Quality

Probiotics improve meat quality by different mechanisms (Dong et al. 2024): (i) Probiotics have antioxidant effects and can reduce ROS production by enhancing superoxide dismutase and glutathione peroxidase (Deng et al. 2012). The phospholipid-containing muscle cell membranes are subsequently restored (Feng and Wang 2020). As a result, fresh meat may retain its myoglobin freshness for a longer period, improving the meat's color and decreasing dripping loss (Bai et al. 2016; Xiang et al. 2019). (ii) *Saccharomyces cerevisiae* increases Ghrelin, promotes the uptake and utilization of fatty acid, and sustains the uptake of glucose (Song et al. 2022) and muscle calcium levels (Chang et al. 2019b; Nidamanuri et al. 2021), which in turn enhance meat tenderness. Another theory was proposed by Huang et al., who suggested that calcium ions in muscle control calpain activity and impact the kind of muscle fiber, influencing meat quality (Huang et al. 2018). (iii) Improvement of fat and protein metabolism. Improvement of muscle fat and protein deposition affects meat's tenderness, water loss, cooking loss, and flavor.

10   Amelioration of Mycotoxins

Biological control of mycotoxins is a new promising approach that uses probiotics, particularly LAB and *Saccharomyces* spp. (Śliżewska et al. 2019; Liu et al. 2022). Decreased levels of Aflatoxin B1 (AFB1) in the liver and kidneys and improved animal performances are observed when Lactobacilli and *Saccharomyces cerevisiae* are added to a broiler diet contaminated with AFB1 (Śliżewska et al. 2019). This effect results from several mechanisms (Fig. 4.5): (i) Adsorption to the bacterial cell

Fig. 4.5 Biological
detoxification of
mycotoxins by probiotics

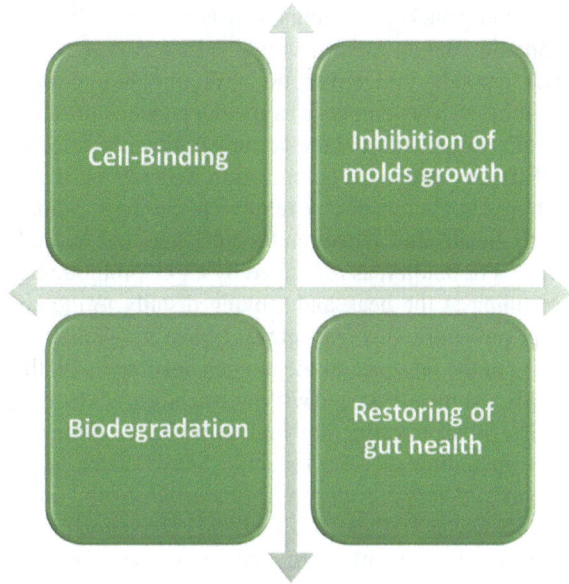

wall through peptidoglycans, exopolysaccharides, and teichoic acids. However, in yeast, adsorption is due to β-D-glucans and mannan oligosaccharides of the cell wall (Karaman et al. 2005). In addition, van der Waals forces, hydrogen bonds, or hydrophobic interactions may also be involved in the absorption process (Pizzolitto et al. 2011). (ii) Inhibition of mold growth. LAB can also inhibit the growth of molds and the production of mycotoxins (Dalié et al. 2010). (iii) Biodegradation. Some *Streptomyces* spp. have shown a strong ability to degrade AFB1 and ZON mycotoxins (Harkai et al. 2016). *Bacillus licheniformis* isolate CFR1 can degrade aflatoxin B1 in liquid culture media with a reduction rate of about 95% (Raksha Rao et al. 2017). It is crucial to address potential risks associated with the biodegradation method. Therefore, to safely use biotechnology in this area, we need to develop methods for monitoring potentially hazardous metabolites and biological effects when degrading mycotoxins.

11   Limitations of Probiotics

Probiotics can possess some disadvantages. (i) Overdose of probiotics may reduce semen quality when used on a reproductive flock of roosters. The number of *Lactobacillus* spp. in the cloaca has a direct effect on semen quality, and prolonged administration of the probiotic could lead to infertility. Therefore, it is important to carefully determine the dose administered when using probiotics on a flock of roosters (Haines et al. 2015; Kiess et al. 2016). (ii) Changes in fatty acid concentrations and bacterial structures of probiotics can occur at high temperatures (Dianawati et al. 2013), particularly when stored in plastic bags. Therefore, properly storing and developing packaging technology is extremely important to retain the right product properties and ensure efficacy. (iii) Inconsistent results about the efficacy of probiotics in poultry and numerous conflicting findings exist on the health and productivity of birds. This emphasizes the dependence of effectiveness on the dose and type of microorganisms and the administration conditions.

Effective probiotic colonization depends on the probiotic strain's survival and stability, strain specificity relative to the host, administration dose and frequency, host's health and nutritional status, age, stress, and host genetics (Mason et al. 2005). For poultry, as measured by CFU, probiotic bacterial colonization increases, beginning at the beak and moving distally to the colon (Simon et al. 2004). The crop, proventriculus, and gizzard have very small numbers of anaerobic bacteria owing to the presence of the oxygen absorbed with the nutrition and the small luminal pH, mainly associated with hydrochloric acid in the proventriculus (Rastall 2004).

The small intestine has high numbers of bacteria consisting of optional anaerobes, including such Lactobacilli, Streptococci, and Enterobacteria and anaerobes like the *Bifidobacterium* spp., *Bacteroides* spp., and *Clostridia* spp. in concentrations between 104 and 108 CFU/ml. The colon and cecum, with a colonization of 1010–1013 CFU/mL, are the most heavily colonized regions of the gut (Gaskins 2003).

One of the major selection criteria for a probiotic candidate is the ability of a probiotic strain to adhere to the mucus and epithelial cell surfaces. An indication of their ability to effect changes in enteric health is the ability of many strains of probiotic bacteria to physically adhere to portions of the GIT microenvironments (Monteagudo-Mera et al. 2019). Attachment to the plasma lemma of the enterocyte is considered the first step in the colonization of surfaces of the host enterocyte. This allows probiotic organisms to resist all other peristalses, mixing with the layer of digesta and mucus and subsequent gut removal.

Nevertheless, adherent probiotic bacteria normally do not permanently colonize the intestinal epithelium and are usually removed from the GI tract a few days after supplementation has ceased (Gusils et al. 1999). Because of the complexity of the intestinal enterocytes and the extensive interaction between intestinal cell types within the intestinal tract, only a few researchers have investigated the adhesion and colonization of probiotic bacteria (Henriksson et al. 1991).

Ensuring the viability and functionality of probiotics is crucial for their positive benefits. The manufacturing process (temperature, oxidation, shear stress, etc.), storage conditions (moisture/low water activity, packing, oxidation, temperature, etc.), and gastrointestinal tract (GIT) conditions (low pH, bile salts, digestive enzymes) are the primary factors that often impact the strains' ability to survive. Therefore, it is important to maintain optimal conditions during production, storage, and ingestion to maximize the effectiveness of probiotics (Blazheva et al. 2022).

Encapsulation protects probiotics by enclosing them in particles with a thin, semi-permeable, and mechanically stable capsule. The capsule must be made of a substance that can release the probiotic cells in a specific location on the intestinal tract. Polysaccharides from plants and animals, resistant starch, oligosaccharides, and fibers from fruits, vegetables, and cereals are commonly used as encapsulating materials. Protein encapsulation ingredients from animals and plants can also be used, but digestive enzymes may degrade them.

Freeze drying, also known as lyophilization, is a process used to remove water from a product that has been frozen and then placed under a vacuum. Cryoprotectants are added to stabilize the product during storage and increase microbial survival. This method is widely used in microencapsulation to enhance the shelf life of probiotic bacteria. Low-molecular-weight cryoprotectants like glucose, mannose, lactose, and trehalose are commonly used. Polysaccharides and proteins like soy protein, milk, gelatin, and maltodextrin are used as high-molecular-weight cryoprotectants. Freeze-drying enhances probiotics' viability and shelf life, although it does not affect how long they survive in the digestive system (Blazheva et al. 2022).

Spray drying. Spray drying is frequently used in feed processing to enhance probiotic stability. The process involves drying probiotics using a spray technique that atomizes both the liquid feed and the probiotics in a hot gas drying chamber. This approach is affordable and easy to scale up, but it can reduce viability due to osmotic stress and high temperatures. Gum Arabic is often used as an encapsulating wall material and has been found to improve viability. *S. cerevisiae* var. boulardii was encapsulated using this method, and gum Arabic and gelatine were the most viable wall materials based on viability tests following spray drying and subsequent tests under simulated GIT conditions (Yoha et al. 2022; Blazheva et al. 2022).

Extrusion. Probiotic cells can be encapsulated through extrusion, which involves suspending bacteria in liquid droplets and hardening them to create gel beads of varying sizes, morphologies, and mechanical strengths. This technique is simple, affordable, and compatible with microbial cells, but its slowness restricts its industrial use. *L. acidophilus* and *B. adolescentis* cells were successfully encapsulated, and alginate/chitosan capsules showed the largest percentage of survivors (Lee et al. 2019). Silk sericin can be used as a wall material and coating layer to protect microbial cells and extend shelf life (Apiwattanasiri et al. 2022).

Emulsion-based system. Two immiscible liquids can create an emulsion when mixed with an emulsifier. The process is easy to scale up but creates particles of different sizes. Zhang et al. (2018) developed a way to contain *Ligilactobacillus salivarius* in secondary emulsions, which improved encapsulation efficiency by up to 90%. El Kadri et al. (2018) used encapsulated *Lacticaseibacillus paracasei* in a milk-based water/oil emulsion, which improved its survival rate. Flaxseed mucilage was used to encapsulate *L. casei* in an alginate matrix, increasing its resistance to the negative effects of the simulated digestive system (Nasiri et al. 2021).

12 Conclusions

The use of probiotics in chicken has grown as an alternative to antibiotics. They have the potential to improve the gut microbiome and benefit the host in a number of aspects. Probiotics improve nutrient utilization, reduce foodborne pathogens, improve meat quality, and lessen the negative impacts of mycotoxins. However, their effects are influenced by the type of species, dosage, administration timing and method, and duration.

References

Abed AH, Radwan SA, Orabi A, Abdelaziz KT (2023) The combined effects of probiotic CLOSTAT® and Aviboost® supplement on growth performance, intestinal morphology, and immune response of broiler chickens. Ger J Vet Res 3(3):7–18. https://doi.org/10.51585/gjvr.2023.3.0058

Adil S, Banday T, Bhat GA, Mir MS, Rehman M (2010) Effect of dietary supplementation of organic acids on performance, intestinal histomorphology, and serum biochemistry of broiler chicken. Vet Med Int 2010:479485. https://doi.org/10.4061/2010/479485

Alakomi HL, Skyttä E, Saarela M, Mattila-Sandholm T, Latva-Kala K, Helander IM (2000) Lactic acid permeabilizes gram-negative bacteria by disrupting the outer membrane. Appl Environ Microbiol 66(5):2001–2005. https://doi.org/10.1128/AEM.66.5.2001-2005.2000

Aljumaah MR, Alkhulaifi MM, Abudabos AM, Aljumaah RS, Alsaleh AN, Stanley D (2020) Bacillus subtilis PB6-based probiotic supplementation plays a role in the recovery after the necrotic enteritis challenge. PLoS One 15(6):e0232781. https://doi.org/10.1371/journal.pone.0232781

Apiwattanasiri P, Charoen R, Rittisak S, Phattayakorn K, Jantrasee S, Savedboworn W (2022) Co-encapsulation efficiency of silk sericin-alginate-prebiotics and the effectiveness of silk sericin coating layer on the survival of probiotic lactobacillus casei. Food Biosci 46:101576. https://doi.org/10.1016/j.fbio.2022.101576

Attia YA, Basiouni S, Abdulsalam NM, Bovera F, Aboshok AA, Shehata AA, Hafez HM (2023) Alternative to antibiotic growth promoters: beneficial effects of Saccharomyces cerevisiae and/or lactobacillus acidophilus supplementation on the growth performance and sustainability of broilers' production. Front Vet Sci 10:1259426. https://doi.org/10.3389/fvets.2023.1259426

Aziz N, Bonavida B (2016) Activation of natural killer cells by probiotics. Forum Immunopathol Dis Ther 7(1–2):41–55. https://doi.org/10.1615/ForumImmunDisTher.2016017095

Bai WK, Zhang FJ, He TJ, Su PW, Ying XZ, Zhang LL, Wang T (2016) Dietary probiotic Bacillus subtilis strain fmbj increases antioxidant capacity and oxidative stability of chicken breast meat during storage. PLoS One 11(12):e0167339. https://doi.org/10.1371/journal.pone.0167339

Beski SSM, Al-Sardary SYT (2014) Effects of dietary supplementation of probiotic and synbiotic on broiler chickens hematology and intestinal integrity. Int J Poult Sci 14(1):31–36. https://doi.org/10.3923/ijps.2015.31.36

Blazheva D, Mihaylova D, Averina OV, Slavchev A, Brazkova M, Poluektova EU, Danilenko VN, Krastanov A (2022) Antioxidant potential of probiotics and postbiotics: a biotechnological approach to improving their stability. Russ J Genet 58(9):1036–1050. https://doi.org/10.1134/S1022795422090058

Brashears MM, Amezquita A, Jaroni D (2005) Lactic acid bacteria and their uses in animal feeding to improve food safety. In: Advances in food and nutrition research. Elsevier, pp 1–31

Cesa-Luna C, Alatorre-Cruz J-M, Carreño-López R, Quintero-Hernández V, Baez A (2021) Emerging applications of bacteriocins as antimicrobials, anticancer drugs, and modulators of the gastrointestinal microbiota. Pol J Microbiol 70(2):143–159. https://doi.org/10.33073/pjm-2021-020

Chang CH, Teng PY, Lee TT, Yu B (2019a) The effects of the supplementation of multi-strain probiotics on intestinal microbiota, metabolites and inflammation of young SPF chickens challenged with salmonella enterica subsp. enterica. Anim Sci J 90(6):737–746. https://doi.org/10.1111/asj.13205

Chang CH, Teng PY, Lee TT, Yu B (2019b) Effects of multi-strain probiotics combined with Gardeniae fructus on intestinal microbiota, metabolites, and morphology in broilers. J Poult Sci 56(1):32–43. https://doi.org/10.2141/jpsa.0170179

Cheng Y, Chen Y, Li X, Yang W, Wen C, Kang Y, Wang A, Zhou Y (2017) Effects of synbiotic supplementation on growth performance, carcass characteristics, meat quality and muscular antioxidant capacity and mineral contents in broilers. J Sci Food Agric 97(11):3699–3705. https://doi.org/10.1002/jsfa.8230

Cheng Y-H, Horng Y-B, Dybus A, Yu Y-H (2021) Bacillus licheniformis-fermented products improve growth performance and intestinal gut morphology in broilers under Clostridium perfringens challenge. J Poult Sci 58(1):30–39. https://doi.org/10.2141/jpsa.0200010

Cummings JH, Macfarlane GT (1997) Collaborative JPEN-clinical nutrition scientific publications role of intestinal bacteria in nutrient metabolism. J Parenter Enter Nutr 21(6):357–365. https://doi.org/10.1177/0148607197021006357

Dalié DKD, Deschamps AM, Richard-Forget F (2010) Lactic acid bacteria—potential for control of mould growth and mycotoxins: a review. Food Control 21(4):370–380. https://doi.org/10.1016/j.foodcont.2009.07.011

de Oliveira JE, van der Hoeven-Hangoor E, van de Linde IB, Montijn RC, van der Vossen JMBM (2014) In ovo inoculation of chicken embryos with probiotic bacteria and its effect on posthatch salmonella susceptibility. Poult Sci 93(4):818–829. https://doi.org/10.3382/ps.2013-03409

Deng Q, Shi H, Luo Y, Zhao H, Liu N (2020) Effect of dietary lactobacilli mixture on listeria monocytogenes infection and virulence property in broilers. Poult Sci 99(7):3655–3662. https://doi.org/10.1016/j.psj.2020.03.058

Deng W, Dong XF, Tong JM, Zhang Q (2012) The probiotic bacillus licheniformis ameliorates heat stress-induced impairment of egg production, gut morphology, and intestinal mucosal immunity in laying hens. Poult Sci 91(3):575–582. https://doi.org/10.3382/ps.2010-01293

Di Gioia D, Biavati B (eds) (2018) Probiotics and prebiotics in animal health and food safety. Springer International Publishing, Cham

Dianawati D, Mishra V, Shah NP (2013) Effect of drying methods of microencapsulated lactobacillus acidophilus and Lactococcus lactis ssp. cremoris on secondary protein structure and glass transition temperature as studied by Fourier transform infrared and differential scanning calorimetry. J Dairy Sci 96(3):1419–1430. https://doi.org/10.3168/jds.2012-6058

Dicks LMT, Dreyer L, Smith C, Van Staden AD (2018) A review: the fate of Bacteriocins in the human gastro-intestinal tract: do they cross the gut–blood barrier? Front Microbiol 9:2297. https://doi.org/10.3389/fmicb.2018.02297

Dong S, Li L, Hao F, Fang Z, Zhong R, Wu J, Fang X (2024) Improving quality of poultry and its meat products with probiotics, prebiotics, and phytoextracts. Poult Sci 103(2):103287

Dreyer L, Smith C, Deane SM, Dicks LMT, Van Staden AD (2019) Migration of bacteriocins across gastrointestinal epithelial and vascular endothelial cells, as determined using in vitro simulations. Sci Rep 9(1):11481. https://doi.org/10.1038/s41598-019-47843-9

El Kadri H, Lalou S, Mantzouridou F, Gkatzionis K (2018) Utilisation of water-in-oil-water (W1/O/W2) double emulsion in a set-type yogurt model for the delivery of probiotic lactobacillus paracasei. Food Res Int Ott Ont 107:325–336. https://doi.org/10.1016/j.foodres.2018.02.049

Elghandour MMY, Khusro A, Adegbeye MJ, Tan Z, Abu Hafsa SH, Greiner R, Ugbogu EA, Anele UY, Salem AZM (2020) Dynamic role of single-celled fungi in ruminal microbial ecology and activities. J Appl Microbiol 128(4):950–965. https://doi.org/10.1111/jam.14427

Feng T, Wang J (2020) Oxidative stress tolerance and antioxidant capacity of lactic acid bacteria as probiotic: a systematic review. Gut Microbes 12(1):1801944. https://doi.org/10.1080/19490976.2020.1801944

Fontana L, Bermudez-Brito M, Plaza-Diaz J, Muñoz-Quezada S, Gil A (2013) Sources, isolation characterisation and evaluation of probiotics. Br J Nutr 109(Suppl 2):S35–S50. https://doi.org/10.1017/S0007114512004011

Fuller R (1992) History and development of probiotics. In: Probiotics. Springer, Netherlands, pp 1–8

Gaskins H (2003) The commensal microbiota and development of mucosal defense in the mammalian intestine. Alberta, Edmonton, Canada

Gordon S (2008) Elie Metchnikoff: father of natural immunity. Eur J Immunol 2008(38):3257–3264. https://doi.org/10.1002/eji.200838855

Guillot JF (1998) Les probiotiques en alimentation animale. Cahiers Agricultures 7(1):49–54

Gusils C, González SN, Oliver G (1999) Some probiotic properties of chicken lactobacilli. Can J Microbiol 45(12):981–987. https://doi.org/10.1139/w99-102

Haines MD, Parker HM, McDaniel CD, Kiess AS (2015) When rooster semen is exposed to lactobacillus fertility is reduced. Int J Poult Sci 14(9):541–547. https://doi.org/10.3923/ijps.2015.541.547

Harimurti S, Hadisaputro W (2015) Probiotics in poultry. In: Liong M-T (ed) Beneficial microorganisms in agriculture, aquaculture and other areas. Springer International Publishing, Cham, pp 1–19

Harkai P, Szabó I, Cserháti M, Krifaton C, Risa A, Radó J, Balázs A, Berta K, Kriszt B (2016) Biodegradation of aflatoxin-B1 and zearalenone by Streptomyces sp. collection. Int Biodeterior Biodegrad 108:48–56. https://doi.org/10.1016/j.ibiod.2015.12.007

Harrington D, Sims M, Kehlet AB (2016) Effect of Bacillus subtilis supplementation in low energy diets on broiler performance. J Appl Poult Res 25(1):29–39. https://doi.org/10.3382/japr/pfv057

Henriksson A, Szewzyk R, Conway PL (1991) Characteristics of the adhesive determinants of lactobacillus fermentum 104. Appl Environ Microbiol 57(2):499–502. https://doi.org/10.1128/aem.57.2.499-502.1991

Higgins SE, Wolfenden AD, Tellez G, Hargis BM, Porter TE (2011) Transcriptional profiling of cecal gene expression in probiotic- and salmonella-challenged neonatal chicks. Poult Sci 90(4):901–913. https://doi.org/10.3382/ps.2010-00907

Hill MJ (1997) Intestinal flora and endogenous vitamin synthesis. Eur J Cancer Prev 6:S43–S45. https://doi.org/10.1097/00008469-199703001-00009

Hong HA, Duc LH, Cutting SM (2005) The use of bacterial spore formers as probiotics. FEMS Microbiol Rev 29(4):813–835. https://doi.org/10.1016/j.femsre.2004.12.001

Huang F, Ding Z, Zhang C, Hu H, Zhang L, Zhang H (2018) Effects of calcium and zinc ions injection on caspase-3 activation and tenderness in post-mortem beef skeletal muscles. Int J Food Sci Technol 53(3):582–589. https://doi.org/10.1111/ijfs.13631

Jazi V, Foroozandeh AD, Toghyani M, Dastar B, Rezaie Koochaksaraie R, Toghyani M (2018) Effects of Pediococcus acidilactici, mannan-oligosaccharide, butyric acid and their combination on growth performance and intestinal health in young broiler chickens challenged with salmonella typhimurium. Poult Sci 97(6):2034–2043. https://doi.org/10.3382/ps/pey035

Jha R, Das R, Oak S, Mishra P (2020) Probiotics (direct-fed Microbials) in poultry nutrition and their effects on nutrient utilization, growth and laying performance, and gut health: a systematic review. Animals 10(10):1863. https://doi.org/10.3390/ani10101863

Jin LZ, Ho YW, Abdullah N, Jalaludin S (2000) Digestive and bacterial enzyme activities in broilers fed diets supplemented with lactobacillus cultures. Poult Sci 79(6):886–891. https://doi.org/10.1093/ps/79.6.886

Kan L, Guo F, Liu Y, Pham VH, Guo Y, Wang Z (2021) Probiotics bacillus licheniformis improves intestinal health of subclinical necrotic enteritis-challenged broilers. Front Microbiol 12:623739. https://doi.org/10.3389/fmicb.2021.623739

Karaman M, Basmacioglu H, Ortatatli M, Oguz H (2005) Evaluation of the detoxifying effect of yeast glucomannan on aflatoxicosis in broilers as assessed by gross examination and histopathology. Br Poult Sci 46(3):394–400. https://doi.org/10.1080/00071660500124487

Kaufmann SHE (2008) Immunology's foundation: the 100-year anniversary of the Nobel prize to Paul Ehrlich and Elie Metchnikoff. Nat Immunol 9(7):705–712. https://doi.org/10.1038/ni0708-705

Kiess AS, Hirai JH, Triplett MD, Parker HM, McDaniel CD (2016) Impact of oral lactobacillus acidophilus gavage on rooster seminal and cloacal lactobacilli concentrations. Poult Sci 95(8):1934–1938. https://doi.org/10.3382/ps/pew112

Latif A, Shehzad A, Niazi S, Zahid A, Ashraf W, Iqbal MW, Rehman A, Riaz T, Aadil RM, Khan IM, Özogul F, Rocha JM, Esatbeyoglu T, Korma SA (2023) Probiotics: mechanism of action, health benefits and their application in food industries. Front Microbiol 14:1216674. https://doi.org/10.3389/fmicb.2023.1216674. eCollection 2023

Laudadio V, Passantino L, Perillo A, Lopresti G, Passantino A, Khan RU, Tufarelli V (2012) Productive performance and histological features of intestinal mucosa of broiler chickens fed different dietary protein levels. Poult Sci 91(1):265–270. https://doi.org/10.3382/ps.2011-01675

Lee Y, Ji YR, Lee S, Choi M-J, Cho Y (2019) Microencapsulation of probiotic lactobacillus acidophilus KBL409 by extrusion technology to enhance survival under simulated intestinal and freeze-drying conditions. J Microbiol Biotechnol 29(5):721–730. https://doi.org/10.4014/jmb.1903.03018

Liu L, Li Q, Yang Y, Guo A (2021) Biological function of short-chain fatty acids and its regulation on intestinal health of poultry. Front Vet Sci 8:736739. https://doi.org/10.3389/fvets.2021.736739

Liu L, Xie M, Wei D (2022) Biological detoxification of mycotoxins: current status and future advances. Int J Mol Sci 23(3):1064. https://doi.org/10.3390/ijms23031064

Liu Q, Yu Z, Tian F, Zhao J, Zhang H, Zhai Q, Chen W (2020) Surface components and metabolites of probiotics for regulation of intestinal epithelial barrier. Microb Cell Factories 19(1):23. https://doi.org/10.1186/s12934-020-1289-4

Lokapirnasari WP, Sahidu AM, Maslachah L, Sabdoningrum EK, Yulianto AB (2020) Effect of lactobacillus casei and lactobacillus acidophilus in laying hens challenged by Escherichia coli infection. Sains Malays 49(6):1237–1244. https://doi.org/10.17576/jsm-2020-4906-03

Manus J, Millette M, Uscanga BRA, Salmieri S, Maherani B, Lacroix M (2021) In vitro protein digestibility and physico-chemical properties of lactic acid bacteria fermented beverages enriched with plant proteins. J Food Sci 86(9):4172–4182. https://doi.org/10.1111/1750-3841.15859

Mao J, Zhang S-Z, Du P, Cheng Z-B, Hu H, Wang S-Y (2020) Probiotics can boost the antitumor immunity of CD8+T cells in BALB/c mice and patients with colorectal carcinoma. J Immunol Res 2020:4092472. https://doi.org/10.1155/2020/4092472

Mason CK, Collins MA, Thompson K (2005) Modified electroporation protocol for lactobacilli isolated from the chicken crop facilitates transformation and the use of a genetic tool. J Microbiol Methods 60(3):353–363. https://doi.org/10.1016/j.mimet.2004.10.013

McFarland LV (2015) From yaks to yogurt: the history, development, and current use of probiotics. Clin Infect Dis Off Publ Infect Dis Soc Am 60(Suppl 2):S85–S90. https://doi.org/10.1093/cid/civ054

Meimandipour A, Hair-bejo M, Shuhaimi M, Azhar K, Soleimani AF, Rasti B, Yazid AM (2010) Gastrointestinal tract morphological alteration by unpleasant physical treatment and modulating role of lactobacillus in broilers. Br Poult Sci 51(1):52–59. https://doi.org/10.1080/00071660903394455

Monteagudo-Mera A, Rastall RA, Gibson GR, Charalampopoulos D, Chatzifragkou A (2019) Adhesion mechanisms mediated by probiotics and prebiotics and their potential impact on human health. Appl Microbiol Biotechnol 103(16):6463–6472. https://doi.org/10.1007/s00253-019-09978-7

Mountzouris KC, Tsitrsikos P, Palamidi I, Arvaniti A, Mohnl M, Schatzmayr G, Fegeros K (2010) Effects of probiotic inclusion levels in broiler nutrition on growth performance, nutrient digestibility, plasma immunoglobulins, and cecal microflora composition. Poult Sci 89(1):58–67. https://doi.org/10.3382/ps.2009-00308

Nagpal R, Kumar A, Kumar M, Behare PV, Jain S, Yadav H (2012) Probiotics, their health benefits and applications for developing healthier foods: a review. FEMS Microbiol Lett 334(1):1–15. https://doi.org/10.1111/j.1574-6968.2012.02593.x

Nasiri H, Golestan L, Shahidi S-A, Darjani P (2021) Encapsulation of lactobacillus casei in sodium alginate microcapsules: improvement of the bacterial viability under simulated gastrointestinal conditions using wild sage seed mucilage. J Food Meas Charact 15(5):4726–4734. https://doi.org/10.1007/s11694-021-01022-5

Neal-McKinney JM, Lu X, Duong T, Larson CL, Call DR, Shah DH, Konkel ME (2012) Production of organic acids by probiotic lactobacilli can be used to reduce pathogen load in poultry. PLoS One 7(9):e43928. https://doi.org/10.1371/journal.pone.0043928

Neupane D, Nepali D, Devkota N, Sharma M, Kadaria I (2019) Effect of probiotics on production and egg quality of dual purpose chicken at Kathmundu in Nepal. Bangladesh J Anim Sci 48(1):29–35. https://doi.org/10.3329/bjas.v48i1.44556

Nidamanuri AL, Leslie Leo Prince L, Yadav SP, Bhattacharya TK, Konadaka SRR, Bhanja SK (2021) Effect of supplementation of fermented yeast culture on hormones and their receptors on exposure to higher temperature and on production performance after exposure in Nicobari chickens. Int J Endocrinol 2021:5539780. https://doi.org/10.1155/2021/5539780

Nurmi E, Rantala M (1973) New aspects of salmonella infection in broiler production. Nature 241:210–211

Onifade AA, Babatunde GM (1996) Supplemental value of dried yeast in a high-fibre diet for broiler chicks. Anim Feed Sci Technol 62(2–4):91–96. https://doi.org/10.1016/S0377-8401(96)00991-1

Pan D, Yu Z (2014) Intestinal microbiome of poultry and its interaction with host and diet. Gut Microbes 5(1):108–119. https://doi.org/10.4161/gmic.26945

Paryad A, Mahmoudi M (2008) Effect of different levels of supplemental yeast Saccharomyces cerevisiae on performance, blood constituents and carcass characteristics of broiler chicks. Afr J Agric Res 3:835–842

Perumalla AVS, Wythe LA, Ricke SC (2023) Probiotics in poultry preharvest food safety: historical developments and current prospects. In: Callaway TR, Ricke SC (eds) Direct-fed Microbials and prebiotics for animals. Springer International Publishing, Cham, pp 127–166

Pizzolitto RP, Bueno DJ, Armando MR, Cavaglieri L, Dalcero AM, Salvano MA (2011) Binding of aflatoxin B1 to lactic acid bacteria and Saccharomyces cerevisiae in vitro: a useful model to determine the most efficient microorganism. In: Guevara-Gonzalez RG (ed) Aflatoxins–biochemistry and molecular biology. InTech

Poudel B, Shterzer N, Sbehat Y, Ben-Porat N, Rakover M, Tovy-Sharon R, Wolicki D, Rahamim S, Bar-Shira E, Mills E (2022) Characterizing the chicken gut colonization ability of a diverse group of bacteria. Poult Sci 101(11):102136. https://doi.org/10.1016/j.psj.2022.102136

Raksha Rao K, Vipin AV, Hariprasad P, Anu Appaiah KA, Venkateswaran G (2017) Biological detoxification of aflatoxin B1 by bacillus licheniformis CFR1. Food Control 71:234–241. https://doi.org/10.1016/j.foodcont.2016.06.040

Rastall RA (2004) Bacteria in the gut: friends and foes and how to alter the balance. J Nutr 134(8 Suppl):2022S–2026S. https://doi.org/10.1093/jn/134.8.2022S

Rousseaux A, Brosseau C, Bodinier M (2023) Immunomodulation of B lymphocytes by prebiotics, probiotics and synbiotics: application in pathologies. Nutrients 15(2):269. https://doi.org/10.3390/nu15020269

Sabry Abd Elraheam Elsayed M, Shehata AA, Mohamed Ammar A, Allam TS, Ali AS, Ahmed RH, Abeer Mohammed AB, Tarabees R (2021) The beneficial effects of a multistrain potential probiotic, formic, and lactic acids with different vaccination regimens on broiler chickens challenged with multidrug-resistant Escherichia coli and salmonella. Saudi J Biol Sci 28(5):2850–2857. https://doi.org/10.1016/j.sjbs.2021.02.017

Samanya M, Yamauchi K (2002) Histological alterations of intestinal villi in chickens fed dried Bacillus subtilis var. natto. Comp Biochem Physiol A Mol Integr Physiol 133(1):95–104. https://doi.org/10.1016/s1095-6433(02)00121-6

Simon O, Vahjen W, Taras D (2004) Interaction of nutrition with intestinal microbial communities. In: Tucker LA, Taylor-Pickard JA (eds) Interfacing immunity, gut health and performance. Nottingham University Press, Nottingham

Simons A, Alhanout K, Duval RE (2020) Bacteriocins, antimicrobial peptides from bacterial origin: overview of their biology and their impact against multidrug-resistant bacteria. Microorganisms 8(5):639. https://doi.org/10.3390/microorganisms8050639

Śliżewska K, Cukrowska B, Smulikowska S, Cielecka-Kuszyk J (2019) The effect of probiotic supplementation on performance and the histopathological changes in liver and kidneys in broiler chickens fed diets with aflatoxin B1. Toxins 11(2):112. https://doi.org/10.3390/toxins11020112

Smith JM (2014) A review of avian probiotics. J Avian Med Surg 28(2):87–94. https://doi.org/10.1647/2012-031

Soltani S, Hammami R, Cotter PD, Rebuffat S, Said LB, Gaudreau H, Bédard F, Biron E, Drider D, Fliss I (2021) Bacteriocins as a new generation of antimicrobials: toxicity aspects and regulations. FEMS Microbiol Rev 45(1):fuaa039. https://doi.org/10.1093/femsre/fuaa039

Song X, Wang M, Jiao H, Zhao J, Wang X, Lin H (2022) Ghrelin is a signal to facilitate the utilization of fatty acids and save glucose by the liver, skeletal muscle, and adipose tissues in chicks. Biochim Biophys Acta BBA–Mol Cell Biol Lipids 1867(2):159081. https://doi.org/10.1016/j.bbalip.2021.159081

Tarabees R, Shehata AA, Allam TS, Dawood A, El-Sayed MSA, Gaber M (2023) Effects of probiotics and prebiotics and their combinations on growth performance and intestinal microbiota of broilers infected with mixed Salmonellae. Alex J Vet Sci 77(1):12. https://doi.org/10.5455/ajvs.143849

Tarabees R, El-Sayed MS, Shehata AA, Diab MS (2020) Effects of the probiotic candidate E. Faecalis-1, the Poulvac E. Coli vaccine, and their combination on growth performance, caecal microbial composition, immune response, and protection against E. Coli O78 challenge in broiler chickens. Probiotics Antimicrob Proteins 12(3):860–872. https://doi.org/10.1007/s12602-019-09588-9

Thoda C, Touraki M (2023) Immunomodulatory properties of probiotics and their derived bioactive compounds. Appl Sci 13(8):4726. https://doi.org/10.3390/app13084726

Tiwari G, Tiwari R, Pandey S, Pandey P (2012) Promising future of probiotics for human health: current scenario. Chron Young Sci 3(1):17. https://doi.org/10.4103/2229-5186.94308

Tsuda H, Miyamoto T (2010) Guidelines for the evaluation of probiotics in food. Report of a joint FAO/WHO working group on drafting guidelines for the evaluation of probiotics in food guidelines for the evaluation of probiotics in food. Report of a joint FAO/WHO working group on drafting guidelines for the evaluation of probiotics in food, 2002. Food Sci Technol Res 1:87–92

Volf J, Polansky O, Varmuzova K, Gerzova L, Sekelova Z, Faldynova M, Babak V, Medvecky M, Smith AL, Kaspers B, Velge P, Rychlik I (2016) Transient and prolonged response of chicken cecum mucosa to colonization with different gut microbiota. PLoS One 11(9):e0163932. https://doi.org/10.1371/journal.pone.0163932

Vreeland RH, Rosenzweig WD, Powers DW (2000) Isolation of a 250 million-year-old halotolerant bacterium from a primary salt crystal. Nature 407(6806):897–900. https://doi.org/10.1038/35038060

Walden KE, Hagele AM, Orr LS, Gross KN, Krieger JM, Jäger R, Kerksick CM (2022) Probiotic BC30 improves amino acid absorption from plant protein concentrate in older women. Probiotics Antimicrob Proteins 16:125. https://doi.org/10.1007/s12602-022-10028-4

Wang H, Ni X, Qing X, Zeng D, Luo M, Liu L, Li G, Pan K, Jing B (2017) Live probiotic lactobacillus johnsonii BS15 promotes growth performance and lowers fat deposition by improving lipid metabolism, intestinal development, and gut microflora in broilers. Front Microbiol 8:1073. https://doi.org/10.3389/fmicb.2017.01073

Wang J, Ji H (2019) Influence of probiotics on dietary protein digestion and utilization in the gastrointestinal tract. Curr Protein Pept Sci 20(2):125–131. https://doi.org/10.2174/1389203719666180517100339

Wang Y, Gu Q (2010) Effect of probiotic on growth performance and digestive enzyme activity of arbor acres broilers. Res Vet Sci 89(2):163–167. https://doi.org/10.1016/j.rvsc.2010.03.009

Xiang Q, Wang C, Zhang H, Lai W, Wei H, Peng J (2019) Effects of different probiotics on laying performance, egg quality, oxidative status, and gut health in laying hens. Anim Open Access J MDPI 9(12):1110. https://doi.org/10.3390/ani9121110

Yoha KS, Nida S, Dutta S, Moses JA, Anandharamakrishnan C (2022) Targeted delivery of probiotics: perspectives on research and commercialization. Probiotics Antimicrob Proteins 14(1):15–48. https://doi.org/10.1007/s12602-021-09791-7

Yousefi B, Eslami M, Ghasemian A, Kokhaei P, Salek Farrokhi A, Darabi N (2019) Probiotics importance and their immunomodulatory properties. J Cell Physiol 234(6):8008–8018. https://doi.org/10.1002/jcp.27559

Zaghari M, Sarani P, Hajati H (2020) Comparison of two probiotic preparations on growth performance, intestinal microbiota, nutrient digestibility and cytokine gene expression in broiler chickens. J Appl Anim Res 48(1):166–175. https://doi.org/10.1080/09712119.2020.1754218

Zhang L, Lou Y, Schutyser MAI (2018) 3D printing of cereal-based food structures containing probiotics. Food Struct 18:14–22. https://doi.org/10.1016/j.foostr.2018.10.002

Prebiotics: An Overview on Their Properties and Mode of Action

5

Awad A. Shehata, Shereen Basiouni, Claudia Huber,
Guillermo Tellez-Isaias, Hafez M. Hafez,
and Wolfgang Eisenreich

Contents

A. A. Shehata · C. Huber · W. Eisenreich (✉)
Structural Membrane Biochemistry, Bavarian NMR Center, Technical University of Munich
(TUM), Garching, Bayern, Germany
e-mail: wolfgang.eisenreich@mytum.de

S. Basiouni
Institute of Molecular Physiology, Johannes-Gutenberg University, Mainz, Germany

G. Tellez-Isaias
Division of Agriculture, Department of Poultry Science, University of Arkansas,
Fayetteville, AR, USA

H. M. Hafez
Institute of Poultry Diseases, Faculty of Veterinary Medicine, Free University of Berlin,
Berlin, Germany

Abstract

Prebiotics, such as inulin and mannan-oligosaccharides, have demonstrated promising outcomes. Prebiotics are non-digestible feed elements that are metabolized by intestinal microbiota members and promote the host's health. Prebiotics are primarily used to modulate the gut microbiota in a way that benefits the host animal, offering benefits to the intestinal environment as well as the body as a whole. Positive gains in productive metrics, including body weight increase, feed conversion ratio, mortality index, and egg production, have been reported. Additionally, prebiotics have been shown to be successful in lowering the colonization of significant infections. Interestingly, prebiotics are less likely to have adverse effects on the host. In this chapter, we will discuss the main prebiotics used in poultry and their mechanism of action.

Keywords

Prebiotics · Inulin · Mannan-oligosaccharides · Intestinal microbiota

1   Introduction

Prebiotics are a relatively recent approach in nutrition that employs indigestible dietary ingredients like oligosaccharides (75). They have the potential to modulate gut microbiota, such as Bifidobacteria and lactic acid bacteria, that can improve the host's health. In the 1990s, Gibson and Roberfroid defined a prebiotic as a non-digestible food ingredient that selectively promotes the growth and/or activity of beneficial bacteria, ultimately leading to health benefits (Gibson and Roberfroid 1995). According to the initial definition, prebiotics must meet the following requirements: first, they cannot be absorbed or hydrolyzed in the upper intestinal tract; second, they act as a specific source of nutrients to support beneficial gut microbiota; and third, they trigger beneficial effects in the gut, or they benefit the host in some way. In this regard, non-digestible carbohydrates such as fructo-oligosaccharides (FOS), galacto-oligosaccharides (GOS), mannan-oligosaccharides (MOS), Chitosan oligosaccharides (COS), and related carbohydrate polymers were classified as prebiotics (Patterson and Burkholder 2003; Ricke 2015, 2018).

In December 2016, a prebiotic was refined by the International Scientific Association of Probiotics and Prebiotics (ISAPP) as "a substrate that is selectively utilized by host microorganisms conferring a health benefit" (Gibson et al. 2017). Even though prebiotics are known to be based on carbohydrates, the word "substrate" refers to any substance that affects only specific bacterial communities. In this regard, other carbohydrates and non-carbohydrates have also been considered as potential prebiotics (Fig. 5.1); thus, they can produce metabolites that modulate the intestinal microbiota (Gibson et al. 2017).

Polyphenols are classified as prebiotics, based on the new definition, since they are transformed into biologically active molecules that ultimately provide health benefits to the host via promoting beneficial gut bacteria. However, Rodríguez-Daza

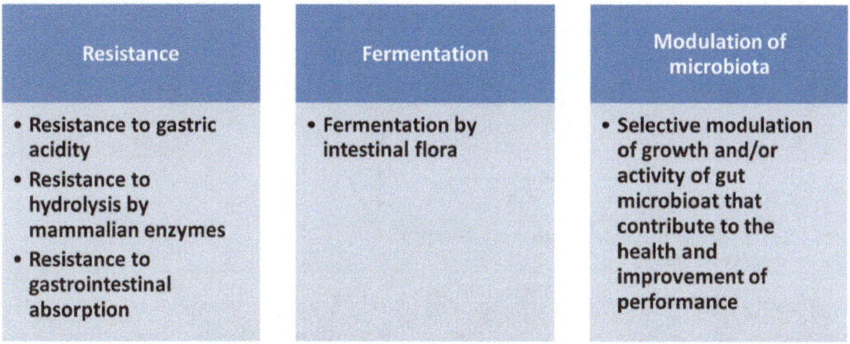

Resistance	Fermentation	Modulation of microbiota
• Resistance to gastric acidity • Resistance to hydrolysis by mammalian enzymes • Resistance to gastrointestinal absorption	• Fermentation by intestinal flora	• Selective modulation of growth and/or activity of gut microbioat that contribute to the health and improvement of performance

Fig. 5.1 Criteria of prebiotic candidate

et al. described polyphenols as duplibiotic, unabsorbed substrates modulating the gut microbiota by both antimicrobial and prebiotic modes of action (Rodríguez-Daza et al. 2021). This review highlights the types of prebiotics and their mode of action in poultry.

Prebiotics are known to strengthen gut microbiota and reduce pathogen colonization in chickens. They are classified as generally accepted as safe (GRAS) (Kocot et al. 2022). The most commonly used prebiotics in poultry are inulin-type fructans (ITF), β-glucan, FOS, MOS, xylooligosaccharides (XOS), GOS, and isomaltooligosaccharides (IMO) (Józefiak et al. 2008; Pourabedin and Zhao 2015; Slawinska et al. 2019)]. FOS, inulin, GOS, and resistant starch are based on 6-carbon sugars such as fructose, glucose, and galactose. However, XOS and AXOS are based on 5-carbon sugars. Prebiotics can be used in feed directly after hatching or even in ovo at a dose of 10 times less than after hatching (Bednarczyk et al. 2016; Siwek et al. 2018).

2 General Mechanisms of Prebiotics

Prebiotics exert beneficial effects through several mechanisms (Solis-Cruz et al. 2020): (i) modulation of intestinal microbiota, (ii) prevention or decreasing the colonization of pathogens, (iii) maintaining gut health, (iv) immunomodulation, and (v) improvement of animal performance (Fig. 5.2).

2.1 Modulation of Gut Microbiota and the Immune System

Prebiotics are beneficial for intestinal health as they help to regulate the gut and promote the growth of "good" bacteria, such as *Lactobacillus* and *Bifidobacterium* species. These bacteria can ferment and metabolize prebiotics, which selectively promotes their growth and activity (Téllez et al. 2015). This may affect the production of short-chain fatty acids (SCFA) and reduce the pH in the intestines. Moreover,

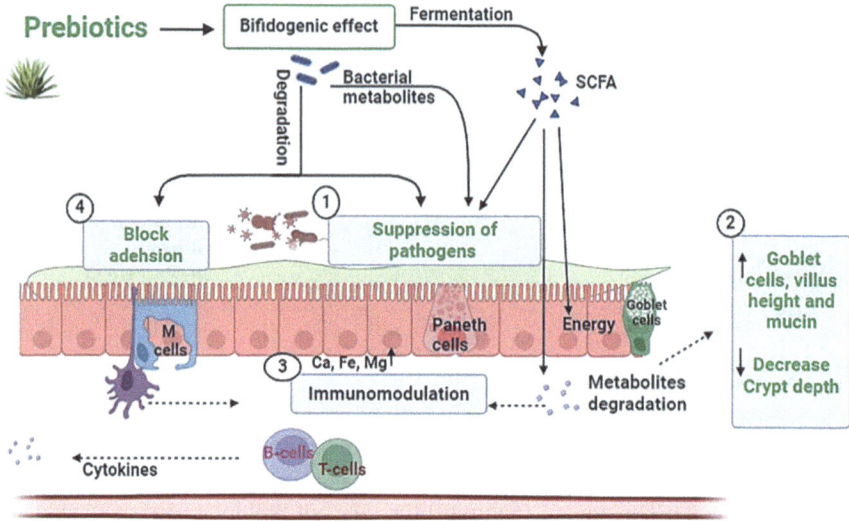

Fig. 5.2 Potential mechanisms of prebiotics. The figure was created by BioRender

they can improve metabolism by increasing the generation of vitamins and digestive enzymes while also lowering levels of cholesterol, triglycerides, and odor compounds. Probiotics can strengthen the immune system, which helps to prevent the growth of harmful bacteria (Patterson and Burkholder 2003).

2.2 Prevention of Pathogen Colonization

Prebiotics decrease the colonization of harmful bacteria in the gastrointestinal system in chickens. The extent of this effect is determined by the nature of the prebiotics being used. Prebiotics prevent pathogen colonization through mechanisms such as increasing SCFA production, stimulating beneficial bacteria for competitive exclusion, and directly binding with pathogen lectins (Solis-Cruz et al. 2020). The reduction of pathogen colonization may be attributed to the stimulating *Lactobacillus* and *Bifidobacterium* which produce SCFAs (Ricke 2018).

Indeed, several findings have explained how SCFAs reduce the colonization of pathogens, but many of the mechanisms are still not fully understood: (i) SCFAs prevent pathogen colonization by lowering intestinal pH, which dissipates the proton motive force across bacterial cell membranes (Russell 1992). (ii) Butyrate has the ability to reduce the expression of invasive genes in *Salmonella* (Van Immerseel et al. 2006). (iii) The increase of SCFAs in the caecum of broiler chickens was also found to be correlated with the reduction of Enterobacteriaceae, while they did not affect *Lactobacillus* (van Der Wielen et al. 2000). (iv) An increase in mucin production can serve as a physical barrier (Willemsen et al. 2003; Tellez et al. 2006). (v) SCFAs support colonocytes and gut integrity by providing a preferable metabolic energy source (Józefiak et al. 2004; Alloui et al. 2013). (vi) Additionally, stimulation of beneficial bacteria such as *Lactobacillus* and *Bifidobacteria* by prebiotics can reduce colonization by the competitive exclusion mechanism (Chen et al. 2007; Muñoz et al. 2012; Bucław 2016). (vii) Certain bacteria, such as *Salmonella* spp. and *E. coli*, attach to intestinal cells using mannose-specific lectins. Prebiotics, such as MOS, can bind to lectins and hinder the colonization process (Gaggìa et al. 2010; Teng and Kim 2018). It was found that mannose at 2.5% (w/v) reduces the colonization of *S. Typhimurium* in broiler chicks (Oyofo et al. 1989).

2.3 Gut Morphology

Several studies have shown that prebiotics can improve gut morphology and function through multiple mechanisms. (i) Prebiotics can improve the height of villi and microvilli (Yusrizal and Chen 2003; Rehman et al. 2007), which leads to an improvement in nutrient utilization. (ii) Prebiotics can increase the number of goblet cells, resulting in increased mucin production (Johansson et al. 2008). The addition of MOS and β-glucans (at 1 and 2 lb./ton) to turkey feed has been shown to enhance villus height, surface area, lamina propria thickness, crypt depth, and goblet cell count (de los Solis Santos et al. 2007).

2.4 Improvement of Bone Health

Prebiotics can enhance the absorption of essential minerals, which is vital for maintaining bone health (Egbu et al. 2022). They can also boost the production of hormones that promote bone growth, such as IGF-1 (Shehata et al. 2022; Jaiswal et al. 2023). By affecting the absorption and deposition of minerals, prebiotics can help to strengthen the skeletal structure of poultry. The prebiotic *Aspergillus* is linked to promoting bone mineralization (Selim et al. 2022; Martin et al. 2023).

3 Commonly Used Prebiotic in Poultry

3.1 Inulin-Type Fructans (ITF): Inulin and Fructo-oligosaccharides (FOS)

Oligosaccharides are made up of different components like maltodextrins and inulin. Although maltodextrins are digestible, in contrast, inulin is indigestible. As a result, inulin reaches the lower gastrointestinal tract where it promotes the growth of beneficial bacteria such as Bifidobacteria. The term "inulin-type" (ITF) is a general term that refers to all $\beta(2 \rightarrow 1)$ linear fructans and includes inulin, short-chain fructo-oligosaccharides, and oligofructose (Roberfroid 2005, 2007), Fig. 5.3.

Several natural sources of inulin are known, including plants, bacteria, and fungi. Chicory root (*Cichorium intybus*) is the most abundant source of inulin, and it is commonly used to produce ITFs that are available commercially. However,

Fig. 5.3 Chemical structure of inulin, glucosidic bond of (2 → 1)-linkages

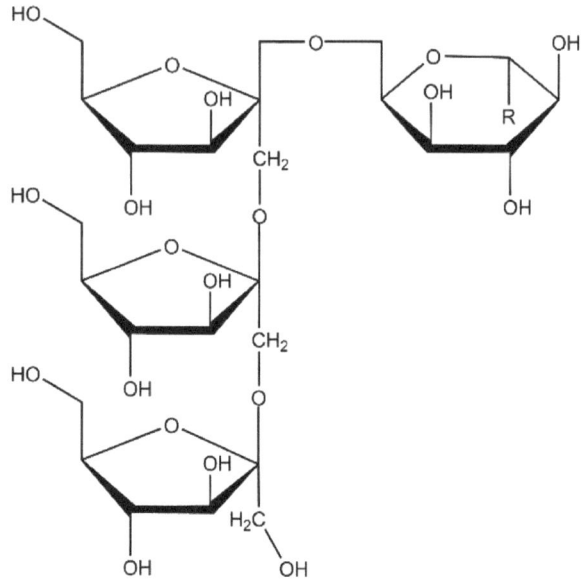

Jerusalem artichoke tubers (*Helianthus tuberosus*), artichoke inflorescence (*Cynara cardunculus*) (Redondo-Cuenca et al. 2021), burdock root (Asteraceae), onions, garlic, and leeks are also rich in inulin.

Based on the production methods, several forms of inulin are known (Niness 1999; Franck 2002): (i) Standard inulin: It can be produced by hot water extraction of inulin-rich plants like chicory roots, followed by purification, evaporation, and spray drying. (ii) High-performance inulin: It can be produced by the same separation procedures followed by eliminating the shorter-chain molecules. It is known as long-chain inulin or high-molecular-weight inulin-type fructan. (iii) Oligofructose: It can be produced by partial enzymatic hydrolysis of the standard inulin into short-chain carbohydrates. Oligofructose that terminates with glucose molecules is known as FOS (Roberfroid 2005; Alexander et al. 2023).

Inulin is a commonly used prebiotic in poultry (Tiengtam et al. 2015). Due to its glucosidic bond of (2 → 1)-linkages, it is indigestible to the digestive enzymes of both monogastric animals and humans (Pool-Zobel et al. 2002) In the large intestine or colon, it is instead fermented by Bifidobacteria and other lactic acid-producing bacteria, increasing their population. Supplementing animal feed with inulin has been demonstrated to boost immunological parameters, adjust gut flora, and improve chicken performance (He et al. 2002; Nabizadeh 2012).

In laying hens, inulin increased Bifidobacterium and decreased coliforms in the cecum (Shang et al. 2010). It was also found that supplementation of broiler chickens with inulin (10 and 15 g/kg feed) upregulated the expression of mucin (Huang et al. 2015). In 2018, Shang and colleagues found that including inulin in the diet of laying hens can enhance their antioxidant status by increasing SOD, CAT, and GSH-Px and decreasing MDA (Shang et al. 2018). A study has found that a combination of inulin and saponin, at a rate of 0.1 and 0.005 g/kg feed, significantly improved egg production in terms of egg, albumen, and shell weights (Lonkar et al. 2020).

FOS, indigestible oligosaccharides (Fig. 5.4), can be produced either by enzymatic synthesis or by hydrolysis of inulin. Enzymatic synthesis requires glucosyltransferases or β-fructofuranosidases found in fungi like *Aureobasidium pullulans*, *Aspergillus* spp., and *Penicillium* spp. (Nobre et al. 2018, 2019; Castro et al. 2019; Karkeszová and Polakovič 2023). Producing short-chain FOS (scFOS) from sucrose is more cost-effective and yields higher purity than extraction from plants.

Supplementing broiler chickens with FOS (4.0 g/kg) resulted in a significant increase in *Lactobacillus* and *Bifidobacterium* in the small intestine, while *E. coli* was significantly reduced (Gibson and Roberfroid 1995). Supplementing ITF to chicken feed has been demonstrated to be an efficient strategy as an alternative to antimicrobials. The beneficial effects of ITF in poultry can be summarized as follows: (i) ITF are not hydrolyzed in the upper GIT (Kumar et al. 2019) and can be fermented by *Bifidobacteria* and *Lactobacillus* spp. (Ricke 2015, 2018). (ii) ITF contributes to the stimulation of immune response (Khodambashi Emami et al. 2012) and maintenance of gut health (Xu et al. 2003); (iii) ITF modulates gut microbiota (Bailey et al. 1991; Yusrizal and Chen 2003). (iv) ITF increases SCFA, which exhibits antibacterial effects against pathogens such as *Clostridium perfringens* (Ricke 2015, 2018; Kumar et al. 2019). (v) ITF reduces the colonization of *Salmonella* spp. (Donalson et al. 2007; Ricke 2015, 2018); (iv) ITF improves animal performance, including BWG and FCR (Ricke 2015, 2018; Shang and Kim 2016; Kim et al. 2019) (Table 5.1).

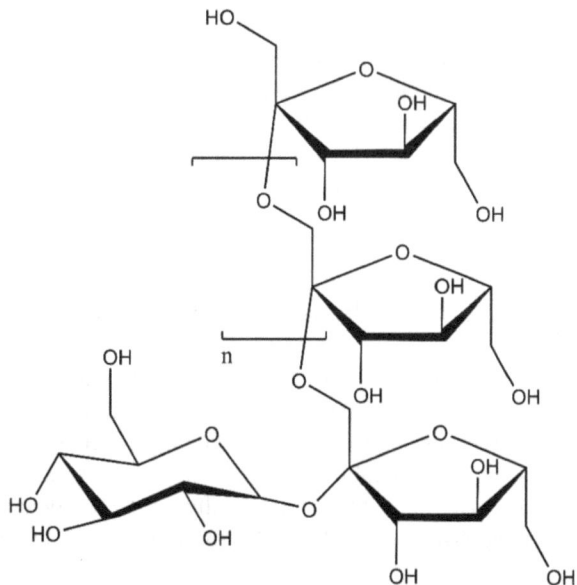

Fig. 5.4 Chemical structure of fructo-oligosaccharides

Table 5.1 Potential positive effects of inulin and FOS in poultry

Species	Age	Dose	Beneficial effects	Reference
Broilers	1 day–5 weeks	200 mg/kg	Improvement of gut health Antioxidant effect No effect on performance	(Yang et al. 2022b)
Broiler	1–35 days	0, and 4 g/kg	Increasing *Lactobacillus* Reduction of coliform Increasing acetic and butyric acid	(Akbaryan et al. 2019)
Layers	60–65 to 63–68 weeks	0, 5, 10 g/kg	Reduction of *Salmonella enteritidis* with 10,000 mg/kg FOS	(Adhikari et al. 2018)
Layer	100–105 to 101–106 week	0, 3750, 7500 mg/kg	Increasing SCFAs: Propionate, butyrate, and lactic acid	(Donalson et al. 2008)
Broilers	0–6 weeks	0.6 g/kg	Leucine aminopeptidase activity improvement of feed conversion ratio	(Williams et al. 2008)
Broiler	28–42 days	0, 4100 mg/kg	Increasing valeric acid	(Cao et al. 2005)
Broiler	1–49 days	0, 2, 4, and 8 g(kg	Increasing of anaerobes and Bifidobacterium with 4000 mg/kg Increasing of *Lactobacillus* with 2000 and 4000 mg/g Reducing *E. coli* with 2000 and 4000 mg/g	(Xu et al. 2003)

Potential Harmful Effects

It was proposed that FOS could also exhibit adverse effects in poultry (Ten Bruggencate et al. 2003) by enhancing the growth of some pathogens. This concern is based on the potentially harmful effect of the excessive production of SCFAs through the rapid fermentation of prebiotics that could decrease the pH and consequently damage the gut mucosa. Additionally, it was postulated that in hosts with healthy microbiota, prebiotics may not provide additional benefits to the hosts (Ten Bruggencate et al. 2003). Based on these concerns, the optimal dose has to be considered. The optimum levels of FOS that may enhance birds' performance and health, especially in stressful conditions, were proposed to be 2.5 and 5.0 g/kg diet. It was found that high doses of FOS (10 g/kg) may cause diarrhea and a reduction in animal performance (Wu et al. 1999).

3.2   Galacto-oligosaccharides (GOS)

Chemical Structure and Technological Production Aspects

GOSs, or galacto-oligosaccharides, are made up of either α-GOS or β-GOS. The main target for prebiotic uses is β-GOS. (Mei et al. 2022), due to its desirable properties, including solubility, low viscosity, stability, and pleasant texture with a sweet flavor (Torres et al. 2010). Chemically, GOS is made up of a glucose molecule at the reducing terminus and 1–7 galactose units connected by glycosidic linkages, such

Fig. 5.5 Chemical structure of galacto-oligosaccharides

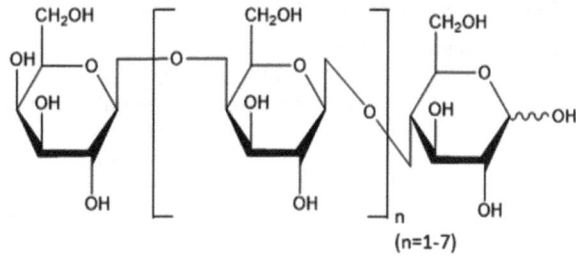

(n=1-7)

as β-(1,2), β-(1,3) β(1,4), and β-(1,6) (De Almeida and Maitan-Alfenas 2021) (Fig. 5.5).

GOS are naturally occurring compounds that can be obtained from biological sources (Kim et al. 2003). However, because of their relatively low concentration, their extraction is not economically advantageous. GOS can also be produced chemically by acidic hydrolysis of lactose. However, the chemical synthesis approach is not recommended since it produces undesired chemicals and lacks product selectivity (Ambrogi et al. 2023). The current recommended approach for commercial GOS production is biosynthesis-utilizing lactose, a byproduct of the dairy industry produced in vast amounts.

GOS is produced by enzymatic hydrolysis (Mei et al. 2022). Transgalactosylation is facilitated by β-galactosidase, which uses lactose as a substrate. The length of GOS can range from 2 to 8 monomers and is influenced by factors like lactose concentration, substrate composition, and reaction (Macfarlane et al. 2008; Ambrogi et al. 2021; Mei et al. 2022).

Beneficial Effects

GOS has the potential to improve chicken performance and enhance the development of beneficial gut flora (Jung et al. 2008; Bednarczyk et al. 2016; Richards et al. 2019). Supplementing chickens with 3.5 mg of a GOS combination in ovo boosted the Bifidobacterium in the intestine; however, Lactobacillus was higher in the control group (Slawinska et al. 2019).

Additionally, GOS may lessen the harmful effects of heat stress (Varasteh et al. 2015). The beneficial effects of GOS can be summarized as follows: (i) Modulation of gut microbiota by increasing *Bifidobacteria* (Tzortzis et al. 2005). Jung et al. reported a significant increase of *Bifidobacteria* and lactobacilli in broilers treated with GOS (Jung et al. 2008). (ii) Increasing the production of SCFAs, including acetic acid and lactic acid (Tzortzis et al. 2005). (iii) Impact on the immunological activities by modulating cytokines. It was found that GOS supplementation increased IL-17A expression and decreased IL-10 and IL-17F in the ileum and cecum (Richards et al. 2019). IL-17A is believed to have a positive impact on gut health, while IL-17F expression is linked to colitis development (Haas et al. 2015; Lee et al. 2018). (iv) Antiadhesive properties by inhibiting the adherence of enteropathogenic *E. coli* to Caco-2 cells and HEP-2 cells by 40–70% and 65%, respectively (Shoaf et al. 2006) (Table 5.2).

Table 5.2 Potential positive effects of GOS in poultry

Species	Age	Dose		Reference
Broiler	1–35	1–11 day (3.5% 12–35 day f1.685%	Increasing *Lactobacillus johnsonii* Upregulation of ileal and cecal interleukin-17A (IL-17A) gene expression Downregulation of IL-10.	(Richards et al. 2019)
Broiler	0–35 days	0, 1000, 2000, 5000	Increased *Lactobacillus* No effect of β-GOS on coliforms and clostridia	(Yousaf et al. 2017)
Broiler	1–70 days	0, 1000	Reduced diversity of microbiota with XOS and GOS; increased Ruminococcaceae, Barnesiellaceae, and Acidaminococcaceae and decreased Bacteroidaceae and Lactobacillaceae Alistipes dominant with GOS	(Yang et al. 2022a)
Layers	1–27 days	1%	GOS can modify both cecal tonsil gene expression and the cecal microbiome	(Hughes et al. 2017)
Ross 308 eggs	In ovo	0.2 mL phosphate buffer saline containing 3.5 mg	No additional benefits to the intestinal transcriptomic response to heat stress	(Jung et al. 2008)

Fig. 5.6 Chemical structure of lactulose

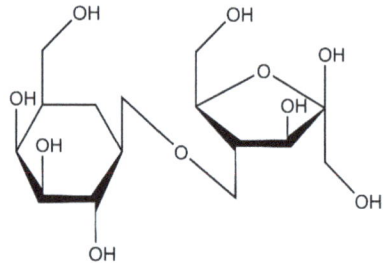

3.3   Lactulose

Lactulose, a synthetic disaccharide, is derived from lactose, similar to GOS (Fig. 5.6). It is mainly used as a laxative; however, at lower doses, lactulose exhibits prebiotic properties by modulating beneficial bacteria like *Lactobacillus* and *Bifidobacterium* (Schumann 2002; Tuohy et al. 2002).

It has been shown that supplementation of a poultry diet with lactulose at 0.1% or 0.2% can enhance the performance and reduce *E. coli* in feces (Cho and Kim 2014). Furthermore, lactulose can enhance gut morphology by favorably modulating the intestinal microflora, which in turn enhances the production of SCFAs in the cecum (Calik and Ergün 2015). The production, weight, and quality of eggs have been reported to be enhanced by the use of lactulose at a dosage of 0.10 mL per kg of body weight. Other beneficial impacts have been reported, including increasing the levels of alkaline phosphatase, lipase, and phosphorus while lowering total bilirubin, total protein, and globulin (Elkomy et al. 2023). Additionally, it has been found that lactulose at this dosage improves intestinal health by increasing the number of goblet cells, mucus secretion, and villus length.

3.4   Isomaltooligosaccharides (IMO)

IMOs, glucose monomers with α-1,6 glucosidic bonds (Fig. 5.7), are found naturally in miso, soy sauce, sake, and honey. Commercially, IMO is produced from starch hydrolysis by α-amylase and pullulanase to produce soluble starch. Maltose, produced by hydrolysis of soluble starch by β-amylase, is a substrate for α-glucosidase (Crittenden and Playne 1996; Rastall and Maitin 2002; Alexander et al. 2023). Chang et al. studied the effect of IMO at a dose of 3 g/kg feed in broiler chickens (1–56 days). IMO provided several beneficial effects, including modulation of intestinal microbiota (increased Bacteroidetes, Tenericutes, Euryarchaeota, Spirochaetae, Ruminocaceae, and Lachnoclostridium) and production of SCFA (butyric acid and valeric acid) (Chang et al. 2022). However, Zhang et al. did not see the effects of IMO on SCFA in broiler chickens supplemented at doses of 3, 6, 9, and 12 g/kg feed (Zhang et al. 2003).

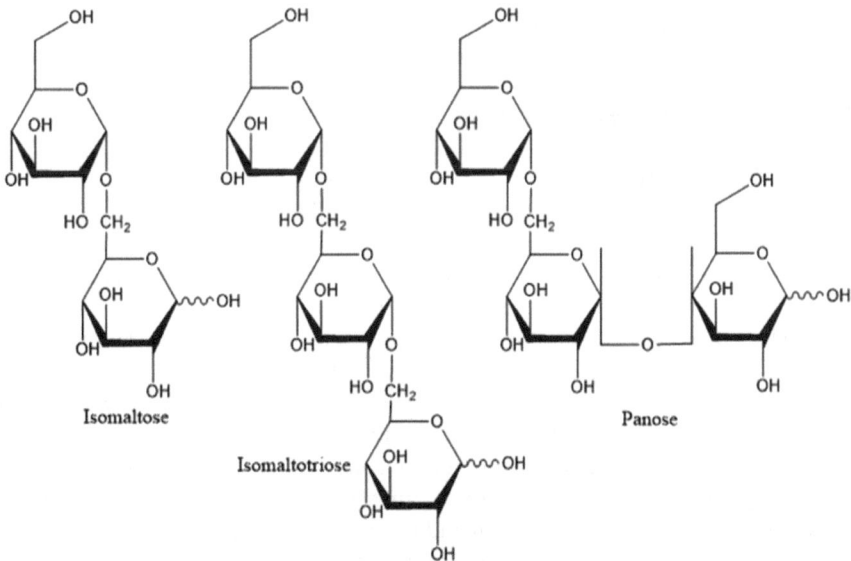

Fig. 5.7 Chemical structures of isomaltooligosaccharides

Fig. 5.8 Chemical
structure of
xylooligosaccharides

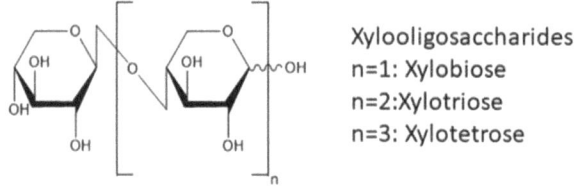

Xylooligosaccharides
n=1: Xylobiose
n=2:Xylotriose
n=3: Xylotetrose

3.5 Xylooligosaccharides (XOS)

XOS, xylose molecules with β-1,4 linkages, are naturally found in honey, bamboo shoots, fruits, vegetables, and milk. Commercially, pXOS can be produced by enzymatic hydrolysis of corn cobs. Other substrates, such as straws, hardwoods, bagasse, hulls, and bran, can be used (Fig. 5.8). Xylanases are produced by *Trichoderma reesei, T. harzianum, T. viride, T. koningii*, and *T. longibrachiatum* (Casci and Rastall 2006). Numerous studies have reported several health benefits of XOS for both humans and animals. These benefits include restoring normal stool consistency, increasing the frequency of bowel movements, reducing serum triglyceride and cholesterol levels, lowering HbA1c and fasting glucose levels, and altering immune cell populations and responsiveness.

Selective modulation of the gut microbiota, including *Lactobacillus* and *Bifidobacterium*, led to an increase in SCFAs, such as butyrate, in poultry (Morgan et al. 2019, 2020; Bautil et al. 2020). Other beneficial effects have been reported, including immunomodulation, increasing immune cells in the GIT, cecal weight, and villi length, improving nutrient utilization, and modulating GIT microbiota composition (Morgan et al. 2019).

3.6 Yeast-Derived Prebiotics

3.6.1 Beta-Glucans

Beta-glucans are important components of cell walls that can be found in various organisms such as bacteria, algae, fungi, and yeast. They are also present in grains like rye, barley, and oats (Vetvicka and Vetvickova 2014). Beta-glucans consist of β-D-glucose molecules, which are linked by $(1 \rightarrow 3)$- or $(1 \rightarrow 4)$-glycosidic linkages. The branching structure of yeast $(1 \rightarrow 3)(1 \rightarrow 6)$-β-glucans is due to the $(1 \rightarrow 6)$-β-links that connect the side chains to the backbone (Fig. 5.9). Fungal ß-glucans possess short branches, determinantal for biological activity, while yeast β-glucans have longer ones (Jacob and Pescatore 2014).

Several studies have demonstrated the potential utility of beta-glucans in both immune system stimulation and the replacement of antibiotics (Chae et al. 2006). (i) It stimulates immunological cells, including macrophages and natural killer cells, and increases the production of antibodies. Similar to mammals, avian macrophages synthesize cytokines and chemokines such as tumor necrosis factor-alpha (TNFα), IL4, IL6, IL8, IL10, and interferon-gamma (IFN-γ) (Cox et al. 2010; Han et al.

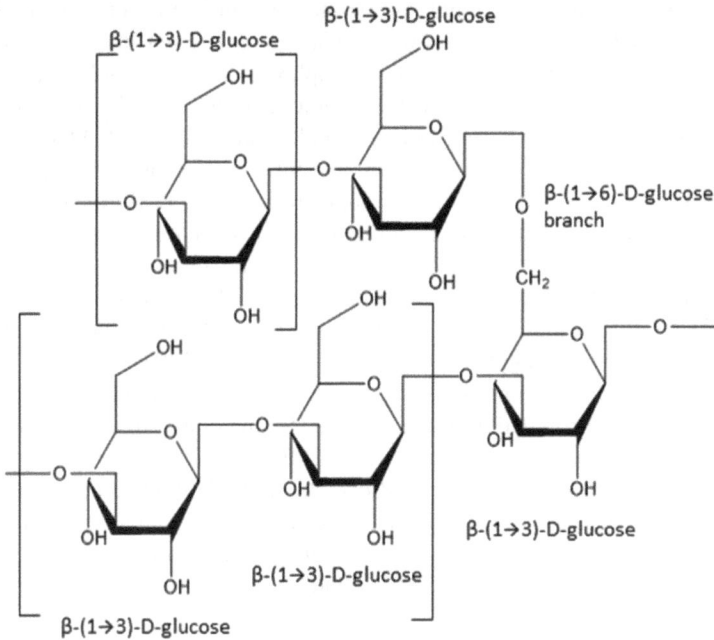

Fig. 5.9 Chemical structure of yeast β-glucans

Fig. 5.10 Chemical structure of mannan-oligosaccharides

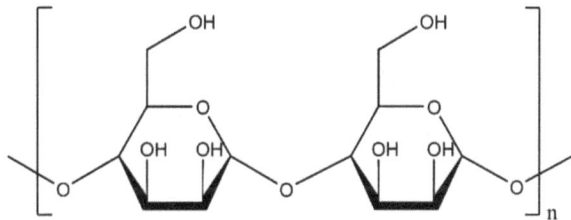

2017). Glucans can also influence the expression of immune-related genes and proteins. The immune stimulation helps fight against enteric infection and immune suppression caused by high-stress rearing conditions (Jacob and Pescatore 2017). Beta-glucans were found to be effective in enhancing the growth of broiler chickens and improving meat quality (Cho et al. 2013). Beta-glucans have been shown to improve gut health in poultry subjected to bacterial challenges and increase the flow of new immune cells into lymphoids (Stier et al. 2014). Beta-glucans can increase macrophage functionality (Jin et al. 2018), affect intestinal morphology, and function as anti-inflammatory immunomodulators (Vetvicka and Vetvickova 2014).

3.6.2 Mannan-oligosaccharides (MOS)

MOS (Fig. 5.10), a structural component of the yeast cell walls, is known to enhance the growth performance of broiler chicks (Hooge 2004; Rosen 2007). It is worth noting that the improvement in gut microbiota is mainly noticed in young birds, as they have a smaller population. However, older birds are not likely to experience any significant improvement due to their complex microbiota population.

MOS is known for its ability to bind pathogenic bacteria that possess type-1 fimbriae (Spring et al. 2000), such as *E. coli* and *Salmonella* spp. By blocking bacterial lectin, MOS can reduce the colonization of these pathogens in the gut (Ofek and Beachey 1978). It has been shown that supplementing MOS from 0.08% to 0.5% can modulate the cecal microbial community by increasing *Lactobacillus* and *Bifidobacterium* while decreasing *Salmonella*, *E. coli*, *Clostridium perfringens*, and *Campylobacter* (Spring et al. 2000; Fernandez et al. 2002; Corrigan et al. 2015).

Additionally, it is important to mention that the gastrointestinal tract of newly hatched chicks usually stabilizes at 2 weeks old (Rosen 2007). The main positive effects of MOS can be summarized as follows: (i) Reduction of pathogenic bacteria such as *Salmonella* population in the broilers (Spring et al. 2000). Degradation of mycotoxins such as aflatoxins (Zaghini et al. 2005). (ii) Immunomodulatory effects by increasing the production of IgA (Ricke 2015, 2018). MOS administration has shown some inconsistencies in animals' performance. Several studies reported no positive effects of MOS on animal performance (Pelicano et al. 2004; Stanczuk et al. 2005). However, Sims et al. (2004) found improvement in the live body weight of turkeys (Sims et al. 2004).

4   Conclusion

Prebiotics for mammalian nutrition must be resistant to stomach acidic pH, fermentable by intestinal microbiota, and selectively boost enteric bacteria. Prebiotics can be an alternative to antibiotics in poultry nutrition, but further research is needed to determine the appropriate dosage and mechanism of action.

References

Adhikari P, Cosby DE, Cox NA, Franca MS, Williams SM, Gogal RM, Ritz CW, Kim WK (2018) Effect of dietary fructooligosaccharide supplementation on internal organs salmonella colonization, immune response, ileal morphology, and ileal immunohistochemistry in laying hens challenged with salmonella enteritidis. Poult Sci 97:2525–2533. https://doi.org/10.3382/ps/pey101

Akbaryan M, Mahdavi A, Jebelli-Javan A, Staji H, Darabighane B (2019) A comparison of the effects of resistant starch, fructooligosaccharide, and zinc bacitracin on cecal short-chain fatty acids, cecal microflora, intestinal morphology, and antibody titer against Newcastle disease virus in broilers. Comp Clin Pathol 28:661–667. https://doi.org/10.1007/s00580-019-02936-9

Alexander C, Lin CY, Boler BMV, Fahey GC Jr, Swanson KS (2023) Prebiotics with plant and microbial origins. In: Callaway TR, Ricke SC (eds) Direct-fed microbials and prebiotics for animals science and mechanisms of action

Alloui MN, Szczurek W, Świątkiewicz S (2013) The Usefulness of Prebiotics and Probiotics in Modern Poultry Nutrition: a Review/Przydatność prebiotyków i probiotyków w nowoczesnym żywieniu drobiu—przegląd. Ann Anim Sci 13:17–32. https://doi.org/10.2478/v10220-012-0055-x

Ambrogi V, Bottacini F, Cao L, Kuipers B, Schoterman M, Van Sinderen D (2023) Galacto-oligosaccharides as infant prebiotics: production, application, bioactive activities and future perspectives. Crit Rev Food Sci Nutr 63:753–766. https://doi.org/10.1080/10408398.2021.1953437

Ambrogi V, Bottacini F, Mac Sharry J, Van Breen J, O'Keeffe E, Walsh D, Schoemaker B, Cao L, Kuipers B, Lindner C, Jimeno ML, Doyagüez EG, Hernandez-Hernandez O, Moreno FJ, Schoterman M, Van Sinderen D (2021) Bifidobacterial β-galactosidase-mediated production of Galacto-oligosaccharides: structural and preliminary functional assessments. Front Microbiol 12:750635. https://doi.org/10.3389/fmicb.2021.750635

Bailey JS, Blankenship LC, Cox NA (1991) Effect of fructooligosaccharide on salmonella colonization of the chicken intestine. Poult Sci 70:2433–2438. https://doi.org/10.3382/ps.0702433

Bautil A, Verspreet J, Buyse J, Goos P, Bedford MR, Courtin CM (2020) Arabinoxylan-oligosaccharides kick-start arabinoxylan digestion in the aging broiler. Poult Sci 99:2555–2565. https://doi.org/10.1016/j.psj.2019.12.041

Bednarczyk M, Stadnicka K, Kozłowska I, Abiuso C, Tavaniello S, Dankowiakowska A, Sławińska A, Maiorano G (2016) Influence of different prebiotics and mode of their administration on broiler chicken performance. Animal 10:1271–1279. https://doi.org/10.1017/S1751731116000173

Bucław M (2016) The use of inulin in poultry feeding: a review. J Anim Physiol Anim Nutr (Berl) 100:1015–1022. https://doi.org/10.1111/jpn.12484

Calik A, Ergün A (2015) Effect of lactulose supplementation on growth performance, intestinal histomorphology, cecal microbial population, and short-chain fatty acid composition of broiler chickens. Poult Sci 94:2173–2182. https://doi.org/10.3382/ps/pev182

Cao BH, Karasawa Y, Guo YM (2005) Effects of green tea polyphenols and Fructo-oligosaccharides in semi-purified diets on broilers` performance and Caecal microflora and their metabolites. Asian Australas J Anim Sci 18:85–89. https://doi.org/10.5713/ajas.2005.85

Casci T, Rastall RA (2006) Manufacture of prebiotic oligosaccharides. In: Gibson GR, Rastall RA (eds) Prebiotics: development & application. Wiley, Chichester

Castro CC, Nobre C, De Weireld G, Hantson A-L (2019) Microbial co-culturing strategies for fructo-oligosaccharide production. New Biotechnol 51:1–7. https://doi.org/10.1016/j.nbt.2019.01.009

Chae BJ, Lohakare JD, Moon WK, Lee SL, Park YH, Hahn T-W (2006) Effects of supplementation of β-glucan on the growth performance and immunity in broilers. Res Vet Sci 80:291–298. https://doi.org/10.1016/j.rvsc.2005.07.008

Chang L, Ding Y, Wang Y, Song Z, Li F, He X, Zhang H (2022) Effects of different oligosaccharides on growth performance and intestinal function in broilers. Front Vet Sci 9:852545. https://doi.org/10.3389/fvets.2022.852545

Chen Y-S, Srionnual S, Onda T, Yanagida F (2007) Effects of prebiotic oligosaccharides and trehalose on growth and production of bacteriocins by lactic acid bacteria. Lett Appl Microbiol 45:190–193. https://doi.org/10.1111/j.1472-765X.2007.02167.x

Cho JH, Kim IH (2014) Effects of lactulose supplementation on performance, blood profiles, excreta microbial shedding of lactobacillus and Escherichia coli, relative organ weight and excreta noxious gas contents in broilers. Anim Physiol Nutr 98:424–430. https://doi.org/10.1111/jpn.12086

Cho JH, Zhang ZF, Kim IH (2013) Effects of single or combined dietary supplementation of β-glucan and kefir on growth performance, blood characteristics and meat quality in broilers. Br Poult Sci 54:216–221. https://doi.org/10.1080/00071668.2013.777691

Corrigan A, De Leeuw M, Penaud-Frézet S, Dimova D, Murphy RA (2015) Phylogenetic and functional alterations in bacterial community compositions in broiler ceca as a result of Mannan oligosaccharide supplementation. Appl Environ Microbiol 81:3460–3470. https://doi.org/10.1128/AEM.04194-14

Cox CM, Sumners LH, Kim S, McElroy AP, Bedford MR, Dalloul RA (2010) Immune responses to dietary β-glucan in broiler chicks during an Eimeria challenge. Poult Sci 89:2597–2607. https://doi.org/10.3382/ps.2010-00987

Crittenden RG, Playne MJ (1996) Production, properties and applications of food-grade oligosaccharides. Trends Food Sci Technol 7:353–361. https://doi.org/10.1016/S0924-2244(96)10038-8

De Almeida MN, Maitan-Alfenas GP (2021) Production of oligosaccharides by fungi or fungal enzymes. In: Encyclopedia of mycology. Elsevier, pp 385–393. https://doi.org/10.1016/B978-0-12-819990-9.00037-8

Donalson LM, Kim WK, Chalova VI, Herrera P, McReynolds JL, Gotcheva VG, Vidanovic D, Woodward CL, Kubena LF, Nisbet DJ, Ricke SC (2008) In vitro fermentation response of laying hen cecal bacteria to combinations of fructooligosaccharide prebiotics with alfalfa or a layer ration. Poult Sci 87:1263–1275. https://doi.org/10.3382/ps.2007-00179

Donalson LM, Kim W-K, Chalova VI, Herrera P, Woodward CL, McReynolds JL, Kubena LF, Nisbet DJ, Ricke SC (2007) In vitro anaerobic incubation of salmonella enterica serotype typhimurium and laying hen cecal bacteria in poultry feed substrates and a fructooligosaccharide prebiotic. Anaerobe 13:208–214. https://doi.org/10.1016/j.anaerobe.2007.05.001

Egbu CF, Motsei LE, Yusuf AO, Mnisi CM (2022) Effect of Moringa oleifera seed extract administered through drinking water on physiological responses, carcass and meat quality traits, and bone parameters in broiler chickens. Appl Sci 12:10330. https://doi.org/10.3390/app122010330

Elkomy HS, Koshich II, Mahmoud SF, Abo-Samaha MI (2023) Use of lactulose as a prebiotic in laying hens: its effect on growth, egg production, egg quality, blood biochemistry, digestive enzymes, gene expression and intestinal morphology. BMC Vet Res 19:207. https://doi.org/10.1186/s12917-023-03741-x

Fernandez F, Hinton M, Gils BV (2002) Dietary mannan-oligosaccharides and their effect on chicken caecal microflora in relation to salmonella Enteritidis colonization. Avian Pathol 31:49–58. https://doi.org/10.1080/03079450120106000

Franck A (2002) Technological functionality of inulin and oligofructose. Br J Nutr 87(Suppl 2):S287–S291. https://doi.org/10.1079/BJNBJN/2002550

Gaggìa F, Mattarelli P, Biavati B (2010) Probiotics and prebiotics in animal feeding for safe food production. Int J Food Microbiol 141:S15–S28. https://doi.org/10.1016/j.ijfoodmicro.2010.02.031

Gibson GR, Hutkins R, Sanders ME, Prescott SL, Reimer RA, Salminen SJ, Scott K, Stanton C, Swanson KS, Cani PD, Verbeke K, Reid G (2017) Expert consensus document: the international scientific Association for Probiotics and Prebiotics (ISAPP) consensus statement on the definition and scope of prebiotics. Nat Rev Gastroenterol Hepatol 14:491–502. https://doi.org/10.1038/nrgastro.2017.75

Gibson GR, Roberfroid MB (1995) Dietary modulation of the human colonic microbiota: introducing the concept of prebiotics. J Nutr 125:1401–1412. https://doi.org/10.1093/jn/125.6.1401

Haas R, Smith J, Rocher-Ros V, Nadkarni S, Montero-Melendez T, D'Acquisto F, Bland EJ, Bombardieri M, Pitzalis C, Perretti M, Marelli-Berg FM, Mauro C (2015) Lactate regulates metabolic and pro-inflammatory circuits in control of T cell migration and effector functions. PLoS Biol 13:e1002202. https://doi.org/10.1371/journal.pbio.1002202

Han D, Lee HT, Lee JB, Kim Y, Lee SJ, Yoon JW (2017) A bioprocessed polysaccharide from Lentinus edodes mycelia cultures with turmeric protects chicks from a lethal challenge of salmonella Gallinarum. J Food Prot 80:245–250. https://doi.org/10.4315/0362-028X.JFP-16-306

He G, Baidoo SK, Yang Q, Golz D, Tungland B (2002) Evaluation of chicory inulin extracts as feed additive for early-weaned pigs. J Anim Sci:80–81

Hooge DM (2004) Meta-analysis of broiler chicken Pen trials evaluating dietary Mannan oligosaccharide, 1993-2003. Int J Poultry Sci 3:163–174. https://doi.org/10.3923/ijps.2004.163.174

Huang Q, Wei Y, Lv Y, Wang Y, Hu T (2015) Effect of dietary inulin supplements on growth performance and intestinal immunological parameters of broiler chickens. Livest Sci 180:172–176. https://doi.org/10.1016/j.livsci.2015.07.015

Hughes R-A, Ali RA, Mendoza MA, Hassan HM, Koci MD (2017) Impact of dietary Galacto-oligosaccharide (GOS) on Chicken's gut microbiota, mucosal gene expression, and salmonella colonization. Front Vet Sci 4:192. https://doi.org/10.3389/fvets.2017.00192

Jacob J, Pescatore A (2017) Glucans and the poultry immune system. Am J Immunol 13:45–49. https://doi.org/10.3844/ajisp.2017.45.49

Jacob JP, Pescatore AJ (2014) Barley β-glucan in poultry diets. Ann Transl Med 2:20. https://doi.org/10.3978/j.issn.2305-5839.2014.01.02

Jaiswal SK, Tomar S, Saxena VK, Uniyal S (2023) Supplementation of lactobacillus reuteri isolated from red jungle fowl along with mannanoligosaccharide improves growth performance, immune response and gut health in broiler chicken. Indian J Anim Sci 93. https://doi.org/10.56093/ijans.v93i6.129352

Jin X, Zhang X, Li J, Yu W, Chen F (2018) Activation of chicken macrophages during in vitro stimulation and expression of immune genes. ajvr 79:1306–1312. https://doi.org/10.2460/ajvr.79.12.1306

Johansson MEV, Phillipson M, Petersson J, Velcich A, Holm L, Hansson GC (2008) The inner of the two Muc2 mucin-dependent mucus layers in colon is devoid of bacteria. Proc Natl Acad Sci USA 105:15064–15069. https://doi.org/10.1073/pnas.0803124105

Józefiak D, Kaczmarek S, Rutkowski A (2008) A note on the effects of selected prebiotics on the performance and ileal microbiota of broiler chickens. J Anim Feed Sci 17:392–397. https://doi.org/10.22358/jafs/66633/2008

Józefiak D, Rutkowski A, Martin SA (2004) Carbohydrate fermentation in the avian ceca: a review. Anim Feed Sci Technol 113:1–15. https://doi.org/10.1016/j.anifeedsci.2003.09.007

Jung SJ, Houde R, Baurhoo B, Zhao X, Lee BH (2008) Effects of galacto-oligosaccharides and a Bifidobacteria lactis-based probiotic strain on the growth performance and fecal microflora of broiler chickens. Poult Sci 87:1694–1699. https://doi.org/10.3382/ps.2007-00489

Karkeszová K, Polakovič M (2023) Production of Fructooligosaccharides using a commercial Heterologously expressed aspergillus sp. Fructosy ltransferase Catalysts 13:843. https://doi.org/10.3390/catal13050843

Khodambashi Emami N, Samie A, Rahmani HR, Ruiz-Feria CA (2012) The effect of peppermint essential oil and fructooligosaccharides, as alternatives to virginiamycin, on growth performance, digestibility, gut morphology and immune response of male broilers. Anim Feed Sci Technol 175:57–64. https://doi.org/10.1016/j.anifeedsci.2012.04.001

Kim S, Kim W, Hwang IK (2003) Optimization of the extraction and purification of oligosaccharides from defatted soybean meal. Int J of Food Sci Tech 38:337–342. https://doi.org/10.1046/j.1365-2621.2003.00679.x

Kim SA, Jang MJ, Kim SY, Yang Y, Pavlidis HO, Ricke SC (2019) Potential for prebiotics as feed additives to limit foodborne campylobacter establishment in the poultry gastrointestinal tract. Front Microbiol 10:91. https://doi.org/10.3389/fmicb.2019.00091

Kocot AM, Jarocka-Cyrta E, Drabińska N (2022) Overview of the importance of biotics in gut barrier integrity. IJMS 23:2896. https://doi.org/10.3390/ijms23052896

Kumar S, Shang Y, Kim WK (2019) Insight into dynamics of gut microbial Community of Broilers fed with Fructooligosaccharides Supplemented low Calcium and Phosphorus Diets. Front Vet Sci 6:95. https://doi.org/10.3389/fvets.2019.00095

Lee Y-S, Kim T-Y, Kim Y, Lee S-H, Kim S, Kang SW, Yang J-Y, Baek I-J, Sung YH, Park Y-Y, Hwang SW, Kim KS, Liu S, Kamada N, Gao N, Kweon M-N (2018) Microbiota-derived lactate accelerates intestinal stem-cell-mediated epithelial development. Cell Host Microbe 24:833–846.e6. https://doi.org/10.1016/j.chom.2018.11.002

Lonkar VD, Yenge G, Ranade A, Doiphode A, Patodkar V, Mote C (2020) Effect of dietary inclusion of combination of inulin and Saponin on egg qualities of laying hens. IJ Vet Sci Bio 15:74–76. https://doi.org/10.21887/ijvsbt.15.3.20

Macfarlane GT, Steed H, Macfarlane S (2008) Bacterial metabolism and health-related effects of galacto-oligosaccharides and other prebiotics. J Appl Microbiol 104:305–344. https://doi.org/10.1111/j.1365-2672.2007.03520.x

Martin K, Laverty L, Filho RLA, Hernandez-Velasco X, Señas-Cuesta R, Gray LS, Marcon RFR, Stein A, Coles ME, Loeza I, Castellanos-Huerta I, El-Ashram S, Al-Olayan E, Tellez-Isaias G, Latorre JD (2023) Evaluation of aspergillus meal prebiotic in productive parameters, bone

mineralization and intestinal integrity in broiler chickens. Ger J Vet Res 3:27–33. https://doi.
org/10.51585/gjvr.2023.3.0061

Mei Z, Yuan J, Li D (2022) Biological activity of galacto-oligosaccharides: a review. Front
Microbiol 13:993052. https://doi.org/10.3389/fmicb.2022.993052

Morgan NK, Keerqin C, Wallace A, Wu S-B, Choct M (2019) Effect of arabinoxylo-oligosaccharides
and arabinoxylans on net energy and nutrient utilization in broilers. Anim Nutr 5:56–62. https://
doi.org/10.1016/j.aninu.2018.05.001

Morgan NK, Wallace A, Bedford MR, Hawking KL, Rodrigues I, Hilliar M, Choct M (2020)
In vitro versus in situ evaluation of xylan hydrolysis into xylo-oligosaccharides in broiler
chicken gastrointestinal tract. Carbohydr Polym 230:115645. https://doi.org/10.1016/j.
carbpol.2019.115645

Muñoz M, Mosquera A, Alméciga-Díaz CJ, Melendez AP, Sánchez OF (2012)
Fructooligosaccharides metabolism and effect on bacteriocin production in lactobacillus strains
isolated from ensiled corn and molasses. Anaerobe 18:321–330. https://doi.org/10.1016/j.
anaerobe.2012.01.007

Nabizadeh A (2012) The effect of inulin on broiler chicken intestinalmicroflora, gut morphology,
and performance. J Anim Feed Sci 21:725–734. https://doi.org/10.22358/jafs/66144/2012

Niness KR (1999) Inulin and oligofructose: what are they? J Nutr 129:1402S–1406S. https://doi.
org/10.1093/jn/129.7.1402S

Nobre C, Alves Filho EG, Fernandes FAN, Brito ES, Rodrigues S, Teixeira JA, Rodrigues LR
(2018) Production of fructo-oligosaccharides by aspergillus ibericus and their chemical characterization. LWT 89:58–64. https://doi.org/10.1016/j.lwt.2017.10.015

Nobre C, Do Nascimento AKC, Silva SP, Coelho E, Coimbra MA, Cavalcanti MTH, Teixeira JA,
Porto ALF (2019) Process development for the production of prebiotic fructo-oligosaccharides
by penicillium citreonigrum. Bioresour Technol 282:464–474. https://doi.org/10.1016/j.
biortech.2019.03.053

Ofek I, Beachey EH (1978) Mannose binding and epithelial cell adherence of Escherichia coli.
Infect Immun 22:247–254. https://doi.org/10.1128/iai.22.1.247-254.1978

Oyofo BA, DeLoach JR, Corrier DE, Norman JO, Ziprin RL, Mollenhauer HH (1989) Prevention
of salmonella typhimurium colonization of broilers with D-mannose. Poult Sci 68:1357–1360.
https://doi.org/10.3382/ps.0681357

Patterson JA, Burkholder KM (2003) Application of prebiotics and probiotics in poultry production. Poult Sci 82:627–631. https://doi.org/10.1093/ps/82.4.627

Pelicano E, de Souza P, de Souza H, Leonel F, Zeola N, Boiago M (2004) Productive traits of broiler
chickens fed diets containing different growth promoters. Rev Bras Cienc Avic 6:177–182.
https://doi.org/10.1590/S1516-635X2004000300008

Pool-Zobel B, Van Loo J, Rowland I, Roberfroid MB (2002) Experimental evidences on the potential of prebiotic fructans to reduce the risk of colon cancer. Br J Nutr 87:273–281. https://doi.
org/10.1079/BJNBJN/2002548

Pourabedin M, Zhao X (2015) Prebiotics and gut microbiota in chickens. FEMS Microbiol Lett
362:fnv122. https://doi.org/10.1093/femsle/fnv122

Rastall RA, Maitin V (2002) Prebiotics and synbiotics: towards the next generation. Curr Opin
Biotechnol 13:490–496. https://doi.org/10.1016/s0958-1669(02)00365-8

Redondo-Cuenca A, Herrera-Vázquez SE, Condezo-Hoyos L, Gómez-Ordóñez E, Rupérez
P (2021) Inulin extraction from common inulin-containing plant sources. Ind Crop Prod
170:113726. https://doi.org/10.1016/j.indcrop.2021.113726

Rehman HU, Vahjen W, Awad WA, Zentek J (2007) Indigenous bacteria and bacterial metabolic
products in the gastrointestinal tract of broiler chickens. Arch Anim Nutr 61:319–335. https://
doi.org/10.1080/17450390701556817

Richards PJ, Flaujac Lafontaine GM, Connerton PL, Liang L, Asiani K, Fish NM, Connerton IF
(2019) Galacto-oligosaccharides modulate the juvenile gut microbiome and innate immunity to
improve broiler chicken performance (preprint). Microbiology. https://doi.org/10.1101/631259

Ricke SC (2018) Impact of prebiotics on poultry production and food safety. Yale J Biol Med
91:151–159

Ricke SC (2015) Potential of fructooligosaccharide prebiotics in alternative and nonconventional poultry production systems. Poult Sci 94:1411–1418. https://doi.org/10.3382/ps/pev049

Roberfroid MB (2007) Inulin-type fructans: functional food ingredients. J Nutr 137:2493S–2502S. https://doi.org/10.1093/jn/137.11.2493S

Roberfroid MB (2005) Introducing inulin-type fructans. Br J Nutr 93(Suppl 1):S13–S25. https://doi.org/10.1079/bjn20041350

Rodríguez-Daza MC, Pulido-Mateos EC, Lupien-Meilleur J, Guyonnet D, Desjardins Y, Roy D (2021) Polyphenol-mediated gut microbiota modulation: toward prebiotics and further. Front Nutr 8:689456. https://doi.org/10.3389/fnut.2021.689456

Rosen GD (2007) Holo-analysis of the efficacy of bio-Mos in broiler nutrition. Br Poult Sci 48:21–26. https://doi.org/10.1080/00071660601050755

Russell JB (1992) Another explanation for the toxicity of fermentation acids at low pH: anion accumulation versus uncoupling. J Appl Bacteriol 73:363–370. https://doi.org/10.1111/j.1365-2672.1992.tb04990.x

Schumann C (2002) Medical, nutritional and technological properties of lactulose. An update. Eur J Nutr 41:1–1. https://doi.org/10.1007/s00394-002-1103-6

Selim S, Abdel-Megeid NS, Khalifa HK, Fakiha KG, Majrashi KA, Hussein E (2022) Efficacy of various feed additives on performance, nutrient digestibility, bone quality, blood constituents, and phosphorus absorption and utilization of broiler chickens fed low phosphorus diet. Animals 12:1742. https://doi.org/10.3390/ani12141742

Shang HM, Hu TM, Lu YJ, Wu HX (2010) Effects of inulin on performance, egg quality, gut microflora and serum and yolk cholesterol in laying hens. Br Poult Sci 51:791–796. https://doi.org/10.1080/00071668.2010.531005

Shang H-M, Zhou H-Z, Yang J-Y, Li R, Song H, Wu H-X (2018) In vitro and in vivo antioxidant activities of inulin. PLoS One 13:e0192273. https://doi.org/10.1371/journal.pone.0192273

Shang Y, Kim WK (2016) Roles of Fructooligosaccharides and Phytase in BroilerChickens: review. Int J of Poultry Sci 16:16–22. https://doi.org/10.3923/ijps.2017.16.22

Shehata AA, Yalçın S, Latorre JD, Basiouni S, Attia YA, Abd El-Wahab A, Visscher C, El-Seedi HR, Huber C, Hafez HM, Eisenreich W, Tellez-Isaias G (2022) Probiotics, prebiotics, and phytogenic substances for optimizing gut health in poultry. Microorganisms 10:395. https://doi.org/10.3390/microorganisms10020395

Shoaf K, Mulvey GL, Armstrong GD, Hutkins RW (2006) Prebiotic galactooligosaccharides reduce adherence of enteropathogenic Escherichia coli to tissue culture cells. Infect Immun 74:6920–6928. https://doi.org/10.1128/IAI.01030-06

Sims MD, Dawson KA, Newman KE, Spring P, Hoogell DM (2004) Effects of dietary mannan oligosaccharide, bacitracin methylene disalicylate, or both on the live performance and intestinal microbiology of turkeys. Poult Sci 83:1148–1154. https://doi.org/10.1093/ps/83.7.1148

Siwek M, Slawinska A, Stadnicka K, Bogucka J, Dunislawska A, Bednarczyk M (2018) Prebiotics and synbiotics—in ovo delivery for improved lifespan condition in chicken. BMC Vet Res 14:402. https://doi.org/10.1186/s12917-018-1738-z

Slawinska A, Dunislawska A, Plowiec A, Radomska M, Lachmanska J, Siwek M, Tavaniello S, Maiorano G (2019) Modulation of microbial communities and mucosal gene expression in chicken intestines after galactooligosaccharides delivery In Ovo. PLoS One 14:e0212318. https://doi.org/10.1371/journal.pone.0212318

de los Solis Santos F, Donoghue AM, Farnell MB, Huff GR, Huff WE, Donoghue DJ (2007) Gastrointestinal maturation is accelerated in Turkey poults supplemented with a mannan-oligosaccharide yeast extract (Alphamune). Poult Sci 86:921–930. https://doi.org/10.1093/ps/86.5.921

Solis-Cruz B, Hernandez-Patlan D, Hargis M, Tellez G (2020) Use of prebiotics as an alternative to antibiotic growth promoters in the poultry industry. In: Franco-Robles E, Ramírez-Emiliano J (eds) Prebiotics and probiotics–potential benefits in nutrition and health. IntechOpen. https://doi.org/10.5772/intechopen.89053

Spring P, Wenk C, Dawson KA, Newman KE (2000) The effects of dietary mannaoligosaccha-rides on cecal parameters and the concentrations of enteric bacteria in the ceca of salmonella-challenged broiler chicks. Poult Sci 79:205–211. https://doi.org/10.1093/ps/79.2.205

Stanczuk J, Zdunczyk Z, Juskiewicz J, Jankowski J (2005) Indices of response of young turkeys to diets containing mannanoligosaccharide or inulin. Vet ir Zootech:98–101

Stier H, Ebbeskotte V, Gruenwald J (2014) Immune-modulatory effects of dietary yeast Beta-1,3/1,6-D-glucan. Nutr J 13:38. https://doi.org/10.1186/1475-2891-13-38

Tellez G, Higgins SE, Donoghue AM, Hargis BM (2006) Digestive physiology and the role of microorganisms. J Appl Poult Res 15:136–144. https://doi.org/10.1093/japr/15.1.136

Téllez G, Lauková A, Latorre JD, Hernandez-Velasco X, Hargis BM, Callaway T (2015) Food-producing animals and their health in relation to human health. Microb Ecol Health Dis 26:25876. https://doi.org/10.3402/mehd.v26.25876

Ten Bruggencate SJM, Bovee-Oudenhoven IMJ, Lettink-Wissink MLG, Van Der Meer R (2003) Dietary Fructo-oligosaccharides dose-dependently increase translocation of salmonella in rats. J Nutr 133:2313–2318. https://doi.org/10.1093/jn/133.7.2313

Teng P-Y, Kim WK (2018) Review: roles of prebiotics in intestinal ecosystem of broilers. Front Vet Sci 5:245. https://doi.org/10.3389/fvets.2018.00245

Tiengtam N, Khempaka S, Paengkoum P, Boonanuntanasarn S (2015) Effects of inulin and Jerusalem artichoke (Helianthus tuberosus) as prebiotic ingredients in the diet of juvenile Nile tilapia (Oreochromis niloticus). Anim Feed Sci Technol 207:120–129. https://doi.org/10.1016/j.anifeedsci.2015.05.008

Torres DPM, Gonçalves M, Teixeira JA, Rodrigues LR (2010) Galacto-oligosaccharides: produc-tion, properties, applications, and significance as prebiotics. Compr Rev Food Sci Food Saf 9:438–454. https://doi.org/10.1111/j.1541-4337.2010.00119.x

Tuohy KM, Ziemer CJ, Klinder A, Knöbel Y, Pool-Zobel BL, Gibson GR (2002) A human volun-teer study to determine the prebiotic effects of lactulose powder on human colonic microbiota. Microb Ecol Health Dis 14:165–173. https://doi.org/10.1080/089106002320644357

Tzortzis G, Goulas AK, Gee JM, Gibson GR (2005) A novel galactooligosaccharide mixture increases the bifidobacterial population numbers in a continuous in vitro fermentation sys-tem and in the proximal colonic contents of pigs in vivo. J Nutr 135:1726–1731. https://doi.org/10.1093/jn/135.7.1726

van Der Wielen PW, Biesterveld S, Notermans S, Hofstra H, Urlings BA, van Knapen F (2000) Role of volatile fatty acids in development of the cecal microflora in broiler chickens during growth. Appl Environ Microbiol 66:2536–2540. https://doi.org/10.1128/AEM.66.6.2536-2540.2000

Van Immerseel F, Russell JB, Flythe MD, Gantois I, Timbermont L, Pasmans F, Haesebrouck F, Ducatelle R (2006) The use of organic acids to combat salmonella in poultry: a mechanistic expla-nation of the efficacy. Avian Pathol 35:182–188. https://doi.org/10.1080/03079450600711045

Varasteh S, Braber S, Akbari P, Garssen J, Fink-Gremmels J (2015) Differences in susceptibility to heat stress along the chicken intestine and the protective effects of Galacto-oligosaccharides. PLoS One 10:e0138975. https://doi.org/10.1371/journal.pone.0138975

Vetvicka V, Vetvickova J (2014) Natural immunomodulators and their stimulation of immune reac-tion: true or false? Anticancer Res 34:2275–2282

Willemsen LEM, Koetsier MA, van Deventer SJH, van Tol E (2003) Short chain fatty acids stimulate epithelial mucin 2 expression through differential effects on prostaglandin E(1) and E(2) production by intestinal myofibroblasts. Gut 52:1442–1447. https://doi.org/10.1136/gut.52.10.1442

Williams J, Mallet S, Leconte M, Lessire M, Gabriel I (2008) The effects of fructo-oligosaccharides or whole wheat on the performance and digestive tract of broiler chickens. Br Poult Sci 49:329–339. https://doi.org/10.1080/00071660802123351

Wu T, Dai X, Wu L (1999) Effect of fructooligsaccharide on the broiler production. Acta Agri Zhejiangensis 11:85–87

Xu ZR, Hu CH, Xia MS, Zhan XA, Wang MQ (2003) Effects of dietary fructooligosaccharide on digestive enzyme activities, intestinal microflora and morphology of male broilers. Poult Sci 82:1030–1036. https://doi.org/10.1093/ps/82.6.1030

Yang C, Qiu M, Zhang Z, Song X, Yang L, Xiong X, Hu C, Pen H, Chen J, Xia B, Du H, Li Q, Jiang X, Yu C (2022a) Galacto-oligosaccharides and xylo-oligosaccharides affect meat flavor by altering the cecal microbiome, metabolome, and transcriptome of chickens. Poult Sci 101:102122. https://doi.org/10.1016/j.psj.2022.102122

Yang C, Tang XW, Liu X, Yang H, Bin DM, Liu HJ, Tang QH, Tang JY (2022b) Effects of dietary oligosaccharides on serum biochemical index, intestinal morphology, and antioxidant status in broilers. Anim Sci J 93:e13679. https://doi.org/10.1111/asj.13679

Yousaf MS, Ahmad I, Ashraf K, Rashid MA, Hafeez A, Ahmad A, Zaneb H, Naseer R, Numan M, Zentek J, Rehman H (2017) Comparative effects of different dietary concentrations of β-galacto-oligosaccharides on serum biochemical metabolites, selected caecel microbiota and immune response in broilers. J Anim Plant Sci:98–105

Yusrizal, Chen TC (2003) Effect of adding chicory Fructans in feed on broiler growth performance, serum cholesterol and intestinal length. Int J Poultry Sci 2:214–219. https://doi.org/10.3923/ijps.2003.214.219

Zaghini A, Martelli G, Roncada P, Simioli M, Rizzi L (2005) Mannanoligosaccharides and aflatoxin B1 in feed for laying hens: effects on egg quality, aflatoxins B1 and M1 residues in eggs, and aflatoxin B1 levels in liver. Poult Sci 84:825–832. https://doi.org/10.1093/ps/84.6.825

Zhang W, Li D, Lu W, Yi G (2003) Effects of isomalto-oligosaccharides on broiler performance and intestinal microflora. Poult Sci 82:657–663. https://doi.org/10.1093/ps/82.4.657

Use of Postbiotics in the Poultry Industry: Current Knowledge and Future Prospects

6

Amr Abd El-Wahab, Christian Visscher,
and Awad A. Shehata

Contents

Abstract

Postbiotics are inactive microbe preparations or their metabolites that are considered health promoters. They are stable during industrial processing and storage, whereas probiotics are vulnerable to heat and oxygen. Postbiotics are considered promising tools for sustainable poultry production because they can improve growth performance, feed efficiency, and health, and lead to a reduction of gut

A. A. El-Wahab
Institute for Animal Nutrition, University of Veterinary Medicine Hannover, Foundation, Hannover, Germany

Department of Nutrition and Nutritional Deficiency Diseases, Faculty of Veterinary Medicine, Mansoura University, Mansoura, Egypt

C. Visscher
Institute for Animal Nutrition, University of Veterinary Medicine Hannover, Foundation, Hannover, Germany

A. A. Shehata (✉)
Structural Membrane Biochemistry, Bavarian NMR Center, Technical University of Munich (TUM), Garching, Bayern, Germany
e-mail: Awad.shehata@tum.de

pathogens. Postbiotcs include numerous significant biological characteristics such as immunomodulatory, antioxidant, and anti-inflammatory responses. Moreover, higher uronic acid concentrations and some antioxidant-producing enzymes, such as glutathione peroxidase, superoxide dismutase, and nicotinamide adenine dinucleotide oxidases and peroxidases, were responsible for the encouraging antioxidant effects of postbiotics. Postbiotics enhance the gut villi, produce more lactic acid, and lower the fecal pH. Collectively, postbiotics thus boost immunity, promote gut health, and improve growth performance parameters. Furthermore, postbiotics enhance the quality of eggs by lowering plasma and yolk cholesterol levels in layers. Several studies have shown that postbiotic substances greatly improve poultry well-being and growth performance. In this chapter, we will shed light on the advantages of postbiotics for the poultry sector as alternatives to antibiotics.

Keywords
Postbiotics · Mechanism of action · Advantages · Applications · Poultry

1   Introduction

A probiotic is defined as "a preparation or a product containing viable, defined microorganisms in sufficient numbers to alter the micro-flora (by implantation or colonization) in a compartment of the host and that exert beneficial health effects in the host" (Schrezenmeir and de Vrese 2001). This is contrasted with postbiotics, which are nonliving products, often made from fungal fermentations, including cellular wreckage and fermentation end products (Loh et al. 2014; Tomasik and Tomasik 2020; Salminen et al. 2021). Prebiotics, in turn, are compounds that are not digested/ degraded by the host animal but are utilized by members of the microbial ecosystem, providing them with a competitive advantage, which in turn can cause a shift in the microbiota composition of the gut of animals and have been termed colonic food (Gibson and Roberfroid 1995; Collins and Gibson 1999; Tannock 1999; Ricke 2018). Synbiotics are combinatorial approaches, often incorporating several modes of action/products, such as a eubiotic coupled with a postbiotic that contains some prebiotic compounds (Collins and Gibson 1999; Schrezenmeir and de Vrese 2001; Salminen et al. 2021).

According to the definition used by The International Scientific Association for Probiotics and Prebiotics (ISAPP), a probiotic is a live, non-pathogenic microorganism that has a positive effect on the host (Salminen et al. 2021). According to Williams (2010), microorganisms classified as probiotics are most often lactic acid bacteria (LAB) (Williams 2010). The latest ISAPP definition says that postbiotics are considered components of bacterial cells, and the by-products produced by post-fermentation such as lactic acid and short-chain fatty acids positively affect the host (Salminen et al. 2021). According to Pandey et al. (2015), a synbiotic is a proper mixing of a probiotic and a prebiotic (Pandey et al. 2015). However, the ISAPP gave a more accurate definition of synbiotics: synbiotics have a positive effect on the host's health, and they are a mixture of microorganisms and nutrients used by the host's microorganisms (Swanson et al. 2020).

Postbiotics may be stored safely and long-lastingly in situations of extreme temperature because they are not affected by acids, pH, or environmental factors. Furthermore, beneficial elements like organic acids and bacteriocins found in postbiotics originating from Lactobacilli increase the counts of LAB (Loh et al. 2014). The study conducted by Cicenia et al. (2016) and Tiptiri-Kourpeti et al. (2016) revealed that various *Lactobacillus* (L.) species, including but not limited to *L. acidophilus*, *L. plantarum*, *L. fermentum*, *L. casei*, *L. rhamnosus*, *L. paracasei*, *L. rhamnosus*, *L. delbrueckil* subsp. Bulgaricus, *L. gasseri*, *L. helveticus*, *L. reuteri*, and *L. johnsonni* were highly effective probiotics. *Lactiplantibacillus plantarum*, formerly known as *Lactobacillus plantarum* (Zheng et al. 2020), is a member of the LAB family, which is well-liked in animal nutrition and has been considered "Generally Recognized As Safe" (GRAS) (Danladi et al. 2022). According to several studies, postbiotics from *L. plantarum* have shown a wide range of antagonistic actions, indicating their capacity to suppress infections from different species (Kareem et al. 2014; Thanh et al. 2010). Chang et al. (2022) found that by enhancing gut health, tight junction permeability, mucin synthesis, beneficial bacteria colonization, and immunology, *L. plantarum* postbiotic might replace growth promoters such as antibiotics. In chickens stressed by heat, postbiotics containing *L. plantarum* show antioxidative properties (Humam et al. 2020).

According to Dunne et al. (1999), bacteria employed to manufacture postbiotics should be non-pathogenic, technologically appropriate for industrial procedures, resistant to acid and bile, and good makers of antimicrobial compounds that influence the GIT's metabolic activities and control immune responses. Additionally, postbiotics have improved the growth performance, intestinal morphology, meat quality, and heat stress tolerance of broilers, laying hens, and pigs (Thanh et al. 2009; Loh et al. 2010; Thu et al. 2011; Choe et al. 2012; Humam et al. 2020). Due to efficient metabolism, absorption, distribution, and excretion, postbiotics can influence host organs and tissues and perform various biological functions (Shenderov 2013). In addition, postbiotics have several interesting properties, including various molecular structures, extended shelf lives, and safe dosages (Shigwedha 2014). Postbiotics are considered more advantageous than probiotics; since they include non-replicating microorganisms and are less likely to result in bacteremia or fungemia (Yelin et al. 2019).

Probiotics are known as "direct-fed microbials" in animal production since they live in the animal's stomach and ideally have a beneficial function. Probiotics will degrade during manufacture and storage, whereas postbiotics will stay stable because most products have a lengthy mean life. Probiotic strains degrade at varying rates depending on their physiological characteristics and storage conditions (temperature, water activity, length, oxygen levels, etc.). As a result, it is challenging to generalize about probiotic product dead cell counts in the final stages of their shelf life (Huber et al. 2005).

Indeed, the most readily available source of animal protein nowadays is thought to be commercial poultry (Zuidhof et al. 2014; Arain et al. 2022). Food animals must be produced safely and effectively to feed the growing population as people switch to diets higher in animal protein (Wu et al. 2014; Nabi et al. 2020). Abd El-Ghany et al. (2022) found that using postbiotics improved immunity, enhanced growth performance, and health status compared to challenged non-treated chickens. Dry or watery postbiotic substances enhance broiler chickens' health, immune system, and growth performance. Inactivated microbes may provide health benefits, but the published literature on postbiotics is unclear and lacks crucial practical variables. However, recent research has been performed in this field (Patel and Denning 2013; Thanh et al. 2009; Zendeboodi et al. 2020; Homayouni Rad et al. 2021). The soluble factors (stabilized bacteria, cellular products, or metabolic by-products) secreted by living bacteria or released following bacterial lysis are referred to as "postbiotics," a term that has recently been used in the poultry industry (Loh et al. 2010; Cicenia et al. 2014; Klemashevich et al. 2014; Blacher et al. 2017; Johnson et al. 2019). Enzymes, peptides, teichoic acids, muropeptides generated from peptidoglycan, polysaccharides, cell surface proteins, and organic acids are some of these compounds (Abd El-Ghany et al. 2022). Postbiotics can be found in the cell walls of Lactobacilli species or cytoplasmic extracts, stable bacteria, cellular products, or fermentation-related metabolic by-products (Johnson et al. 2019).

2 Mechanisms of Action

Postbiotics have five distinct mechanisms of action (Salminen et al. 2021):

i. Postbiotics have the potential to indirectly alter the microbiota through mechanisms like quorum quenching or carrying quorum sensing molecules (Grandclément et al. 2016), or having lactic acid, which certain microorganisms use to produce butyrate and short-chain fatty acids (SCFA) that are beneficial to the microbiota (Laverde Gomez et al. 2019). Postbiotic adhesions, such as lectins and fimbriae (Tytgat et al. 2016), can also compete with resident microorganisms for adhesion sites;

ii. Enhancement of the intestinal barrier's functionality. If a postbiotic contains a sufficient amount of SCFA, it may guard against lipopolysaccharide-induced disturbances and change the functions of epithelial barriers (Feng et al. 2018).

iii. Modification by immune responses. Generally, at the local and systemic levels, immune-modulating activities are initiated by molecular patterns linked to microorganisms that engage with specific pattern-recognition receptors of immune cells. These receptors control cytokines and immunological responses, such as receptors of the nucleotide-binding oligomerization domain, C-type lectins, and Toll-like receptors (Lebeer et al. 2010).

iv. Modification of the metabolic response system. The enzymes and metabolites may directly impact systemic metabolic reactions in postbiotics on and inside the surface of inactivated microorganisms. Bile acids have several downstream

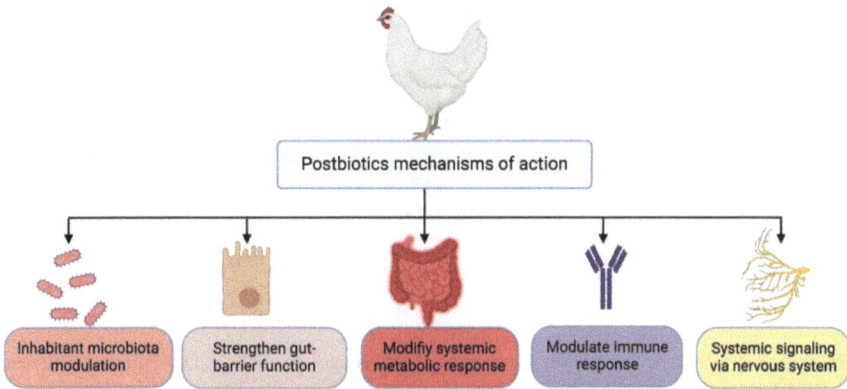

Fig. 6.1 Mechanisms of action of postbiotics (The Figure was created by Biorender.com)

impacts on the host's metabolic processes such as lipids, xenobiotics, glucose, and energy metabolism (Long et al. 2017).

v. Neurological system stimulation. When microbe metabolites, including SCFA, are accessible in adequate concentrations for postbiotic production, they stimulate enterochromaffin cells, which produce serotonin and enter the bloodstream (Iwasaki et al. 2019). A summary of some mechanisms of action of postbiotics is given in Fig. 6.1.

3   General Functions and Advantages of Postbiotics

Different diseases and problems with gut health are caused by disruptions to the gut microbiota, such as pathogen colonization and the growth of native pathobionts. Postbiotics have components that from a therapeutic perspective prohibit pathogenic microbes in the intestine by competing with them to adhere to the mucosa and epithelium (Mantziari et al. 2020). A small antimicrobial peptide called bacteriocin exhibits inhibitory activity against harmful pathogens. According to Yang et al. (2014) and Simons et al. (2020), it might be a good tool to use as an antibacterial agent in food and other medicinal applications. Postbiotics have antibacterial, antioxidant, immunomodulatory, anti-inflammatory, antiproliferative, hypocholesterolemic, and hepatoprotective properties (Aguilar-Toalá et al. 2018). They also improve the health status of the host (see Fig. 6.2). Postbiotic supplementation enhances body weight and health in broiler chickens under normal conditions by boosting immunity and gut health via improved intestinal villus, increased LAB population, decreased Enterobacteriaceae population, and decreased fecal pH (Thanh et al. 2009; Loh et al. 2010; Rosyidah et al. 2011; Kareem et al. 2016, 2017).

Strains of *Lactobacillus plantarum* are well-known postbiotic producers, either alone or in combination (Thanh et al. 2009). The effectiveness of *L. plantarum* strains were shown in broilers (Thanh et al. 2009; Loh et al. 2010; Faseleh Jahromi et al. 2016; Kareem et al. 2016, 2017).

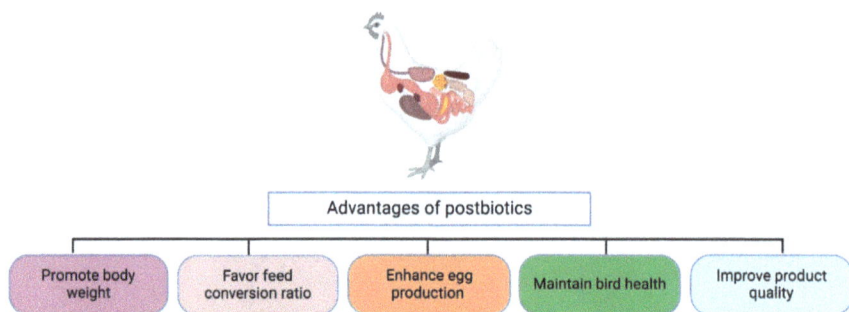

Fig. 6.2 Postbiotics perform many functions and advantages in poultry (The Figure was created by Biorender.com)

Apart from enhancing productivity and gut well-being and managing intestinal infections, postbiotics have also been employed to alleviate antibiotic-related issues (Faseleh Jahromi et al. 2016). Many probiotics are sensitive to heat and oxygen, making it difficult to maintain their stability. Nevertheless, inert bacteria can be stored for a long time. The primary characteristic of postbiotics is their innate stability during storage and industrial operations. Postbiotics may be a better option for preserving live microorganisms than probiotics in areas lacking consistent cold chains or high ambient temperatures. It is challenging to evaluate the percentage of dead cells in probiotics at the end of their shelf life because the rate of death differs based on storage circumstances (oxygen levels, water activity, time, temperature, etc.) and the physiological properties of the strain (Huber et al. 2005).

4 Immunological Response of Postbiotics

Numerous studies have revealed that postbiotics have immunomodulatory effects comparable to those of probiotics. The production of interleukin-10 (IL-10), a post-biotic cytokine with anti-inflammatory characteristics, was found to be upregulated in the cell-free supernatant obtained from *L. reuteri* DSM 17938 (Haileselassie et al. 2016). The IL-10 is essential for immune system modulation because it affects mucosa-like dendritic cells that are driven by retinoic acid. The T-regulatory cells subsequently benefited from this elevation of IL-10 production (Haileselassie et al. 2016). As postbiotics maintain the intestinal mucosal barrier, assist the innate and adaptive immune systems, and use antimicrobial substances to stop the growth of infections, they have been linked to immunomodulatory actions (De Marco et al. 2018). It has been demonstrated that postbiotics, which include pili and protein p40/p75, preserve the intestinal barrier, promote the synthesis of bacteriocins, aggregation factor, and S-layer proteins, and assist in the elimination of pathogens (Teame et al. 2020). Bacterial cell walls contain peptidoglycan and lipoteichoic acid components, which can vary in quantity and affect the immunostimulant activity of the bacteria. By modulating the synthesis of IL-8, postbiotics derived from *Streptococcus thermophilus* may protect the stomach mucosa and enhance the body's natural anti-inflammatory response (Marcial et al. 2017).

Years ago, it was reported that postbiotics, which are produced after probiotics are inactivated, had a far stronger immunomodulatory impact than probiotics (de Almada et al. 2016). The findings of Abd El-Ghany et al. (2022) demonstrated that the postbiotic substance added to feed and water treatments markedly improved the profitability conditions for health and performance, immunological response, bursa of Fabricius/body weight ratio, and intestinal coliform count in *E. coli*-challenged chickens. By improving the humoral immune response and the hepatic health of experimentally infected broiler chickens with *Clostridium perfringens*, the combined feed and water treatment of a postbiotic called "Culbac" showed more promising results for controlling necrotic enteritis than probiotics and antibiotics (Abd El-Ghany et al. 2022).

Postbiotics boosted the growth hormone receptor and insulin-like growth factor 1 (Danladi et al. 2022). During the finisher phase of broiler chickens, there were marked changes in the plasma immunoglobulin IgM level (Danladi et al. 2022). A unique *Saccharomyces cerevisiae* (*S. cerevisiae*) fermentation process is used to produce functional metabolites, which comprise a postbiotic product called *Saccharomyces cerevisiae* fermentation-based postbiotic. This is achieved by raising feed efficiency, growth rate, egg production, improving immune function, and decreasing intestinal lesions during *Eimeria maxima* and *Eimeria tenella* infections, and potentially lowering colonization by foodborne pathogens (Gingerich et al. 2021). Johnson et al. (2019) concluded that the postbiotic studied here imparts an immunomodulatory effect on jejunal tissue in broilers. When paired with a pathogenic challenge of *C. perfringens*, the postbiotic alone reduces the activation of the typical immune response to *C. perfringens* while simultaneously modulating the activation of the innate immune response. The gut health may benefit from this intervention, particularly in light of the limited use of in-feed antibiotics, which are thought to have an anti-inflammatory effect on the gut.

Immunomodulatory effects of postbiotics may be due to a variety of substances found in their cell walls, such as peptides and/or proteins, small molecules such as short-chain fatty acids, or a combination of these (Vinolo et al. 2011; Sun et al. 2018). However, more research is needed to fully understand the processes behind the immunomodulatory properties displayed by postbiotics. Cross-talk between immunity and microbiota is essential for the healthy development and operation of immunity. Microbial metabolites significantly influence immune system regulation (Blacher et al. 2017). Determining the mode of action and efficacy of feed additives is difficult in feed additive research. This is particularly true for products that influence the immune system in response to stress or challenge or may promote growth in specific production conditions.

5   Promising Advantages in the Performance and Production

Postbiotics can be added to feed to help broilers (Kareem et al. 2017) and layers (Choe et al. 2012; Loh et al. 2014) perform better in terms of health, growth, and production. According to Kareem et al. (2016), broiler chicken rations that included prebiotics and postbiotics, increased the birds' overall body weight, feed efficiency, and sustained growth factor 1 as well as the mRNA expression of growth hormones and intestinal mucosal structure. Postbiotic metabolites also decrease the cholesterol levels in the yolk and plasma of the egg in layers and enhance the quality of eggs (Loh et al. 2014).

Including postbiotic (*L. plantarum*) in diets of broiler chickens demonstrated improved maintenance of gut microbiota, growth performance, and intestinal morphology, even in heat stress situations (Humam et al. 2020). Abd El-Ghany et al. (2022) showed that feeding stabilized non-viable Lactobacilli postbiotic affected the intestinal health and growth performance of broilers with colisepticemia. Compared to untreated chickens, postbiotic chemicals dramatically improved the recovery from illness, improved growth performance, improved bursa-to-body weight ratio, and decreased the counts of coliform in the chickens' gut (Abd El-Ghany et al. 2022). Moreover, Johnson's study reported that postbiotic-treated broilers significantly gained weight and had lower lesion ratings than challenged groups (Johnson et al. 2019). The outcomes of Kalavathy's research showed that adding a variety of 12 *Lactobacillus* strains to broiler chicks had beneficial effects such as enhancing growth performance, favoring feed conversion ratio, lowering levels of serum total cholesterol, low-density lipoprotein cholesterol, and triglycerides (Kalavathy et al. 2003). Kareem et al. (2015) hypothesized that the meat color determination in accordance with the Commission International de l'Eclairage (CIE) system (lightness (L*), redness (a*), and yellowness (b*) of breast muscle would increase, and drip loss would decrease in control birds compared to those treated with postbiotics and inulin.

Postbiotics are novel additives in quails' diets that are thought to be the most successful substitute for growth performance promoters. *Lactobacillus animalis* obtained from postbiotics can enhance quails' health status and growth performance by modifying the microbiota in the gut. Administering quails postbiotic fluids rather than antibiotics makes it feasible to balance their gut flora (Kareem 2020). According to Zhang and Kim (2014), postbiotics increase feed efficiency and lower the illness and mortality rate, lessening chicken production's environmental effects. Postbiotics and prebiotics, through increasing feed efficiency and decreasing excrement output, can mitigate the environmental effects of chicken production. The largest energy consumer and emission in the production of chickens is feed. Probiotic-supplemented broilers exhibit lower ammonia emissions, a disagreeable contaminant that can harm the environment (Jeong and Kim 2014; Zhang and Kim 2014).

It is necessary to observe that the Insulin-like growth factor-1 (IGF-1) level, feeding level, and growth rate are concurrent (Beckman et al. 2004). The dependence of nutritional and growth hormones on hepatic IGF1 production has been demonstrated (Moriyama 1995; Beckman 2011). Moreover, among the genes influencing growth, IGF1 has been demonstrated as an indicator of growth rate in chickens by several authors (Jones and Clemmons 1995; Beccavin et al. 2001). The pituitary releases the growth hormone, which stimulates the hepatic production of IGF1 through the actions of growth hormone-activated growth hormone receptors. However, the overall nutritional status of the animal modulated the ability of hepatic tissue to react to growth hormones (Beckman 2011). The IGF1 level can be influenced by factors and situations that affect primary processes and control the IGF1 production. These results might provide a basis for developing IGF1 as a growth index. Nonetheless, some studies have not shown any relation between IGF1 and growth (Silverstein et al. 1998), which can lead to uncertainties about the consistency of IGF1 growth relationships.

The effects of adding a proprietary saponin-based feed additive to a well-established postbiotic feed additive (Original XPC, Diamond V) on turkey performance were studied (Chaney et al. 2023). At weeks 12 and 15, there were noticeable variations in the birds' body weight; at both times, the postbiotic plus saponin therapy group had heavier birds. Additionally, Chaney and his colleagues discovered noteworthy variations in the feed conversion ratio between the ages of the animals from 0 to 18 weeks, with the postbiotic alone showing an improvement in FCR over the control group. There were no discernible variations in terms of livability or feed intake. Chaney et al. (2023) concluded that a combination of a postbiotic plus saponin may exert additive effects on turkey growth. Postbiotics have been found to have a variety of benefits and uses in poultry. As displayed in Table 6.1, various authors conducted tests with varying postbiotic inclusion levels. Dietary interventions had no discernible effects on the growth performance measures (feed conversion ratio, cumulative weight gain, and final body weight), according to Danladi and his co-authors (2022). However, feed intake recorded a significant ($p < 0.05$) change in the starter and finisher phases across the dietary treatments.

6 Promising Effects on Gut Morphology, Metabolites, and Microbiome

Broiler chickens given metabolite combinations produced from strains of *Lactobacillus plantarum* in their diet exhibited an improvement in the height of intestinal villi (Thanh et al. 2009; Loh et al. 2010). Additionally, adding postbiotics and inulin mixtures to broilers' diets improved their intestinal mucosal morphology by resulting in longer villi (Kareem et al. 2016). A combination of *L. plantarum* metabolites significantly raised the intestinal villus height to crypt depth ratio, as Humam et al. (2019) recently showed. Villus height was significantly increased, while the crypt depth decreased significantly due to dietary treatments. Villus height and crypt depth are reliable indicators of gut function and health (Danladi et al. 2022). According to UNI et al. (1995), the health status of the gastrointestinal tract of an animal is a true reflection of intestinal morphology. The postbiotic altered the

Table 6.1 Effects of postbiotics on poultry health and performance (examples)

Postbiotic strain	Species/age	Results	References
0.1% *Lactiplantibacillus plantarum* Postbiotics	Six-week-old broiler chicks	– *L. plantarum* postbiotic increased growth performance and mucin production, postbiotic ameliorated immune status and tight junction permeability – It improved beneficial bacteria Colonization and gut health	(Chang et al. 2022)
1.0% Inulin+different levels of postbiotic (0.15%, 0.30%, 0.45%, 0.60%)		– The level of 1.0% inulin+0.15% postbiotic had the optimal level – Combinations of inulin and postbiotic increased body weight and immune response	
0.2% Postbiotic	Five-week-old broiler chicks	– Paraprobiotics and postbiotics revealed a positive influence on the microbiota by supporting the decrease in harmful microbes like the Proteobacteria while increasing beneficial microbes like the Firmicutes – Paraprobiotics or postbiotics can positively affect the colon mucosa microbiota	(Danladi et al. 2022)
Postbiotic Culbac® (fermentation product produced by *Lactobacillus* acidophilus) Starter diet: 1 kg/ton, grower and finisher diet: 0.5 kg/ton in dry form and in the aqueous form, 4 mL/L drinking water	Five-week-old broiler chicks	– Postbiotic compound either in a dry and/or an aqueous form improved the health, performance, and immunity of colisepticemic broiler chickens	(Abd El-Ghany et al. 2022)
Saccharomyces cerevisiae fermentation-based postbiotic (SCFP) at 1.5 kg/ton (0–21 d) and 1.0 kg/ton (22–32 d)	Layer pullets	– SCFP decreased Salmonella Enteritidis in the layer pullet's ceca – SCFP decreased the proportion of ceca with enumerable S. Enteritidis	(Gingerich et al. 2021)
0.2%, 0.4%, and 0.6% postbiotic derived from *Lactobacillus animalis*	Five-week-old quails	– Postbiotic 0.4% increased body weight and body weight gain – Postbiotics promoted the health of quails by modulating gut microbiota	(Kareem 2020)

(continued)

Table 6.1 (continued)

Postbiotic strain	Species/age	Results	References
Postbiotic product (1 ounce/gallon of freshwater)	Three-week-old broiler chicks	– Postbiotic administration boosts immunomodulatory responses in the gut – Postbiotic reduces disease pathogenesis following challenge	(Johnson et al. 2019)
0.2%, 0.4%, 0.6%, and 0.8% cell free supernatant (postbiotic: *L. plantarum* RI11)	Six-week-old broiler chicks	– Postbiotic RI11 augmented plasma glutathione, catalase, and glutathione peroxidase, and boosted zonula occludens-1, mucin 2, IL-10, and mRNA expression – Postbiotic RI11 declined heat shock protein 70 mRNA expression and plasma tumor necrosis factor alpha, IL8, alpha-1-acid glycoprotein	(Humam et al. 2019)
L. plantarum RG14 and RI11 at 0.3% with 1.0% inulin rom d 1 till d 42	Broilers/d 1–d 42	– Positively enhanced immune response – Reduced the proinflammatory responses – Reduced ($p < 0.05$) Enterobacteriaceae count	(Kareem et al. 2016)
L. plantarum RI11, RG14, and RG11 strains/cell-free supernatant (postbiotic component)	Laying hens	– Reduced plasma and yolk cholesterol concentrations	(Choe et al. 2012)
L. plantarum strains/cell-free supernatant (postbiotic component)	Laying hen and broilers	– Reduced fecal Enterobacteriaceae	
L. plantarum strains/cell-free supernatant (postbiotic component)	Broilers	– Higher growth hormone receptor messenger RNA	(Thanh et al. 2009)

mucosal architecture regarding longer villi and increased birds' performance (Yang et al. 2007; Thanh et al. 2009).

The GIT of chickens is considered home to a diverse range of microorganisms, and GIT health is crucial for immune development, nutrient absorption, and disease resistance. When changes occur in the GIT microbiome, the birds' feed efficiency, productivity, and health can be influenced (Gaskins et al. 2002; Jeurissen et al. 2002; Kohl 2012). The different sections of the GIT of chickens are heavily populated with complex microbiomes (bacteria, fungi, Archaea, protozoa, and viruses) dominated by bacteria (Wei et al. 2013). By adhering to the enterocytes' epithelial walls, the gut microbiota can form a barrier that prevents harmful bacteria from colonizing, as shown by Yegani and Korver (2008).

Furthermore, they can promote the growth of the immune system, which includes the lamina propria, intestinal immune cells (such as cytotoxic and helper-T cells, immunoglobulin-producing cells, and phagocytic cells), mucus layer, and epithelial monolayer (Dibner and Richards 2005; Shakouri et al. 2009; Oakley et al. 2014). Under the influence of the commensal microbiota, the gut mucus layer stimulates mucin secretion and epithelial cell turnover. This keeps the GIT lubricated and stops microbes from colonizing the host's intestinal epithelial cells (Shang et al. 2018). From the undigested diet, the microbiota in the hindgut (ceca and colon) can produce nutrients, energy, and SCFA that are subsequently available to the host (Gaskins et al. 2002; Dibner and Richards 2005). *Salmonella* sp. and other foodborne pathogens can be eliminated by SCFA's bacteriostatic properties (Ricke 2003). According to earlier studies, the formation of SCFA lowers the colonic pH, preventing bile from being converted into secondary biliary products (Christl et al. 1997).

The bacteria use feed ingredients that the host does not absorb or digest as a growth substrate (Pan and Yu 2014). Diets influence the variety and makeup of the gut microbiota in chickens (Borda-Molina et al. 2018). The composition of the mucosa-attached microbiota is influenced by several host factors, such as the expression of specific adhesion sites on the enterocyte membrane, secretion of secretory immunoglobulins, and mucus production rate (Shang et al. 2018). It is commonly known that feed can affect the mucosa-attached and luminal microbiota, which in turn can affect gut health (Jeurissen et al. 2002).

Postbiotics can be effectively used in place of antibiotics in broiler chicken feed, by improving animal performance (Thanh et al. 2009; Loh et al. 2010; Kareem et al. 2017; Humam et al. 2019). Any elements/molecules arising from a probiotic's metabolic activity or any released chemicals that have the potential to directly or indirectly benefit the host are referred to as postbiotics (Tsilingiri et al. 2012). Postbiotics have several advantages, such as preventing the growth of harmful bacteria, which promotes effective nutrient absorption and improved growth (Thanh et al. 2010; Fujiki et al. 2012; Kareem et al. 2014).

Indeed, the GIT of chickens harbors a large microbial community that contributes significantly to their growth and well-being by improving their ability to absorb nutrients and fortifying their immune system (Choi et al. 2015). A diversified microbiota that acts as a second line of defense against pathogen colonization controls immunological development and maturation and produces metabolites for host nutrition can flourish on the GIT (Sergeant et al. 2014; Roberts et al. 2015). Butyrate is one SCFA that gives intestinal lining epithelial cells energy and suppresses the development of dangerous bacterial virulence factors (Rinttilä and Apajalahti 2013; Sergeant et al. 2014; Roto et al. 2015; Polansky et al. 2016).

There is no doubt that cecal microbiota has the greatest diversity in the GIT. It is the source for most, if not all, mucosa-associated microbiota of the proximal and distal colon (Wang et al. 2010). Until now, most studies on chicken intestinal microbiota have focused more on the cecal section of the hindgut. The implication of an increased abundance of Firmicutes due to dietary supplementation with postbiotics is more butyrate production in broiler chickens (Danladi et al. 2022). Firmicutes were reported as the phylum with a larger number of taxa encoding enzymes required for butyrate production (Segura-Wang et al. 2021).

It has previously been documented that a combination of postbiotic metabolites raises the levels of butyric acid in the feces of broiler chickens (Thanh et al. 2009). Moreover, butyrate serves as the primary energy source of enterocytes and controls cellular proliferation and differentiation in the intestinal mucosa, which raises the weight of the intestinal tissue (Rinttilä and Apajalahti 2013; Koh et al. 2016; Beauclercq et al. 2018). One of the main ways that butyrate promotes the formation of intestinal mucosa is by stimulating the production of gastrointestinal peptides and growth factors, which in turn causes cell proliferation (Guilloteau et al. 2010).

Producing propionate and succinate was associated with *Bacteroides* as terminal products of metabolism, as Adamberg et al. (2014) reported. Propionate is a less preferred substrate of colonocytes but is transported to the liver and used as an important energy source for the host (Koh et al. 2016). The strain *Faecalibacterium* was reported to be a carrier of the enzymes necessary for butyrate production and present from the early stages of development. Therefore, the strain will actively participate in future intervention and modulation of the gut microbiota by improving poultry's overall health and growth performance (Segura-Wang et al. 2021). *Lactobacillus*, an important probiotic bacterium promoting a healthy gut, was the fourth most predominant genus. A recent study with postbiotics also revealed a significant (p < 0.05) increase in the population of *Lactobacillus* in the cecum of broiler chickens (Humam et al. 2019). *Lactobacillus* is a beneficial microbe that can produce bacteriocins, a natural antimicrobial compound capable of inhibiting the growth of pathogens at molecular and cellular levels (Drider et al. 2006). Danladi et al. (2022) used the postbiotics from the following strains of *L. plantarum* (RI11, RS5, and TL1) and incorporated them into the broiler starter and finisher diets at a 0.2% level. Danladi and his co-authors noted that supplementation of postbiotics in the diets of broilers demonstrated positive effects on the microbiota by supporting the increase of beneficial microbes like *Firmicutes* while decreasing harmful microbes like *Proteobacteria*.

Loh et al. (2014) found that in laying hens, the dietary postbiotics raised the fecal LAB and decreased the pH and Enterobacteriaceae in the feces. Also, Rosyidah et al. (2011) observed that broiler chickens fed a diet with a metabolite produced by *L. plantarum* resulted in an increase in LAB count and a decrease in fecal Enterobacteriaceae counts. Nabizadeh (2012) noted that feeding broiler chickens with 1% inulin decreased the pH and *E. coli* numbers in their cecum. Postbiotics derived from *Lactobacillus* contain useful substances, including bacteriocin and organic acids that promote the growth of lactic acid bacteria (Loh et al. 2014).

Moreover, van der Wielen et al. (2000) reported that during the growth of broiler chickens, volatile fatty acids are responsible for reducing the number of Enterobacteriaceae in the ceca. An important and well-understood function of the gut microbiota is metabolizing host-indigestible dietary components. The microbiome's capacity greatly increases the host's ability to use the feed's energy content. The microbiota and the metabolites they generate also significantly impact the host's signal transduction and are frequently host response regulators (Loh et al. 2014; Blacher et al. 2017).

7   Positive Impact of Postbiotics on the Quality of Meat and Eggs

Due to the use of postbiotics, it is now possible to raise animals, including poultry, without the use of antibiotics, which has resulted in the production of chicken products such as meat and eggs that are both safe and of high-quality worldwide. Poultry is raised specifically for producing meat worldwide because it is tender, low in fat content, and has a relatively quick production cycle (Haque et al. 2020). Postbiotics (specifically RI11 strain isolated from L. *plantarum*) have the potential to be used as sources of antioxidants in heat stressed broilers. Postbiotics have also been shown to raise breast meat pH while decreasing shear force and CIE L* (Humam et al. 2019). In broiler chickens, improved meat quality and decreased plasma cholesterol levels were reported (Choe et al. 2012; Loh et al. 2014). The effects of postbiotics and inulin on the meat's quality were beneficial (Kareem et al. 2015). Poultry meat plays a role in boosting food security, protein supply, and employment (Reuben et al. 2021). This study showed that laying hens benefit from postbiotic metabolite combinations from L. *plantarum* strains. All metabolite combinations increased hen-day egg output (Loh et al. 2014). As the world's population grows, so does the demand for meat and eggs. Probiotics/postbiotics may improve their quality. Poultry farming focuses on safe and healthy products. Probiotics improve animal productivity and quality (Hussein et al. 2020).

8   Conclusions

This chapter makes clear that postbiotics can be employed in feed, medicinal techniques, and as antibiotic growth promoter substitutes in poultry when given in sufficient levels because of their positive microbial influence on health. Postbiotics for poultry enhance production, health, and nutrition. They might take the place of other synthetic compounds and antibiotic growth boosters for poultry. Gut microbiota and modulation of the immune system will also significantly lower the cost of treating illnesses and avoid bird loss. Postbiotic-based sustainable chicken farming will ensure the safety and security of food. An additional research study is needed to restrict the presence of resistant effects among pathogenic bacteria by employing postbiotics and to prevent the use of antibiotics for the prevention of disease. Subsequent investigations into the prebiotic–postbiotic relationship could yield new insights or enhance health, body weight, and meat quality.

References

Abd El-Ghany WA, Fouad H, Quesnell R, Sakai L (2022) The effect of a postbiotic produced by stabilized non-viable lactobacilli on the health, growth performance, immunity, and gut status of colisepticaemic broiler chickens. Trop Anim Health Prod 54(5):286

Adamberg S, Tomson K, Vija H, Puurand M, Kabanova N, Visnapuu T, Jõgi E, Alamäe T, Adamberg K (2014) Degradation of fructans and production of propionic acid by Bacteroides thetaiotaomicron are enhanced by the shortage of amino acids. Front Nutr 1:21

Aguilar-Toalá J, Garcia-Varela R, Garcia H, Mata-Haro V, González-Córdova A, Vallejo-Cordoba B, Hernández-Mendoza A (2018) Postbiotics: an evolving term within the functional foods field. Trends Food Sci Technol 75:105–114

Arain MA, Nabi F, Shah QA, Alagawany M, Fazlani SA, Khalid M, Soomro F, Khand FM, Farag MR (2022) The role of early feeding in improving performance and health of poultry: herbs and their derivatives. Worlds Poult Sci J 78(2):499–513

Beauclercq S, Nadal-Desbarats L, Hennequet-Antier C, Gabriel I, Tesseraud S, Calenge F, Le Bihan-Duval E, Mignon-Grasteau S (2018) Relationships between digestive efficiency and metabolomic profiles of serum and intestinal contents in chickens. Sci Rep 8(1):6678

Beccavin C, Chevalier B, Cogburn L, Simon J, Duclos MJ (2001) Insulin-like growth factors and body growth in chickens divergently selected for high or low growth rate. J Endocrinol 168(2):297–306

Beckman BR (2011) Perspectives on concordant and discordant relations between insulin-like growth factor 1 (IGF1) and growth in fishes. Gen Comp Endocrinol 170(2):233–252

Beckman BR, Shimizu M, Gadberry BA, Parkins PJ, Cooper KA (2004) The effect of temperature change on the relations among plasma IGF-I, 41-kDa IGFBP, and growth rate in postsmolt coho salmon. Aquaculture 241(1–4):601–619

Blacher E, Levy M, Tatirovsky E, Elinav E (2017) Microbiome-modulated metabolites at the interface of host immunity. J Immunol 198(2):572–580

Borda-Molina D, Seifert J, Camarinha-Silva A (2018) Current perspectives of the chicken gastrointestinal tract and its microbiome. Comput Struct Biotechnol J 16:131–139

Chaney E, Miller EA, Firman J, Binnebose A, Kuttappan V, Johnson TJ (2023) Effects of a postbiotic, with and without a saponin-based product, on Turkey performance. Poult Sci 102(5):102607

Chang HM, Loh TC, Foo HL, Lim ETC (2022) Lactiplantibacillus plantarum Postbiotics: alternative of antibiotic growth promoter to ameliorate gut health in broiler chickens. Front Vet Sci 9:883324

Choe D, Loh T, Foo H, Hair-Bejo M, Awis Q (2012) Egg production, faecal pH and microbial population, small intestine morphology, and plasma and yolk cholesterol in laying hens given liquid metabolites produced by lactobacillus plantarum strains. Br Poult Sci 53(1):106–115

Choi KY, Lee TK, Sul WJ (2015) Metagenomic analysis of chicken gut microbiota for improving metabolism and health of chickens—a review. Asian Australas J Anim Sci 28(9):1217

Christl SU, Bartram P, Paul A, Kelber E, Scheppach W, Kasper H (1997) Bile acid metabolism by colonic bacteria in continuous culture: effects of starch and pH. Ann Nutr Metab 41(1):45–51

Cicenia A, Santangelo F, Gambardella L, Pallotta L, Iebba V, Scirocco A, Marignani M, Tellan G, Carabotti M, Corazziari ES (2016) Protective role of postbiotic mediators secreted by lactobacillus rhamnosus GG versus lipopolysaccharide-induced damage in human colonic smooth muscle cells. J Clin Gastroenterol 50:S140–S144

Cicenia A, Scirocco A, Carabotti M, Pallotta L, Marignani M, Severi C (2014) Postbiotic activities of lactobacilli-derived factors. J Clin Gastroenterol 48:S18–S22

Collins MD, Gibson GR (1999) Probiotics, prebiotics, and synbiotics: approaches for modulating the microbial ecology of the gut. Am J Clin Nutr 69(5):1052s–1057s

Danladi Y, Loh TC, Foo HL, Akit H, Tamrin NA, Naeem Azizi M (2022) Effects of postbiotics and paraprobiotics as replacements for antibiotics on growth performance, carcass characteristics, small intestine histomorphology, immune status and hepatic growth gene expression in broiler chickens. Animals 12(7):917

de Almada CN, Almada CN, Martinez RC, Sant'Ana AS (2016) Paraprobiotics: evidences on their ability to modify biological responses, inactivation methods and perspectives on their application in foods. Trends Food Sci Technol 58:96–114

De Marco S, Sichetti M, Muradyan D, Piccioni M, Traina G, Pagiotti R, Pietrella D (2018) Probiotic cell-free supernatants exhibited anti-inflammatory and antioxidant activity on human gut epithelial cells and macrophages stimulated with LPS. Evid Based Complement Alternat Med 2018:1756308

Dibner J, Richards JD (2005) Antibiotic growth promoters in agriculture: history and mode of action. Poult Sci 84(4):634–643

Drider D, Fimland G, Héchard Y, McMullen LM, Prévost H (2006) The continuing story of class IIa bacteriocins. Microbiol Mol Biol Rev 70(2):564–582

Dunne C, Murphy L, Flynn S, O'Mahony L, O'Halloran S, Feeney M, Morrissey D, Thornton G, Fitzgerald G, Daly C (1999) Probiotics: from myth to reality. Demonstration of functionality in animal models of disease and in human clinical trials. Lactic acid bacteria: genetics, metabolism and applications: proceedings of the Sixth Symposium on lactic acid bacteria: genetics, metabolism and applications, 19–23 September 1999, Veldhoven, The Netherlands

Faseleh Jahromi M, Wesam Altaher Y, Shokryazdan P, Ebrahimi R, Ebrahimi M, Idrus Z, Tufarelli V, Liang JB (2016) Dietary supplementation of a mixture of lactobacillus strains enhances performance of broiler chickens raised under heat stress conditions. Int J Biometeorol 60:1099–1110

Feng Y, Wang Y, Wang P, Huang Y, Wang F (2018) Short-chain fatty acids manifest stimulative and protective effects on intestinal barrier function through the inhibition of NLRP3 inflammasome and autophagy. Cell Physiol Biochem 49(1):190–205

Fujiki T, Hirose Y, Yamamoto Y, Murosaki S (2012) Enhanced immunomodulatory activity and stability in simulated digestive juices of lactobacillus plantarum L-137 by heat treatment. Biosci Biotechnol Biochem 76(5):918–922

Gaskins H, Collier C, Anderson D (2002) Antibiotics as growth promotants: mode of action. Anim Biotechnol 13(1):29–42

Gibson GR, Roberfroid MB (1995) Dietary modulation of the human colonic microbiota: introducing the concept of prebiotics. J Nutr 125(6):1401–1412

Gingerich E, Frana T, Logue C, Smith D, Pavlidis H, Chaney W (2021) Effect of feeding a postbiotic derived from saccharomyces cerevisiae fermentation as a preharvest food safety hurdle for reducing salmonella Enteritidis in the ceca of layer pullets. J Food Prot 84(2):275–280

Grandclément C, Tannières M, Moréra S, Dessaux Y, Faure D (2016) Quorum quenching: role in nature and applied developments. FEMS Microbiol Rev 40(1):86–116

Guilloteau P, Martin L, Eeckhaut V, Ducatelle R, Zabielski R, Van Immerseel F (2010) From the gut to the peripheral tissues: the multiple effects of butyrate. Nutr Res Rev 23(2):366–384

Haileselassie Y, Navis M, Vu N, Qazi KR, Rethi B, Sverremark-Ekström E (2016) Postbiotic modulation of retinoic acid imprinted mucosal-like dendritic cells by probiotic lactobacillus reuteri 17938 in vitro. Front Immunol 7:96

Haque MH, Sarker S, Islam MS, Islam MA, Karim MR, Kayesh MEH, Shiddiky MJ, Anwer MS (2020) Sustainable antibiotic-free broiler meat production: current trends, challenges, and possibilities in a developing country perspective. Biology 9(11):411

Homayouni Rad A, Aghebati Maleki L, Samadi Kafil H, Abbasi A (2021) Postbiotics: A novel strategy in food allergy treatment. Crit Rev Food Sci Nutr 61(3):492–499

Huber M, Mossmann H, Bessler W (2005) Th1-orientated immunological properties of the bacterial extract OM-85-BV. Eur J Med Res 10(5):209–217

Humam AM, Loh TC, Foo HL, Izuddin WI, Awad EA, Idrus Z, Samsudin AA, Mustapha NM (2020) Dietary supplementation of postbiotics mitigates adverse impacts of heat stress on antioxidant enzyme activity, total antioxidant, lipid peroxidation, physiological stress indicators, lipid profile and meat quality in broilers. Animals 10(6):982

Humam AM, Loh TC, Foo HL, Samsudin AA, Mustapha NM, Zulkifli I, Izuddin WI (2019) Effects of feeding different postbiotics produced by lactobacillus plantarum on growth performance, carcass yield, intestinal morphology, gut microbiota composition, immune status, and growth gene expression in broilers under heat stress. Animals 9(9):644

Hussein EO, Ahmed SH, Abudabos AM, Suliman GM, Abd El-Hack ME, Swelum AA, Alowaimer N, A. (2020) Ameliorative effects of antibiotic-, probiotic- and phytobiotic-supplemented diets on the performance, intestinal health, carcass traits, and meat quality of Clostridium perfringens-infected broilers. Animals 10(4):669

Iwasaki M, Akiba Y, Kaunitz JD (2019) Duodenal chemosensing of short-chain fatty acids: implications for GI diseases. Curr Gastroenterol Rep 21:1–8

Jeong J, Kim I (2014) Effect of Bacillus subtilis C-3102 spores as a probiotic feed supplement on growth performance, noxious gas emission, and intestinal microflora in broilers. Poult Sci 93(12):3097–3103

Jeurissen S, Lewis F, van der Klis JD, Mroz Z, Rebel J, Ter Huurne A (2002) Parameters and techniques to determine intestinal health of poultry as constituted by immunity, integrity, and functionality. Curr Issues Intest Microbiol 3(1):1–14

Johnson CN, Kogut MH, Genovese K, He H, Kazemi S, Arsenault RJ (2019) Administration of a postbiotic causes immunomodulatory responses in broiler gut and reduces disease pathogenesis following challenge. Microorganisms 7(8):268

Jones JI, Clemmons DR (1995) Insulin-like growth factors and their binding proteins: biological actions. Endocr Rev 16(1):3–34

Kalavathy R, Abdullah N, Jalaludin S, Ho Y (2003) Effects of lactobacillus cultures on growth performance, abdominal fat deposition, serum lipids and weight of organs of broiler chickens. Br Poult Sci 44(1):139–144

Kareem K, Loh T, Foo H, Asmara S, Akit H (2017) Influence of postbiotic RG14 and inulin combination on cecal microbiota, organic acid concentration, and cytokine expression in broiler chickens. Poult Sci 96(4):966–975

Kareem KY (2020) Effect of different levels of postbiotic on growth performance, intestinal microbiota count and volatile fatty acids on quail. Plant Archives 20(2):2885–2887

Kareem KY, Hooi Ling F, Teck Chwen L, May Foong O, Anjas Asmara S (2014) Inhibitory activity of postbiotic produced by strains of lactobacillus plantarum using reconstituted media supplemented with inulin. Gut Pathog 6(1):1–7

Kareem KY, Loh TC, Foo HL, Akit H, Samsudin AA (2016) Effects of dietary postbiotic and inulin on growth performance, IGF1 and GHR mRNA expression, faecal microbiota and volatile fatty acids in broilers. BMC Vet Res 12:1–10

Kareem KY, Loh TC, Foo HL, Asmara SA, Akit H, Abdulla NR, Ooi MF (2015) Carcass, meat and bone quality of broiler chickens fed with postbiotic and prebiotic combinations. Int J Probiotics Prebiotics 10(1):23

Klemashevich C, Wu C, Howsmon D, Alaniz RC, Lee K, Jayaraman A (2014) Rational identification of diet-derived postbiotics for improving intestinal microbiota function. Curr Opin Biotechnol 26:85–90

Koh A, De Vadder F, Kovatcheva-Datchary P, Bäckhed F (2016) From dietary fiber to host physiology: short-chain fatty acids as key bacterial metabolites. Cell 165(6):1332–1345

Kohl KD (2012) Diversity and function of the avian gut microbiota. J Comp Physiol B 182:591–602

Laverde Gomez JA, Mukhopadhya I, Duncan SH, Louis P, Shaw S, Collie-Duguid E, Crost E, Juge N, Flint HJ (2019) Formate cross-feeding and cooperative metabolic interactions revealed by transcriptomics in co-cultures of acetogenic and amylolytic human colonic bacteria. Environ Microbiol 21(1):259–271

Lebeer S, Vanderleyden J, De Keersmaecker SC (2010) Host interactions of probiotic bacterial surface molecules: comparison with commensals and pathogens. Nat Rev Microbiol 8(3):171–184

Loh TC, Choe DW, Foo HL, Sazili AQ, Bejo MH (2014) Effects of feeding different postbiotic metabolite combinations produced by lactobacillus plantarum strains on egg quality and production performance, faecal parameters and plasma cholesterol in laying hens. BMC Vet Res 10:1–9

Loh TC, Thanh NT, Foo HL, Hair-Bejo M, Azhar BK (2010) Feeding of different levels of metabolite combinations produced by lactobacillus plantarum on growth performance, fecal microflora, volatile fatty acids and villi height in broilers. Anim Sci J 81(2):205–214

Long SL, Gahan CG, Joyce SA (2017) Interactions between gut bacteria and bile in health and disease. Mol Asp Med 56:54–65

Mantziari A, Salminen S, Szajewska H, Malagón-Rojas JN (2020) Postbiotics against pathogens commonly involved in pediatric infectious diseases. Microorganisms 8(10):1510

Marcial G, Villena J, Faller G, Hensel A, de Valdéz GF (2017) Exopolysaccharide-producing Streptococcus thermophilus CRL1190 reduces the inflammatory response caused by helicobacter pylori. Benefic Microbes 8(3):451–461

Moriyama S (1995) Increased plasma insulin-like growth factor-I (IGF-I) following oral and intraperitoneal administration of growth hormone to rainbow trout, Oncorhynchus mykiss. Growth Regul 5(3):164–167

Nabi F, Arain MA, Rajput N, Alagawany M, Soomro J, Umer M, Soomro F, Wang Z, Ye R, Liu J (2020) Health benefits of carotenoids and potential application in poultry industry: A review. J Anim Physiol Anim Nutr 104(6):1809–1818

Nabizadeh A (2012) The effect of inulin on broiler chicken intestinal microflora, gut morphology, and performance. J Anim Feed Sci 21(4):725–734

Oakley BB, Lillehoj HS, Kogut MH, Kim WK, Maurer JJ, Pedroso A, Lee MD, Collett SR, Johnson TJ, Cox NA (2014) The chicken gastrointestinal microbiome. FEMS Microbiol Lett 360(2):100–112

Pan D, Yu Z (2014) Intestinal microbiome of poultry and its interaction with host and diet. Gut Microbes 5(1):108–119

Pandey KR, Naik SR, Vakil BV (2015) Probiotics, prebiotics and synbiotics-a review. J Food Sci Technol 52:7577–7587

Patel RM, Denning PW (2013) Therapeutic use of prebiotics, probiotics, and postbiotics to prevent necrotizing enterocolitis: what is the current evidence? Clin Perinatol 40(1):11–25

Polansky O, Sekelova Z, Faldynova M, Sebkova A, Sisak F, Rychlik I (2016) Important metabolic pathways and biological processes expressed by chicken cecal microbiota. Appl Environ Microbiol 82(5):1569–1576

Reuben RC, Sarkar SL, Roy PC, Anwar A, Hossain MA, Jahid IK (2021) Prebiotics, probiotics and postbiotics for sustainable poultry production. Worlds Poult Sci J 77(4):825–882

Ricke S (2003) Perspectives on the use of organic acids and short chain fatty acids as antimicrobials. Poult Sci 82(4):632–639

Ricke SC (2018) Focus: nutrition and food science: impact of prebiotics on poultry production and food safety. Yale J Biol Med 91(2):151

Rinttilä T, Apajalahti J (2013) Intestinal microbiota and metabolites—implications for broiler chicken health and performance. J Appl Poult Res 22(3):647–658

Roberts T, Wilson J, Guthrie A, Cookson K, Vancraeynest D, Schaeffer J, Moody R, Clark S (2015) New issues and science in broiler chicken intestinal health: emerging technology and alternative interventions. J Appl Poult Res 24(2):257–266

Rosyidah M, Loh T, Foo H, Cheng X, Bejo M (2011) Effect of feeding metabolites and acidifier on growth performance, faecal characteristics and microflora in broiler chickens. J Anim Vet Adv 10(21):2758–2764

Roto SM, Rubinelli PM, Ricke SC (2015) An introduction to the avian gut microbiota and the effects of yeast-based prebiotic-type compounds as potential feed additives. Front Vet Sci 2:28

Salminen S, Collado MC, Endo A, Hill C, Lebeer S, Quigley EM, Sanders ME, Shamir R, Swann JR, Szajewska H (2021) The international scientific Association of Probiotics and Prebiotics (ISAPP) consensus statement on the definition and scope of postbiotics. Nat Rev Gastroenterol Hepatol 18(9):649–667

Schrezenmeir J, de Vrese M (2001) Probiotics, prebiotics, and synbiotics—approaching a definition. Am J Clin Nutr 73(2):361s–364s

Segura-Wang M, Grabner N, Koestelbauer A, Klose V, Ghanbari M (2021) Genome-resolved metagenomics of the chicken gut microbiome. Front Microbiol 12:726923

Sergeant MJ, Constantinidou C, Cogan TA, Bedford MR, Penn CW, Pallen MJ (2014) Extensive microbial and functional diversity within the chicken cecal microbiome. PLoS One 9(3):e91941

Shakouri M, Iji P, Mikkelsen LL, Cowieson A (2009) Intestinal function and gut microflora of broiler chickens as influenced by cereal grains and microbial enzyme supplementation. J Anim Physiol Anim Nutr 93(5):647–658

Shang Y, Kumar S, Oakley B, Kim WK (2018) Chicken gut microbiota: importance and detection technology. Front Vet Sci 5:254

Shenderov BA (2013) Metabiotics: novel idea or natural development of probiotic conception. Microb Ecol Health Dis 24(1):20399

Shigwedha N (2014) Probiotical cell fragments (PCFs) as "novel nutraceutical ingredients". J Biosci Med 2(03):43

Silverstein JT, Shearer KD, Dickhoff WW, Plisetskaya EM (1998) Effects of growth and fatness on sexual development of Chinook salmon (Oncorhynchus tshawytscha) parr. Can J Fish Aquat Sci 55(11):2376–2382

Simons A, Alhanout K, Duval RE (2020) Bacteriocins, antimicrobial peptides from bacterial origin: overview of their biology and their impact against multidrug-resistant bacteria. Microorganisms 8(5):639

Sun M, Wu W, Chen L, Yang W, Huang X, Ma C, Chen F, Xiao Y, Zhao Y, Ma C (2018) Microbiota-derived short-chain fatty acids promote Th1 cell IL-10 production to maintain intestinal homeostasis. Nat Commun 9(1):3555

Swanson KS, Gibson GR, Hutkins R, Reimer RA, Reid G, Verbeke K, Scott KP, Holscher HD, Azad MB, Delzenne NM (2020) The international scientific Association for Probiotics and Prebiotics (ISAPP) consensus statement on the definition and scope of synbiotics. Nat Rev Gastroenterol Hepatol 17(11):687–701

Tannock G (1999) Probiotics: a critical review. J Antimicrob Chemother 43(1):849–852

Teame T, Wang A, Xie M, Zhang Z, Yang Y, Ding Q, Gao C, Olsen RE, Ran C, Zhou Z (2020) Paraprobiotics and postbiotics of probiotic lactobacilli, their positive effects on the host and action mechanisms: A review. Front Nutr 7:570344

Thanh N, Loh T, Foo H, Hair-Bejo M, Azhar B (2009) Effects of feeding metabolite combinations produced by lactobacillus plantarum on growth performance, faecal microbial population, small intestine villus height and faecal volatile fatty acids in broilers. Br Poult Sci 50(3):298–306

Thanh NT, Chwen LT, Foo HL, Hair-Bejo M, Kasim AB (2010) Inhibitory activity of metabolites produced by strains of lactobacillus plantarum isolated from Malaysian fermented food. Int J Probiotics Prebiotics 5(1):37

Thu T, Loh TC, Foo H, Yaakub H, Bejo M (2011) Effects of liquid metabolite combinations produced by lactobacillus plantarum on growth performance, faeces characteristics, intestinal morphology and diarrhoea incidence in postweaning piglets. Trop Anim Health Prod 43:69–75

Tiptiri-Kourpeti A, Spyridopoulou K, Santarmaki V, Aindelis G, Tompoulidou E, Lamprianidou EE, Saxami G, Ypsilantis P, Lampri ES, Simopoulos C (2016) Lactobacillus casei exerts antiproliferative effects accompanied by apoptotic cell death and up-regulation of TRAIL in colon carcinoma cells. PLoS One 11(2):e0147960

Tomasik P, Tomasik P (2020) Probiotics, non-dairy prebiotics and postbiotics in nutrition. Appl Sci 10(4):1470

Tsilingiri K, Barbosa T, Penna G, Caprioli F, Sonzogni A, Viale G, Rescigno M (2012) Probiotic and postbiotic activity in health and disease: comparison on a novel polarised ex-vivo organ culture model. Gut 61(7):1007–1015

Tytgat HL, Douillard FP, Reunanen J, Rasinkangas P, Hendrickx AP, Laine PK, Paulin L, Satokari R, de Vos WM (2016) Lactobacillus rhamnosus GG outcompetes enterococcus faecium via mucus-binding pili: evidence for a novel and heterospecific probiotic mechanism. Appl Environ Microbiol 82(19):5756–5762

Uni Z, Noy Y, Sklan D (1995) Posthatch changes in morphology and function of the small intestines in heavy-and light-strain chicks. Poult Sci 74(10):1622–1629

van der Wielen PW, Biesterveld S, Notermans S, Hofstra H, Urlings BA, van Knapen F (2000) Role of volatile fatty acids in development of the cecal microflora in broiler chickens during growth. Appl Environ Microbiol 66(6):2536–2540

Vinolo MA, Rodrigues HG, Nachbar RT, Curi R (2011) Regulation of inflammation by short chain fatty acids. Nutrients 3(10):858–876

Wang Y, Devkota S, Musch MW, Jabri B, Nagler C, Antonopoulos DA, Chervonsky A, Chang EB (2010) Regional mucosa-associated microbiota determine physiological expression of TLR2 and TLR4 in murine colon. PLoS One 5(10):e13607

Wei S, Morrison M, Yu Z (2013) Bacterial census of poultry intestinal microbiome. Poult Sci 92(3):671–683

Williams NT (2010) Probiotics. Am J Health Syst Pharm 67(6):449–458

Wu G, Fanzo J, Miller DD, Pingali P, Post M, Steiner JL, Thalacker-Mercer AE (2014) Production and supply of high-quality food protein for human consumption: sustainability, challenges, and innovations. Ann N Y Acad Sci 1321(1):1–19

Yang S-C, Lin C-H, Sung CT, Fang J-Y (2014) Antibacterial activities of bacteriocins: application in foods and pharmaceuticals. Front Microbiol 5:241

Yang Y, Iji P, Choct M (2007) Effects of different dietary levels of mannanoligosaccharide on growth performance and gut development of broiler chickens. Asian Australas J Anim Sci 20(7):1084–1091

Yegani M, Korver D (2008) Factors affecting intestinal health in poultry. Poult Sci 87(10):2052–2063

Yelin I, Flett KB, Merakou C, Mehrotra P, Stam J, Snesrud E, Hinkle M, Lesho E, McGann P, McAdam AJ (2019) Genomic and epidemiological evidence of bacterial transmission from probiotic capsule to blood in ICU patients. Nat Med 25(11):1728–1732

Zendeboodi F, Khorshidian N, Mortazavian AM, da Cruz AG (2020) Probiotic: conceptualization from a new approach. Curr Opin Food Sci 32:103–123

Zhang Z, Kim I (2014) Effects of multistrain probiotics on growth performance, apparent ileal nutrient digestibility, blood characteristics, cecal microbial shedding, and excreta odor contents in broilers. Poult Sci 93(2):364–370

Zheng J, Wittouck S, Salvetti E, Franz CM, Harris HM, Mattarelli P, O'toole PW, Pot B, Vandamme P, Walter J (2020) A taxonomic note on the genus lactobacillus: description of 23 novel genera, emended description of the genus lactobacillus Beijerinck 1901, and union of Lactobacillaceae and Leuconostocaceae. Int J Syst Evol Microbiol 70(4):2782–2858

Zuidhof M, Schneider B, Carney V, Korver D, Robinson F (2014) Growth, efficiency, and yield of commercial broilers from 1957, 1978, and 2005. Poult Sci 93(12):2970–2982

The Use of Enzymes in Poultry Nutrition: An Opportunity for Feed Enzymes

Amr Abd El-Wahab, Claudia Huber, Hesham R. El-Seedi,
Christian Visscher, and Awad A. Shehata

Contents

1   Introduction.. 123
2   Phosphorus Availability in Poultry Diets.................................... 123
   2.1   Phytic Acid and Phytate... 124
   2.2   Factors Affecting Phytate-P Hydrolysis.............................. 125
      2.2.1   Effect of Calcium.. 125
      2.2.2   Effect of Other Nutrients.................................. 126
3   Anti-nutritional Factors... 126
4   Enzymes in Poultry Diets: Key to Unlocking Hidden Nutrients........... 127
   4.1   Phytate-Degrading Enzymes (Phytase)............................... 130
      4.1.1   Phytase Discovery and General Description.............. 130
      4.1.2   Intrinsic Plant Phytase.................................... 130
      4.1.3   Exogenous Phytase... 131
      4.1.4   Effect of Phytase on Phosphorus......................... 132
      4.1.5   Effect of Phytase on Protein.............................. 133
      4.1.6   Effect of Phytase on Energy.............................. 133

A. A. El-Wahab (✉)
Institute for Animal Nutrition, University of Veterinary Medicine Hannover Foundation, Hannover, Germany

Department of Nutrition and Nutritional Deficiency Diseases, Faculty of Veterinary Medicine, Mansoura University, Mansoura, Egypt
e-mail: amr.abd.el-wahab@tiho-hannover.de

C. Huber · A. A. Shehata
Structural Membrane Biochemistry, Bavarian NMR Center, Technical University of Munich (TUM), Garching, Bayern, Germany

H. R. El-Seedi
Pharmacognosy Group, Department of Pharmaceutical Biosciences, Uppsala University, SE, Uppsala, Sweden

C. Visscher
Institute for Animal Nutrition, University of Veterinary Medicine Hannover Foundation, Hannover, Germany

© The Author(s), under exclusive license to Springer Nature Switzerland AG 2024
A. A. Shehata et al. (eds.), *Alternatives to Antibiotics against Pathogens in Poultry*, https://doi.org/10.1007/978-3-031-70480-2_7

Abstract

The goal of animal nutrition is to increase the digestibility of nutrients (such as minerals) for example, by the use of diverse feed treatment techniques or the application of exogenous enzymes. For example, phosphorus (P) is primarily stored in the form of phytates in plant seeds, thus being poorly available for monogastric livestock, such as poultry. As phytate is a polyanionic molecule, it has the capacity to chelate positively charged cations, especially calcium (Ca), iron (Fe), and zinc (Zn). Furthermore, it probably compromises the utilization of other dietary nutrients, including protein, starch, and lipids. Reduced efficiency of utilization implies both higher levels of supplementation and increased discharge of the undigested nutrients to the environment. The enzyme phytase catalyzes the stepwise hydrolysis of phytate. Moreover, poultry are often fed with cereal grains, which provide them with the necessary energy. However, these grains contain anti-nutritional factors such as non-starch polysaccharides (NSP), which birds cannot digest due to the lack of vital enzymes in their intestines. This may also lead to an increase in intestinal viscosity, slowing down the migration and absorption of nutrients and causing competition within gut microbiota for digestible nutrients, which in turn affects the overall health of the birds and may increase production costs. Additionally, carbohydrases supplementation plays an important role in poultry diets with high NSP contents. This chapter will discuss the main exogenous enzymes used in poultry to improve feed digestibility and utilization and enhance growth performance.

Keywords

Enzymes · Poultry · Feed utilization · Health · Growth performance · Applications

1   Introduction

In poultry farming, feed constitutes 70–75% of total production costs. Poultry feed is based primarily on cereal grains, mainly corn, wheat, sorghum, etc., which are supplied to meet most of the energy and protein requirements in the poultry diet. The more affordable ingredients, including barley, triticale, rye, olive cake, and sunflower meal (Teymouri et al. 2018; Waititu et al. 2018), could play a role in the substitution of corn, wheat, and soybean, but have some anti-nutritional factors (ANF) which may affect growth performance and intestinal health of the birds.

The United States Department of Agriculture (USDA) predicts that by 2031, the world population will be 8.5 billion, a 0.9% increase from 2022 (USDA 2022). To keep pace with the continually growing population, farmers will need to find a way to produce quality livestock and poultry products without competing with raw materials needed for human consumption. The feed-to-price ratio of broilers is expected to rise steadily over the next 10 years. The USDA estimates that by 2031, the feed-to-price ratio of broilers will rise over 34% (USDA 2022). Therefore, there is a huge demand for products that can be added to poultry diets to increase the nutritional value of feed ingredients and consequently reduce feed costs.

Enzymes have become prevalent feed additives due to their ability to improve feed efficiency and reduce feed costs. The addition of enzymes improves nutrient availability, which saves on feed ingredient costs. The supplementation of exogenous enzymes is estimated to save the global feed market about 3–5 billion USD every year (Lourenco et al. 2020). For example, the estimated Feed Market Size for phytase alone was valued at 592.5 million USD in 2022 (Walker et al. 2024). This chapter sheds light on the main challenges related to nutrient availability and the role of enzymes in improving nutrient digestibility and animal performance.

2   Phosphorus Availability in Poultry Diets

Phosphorus is crucial for good growth, bone development, energy metabolism, and other physiological processes; all living organisms require a constant supply of P. It is found mainly in bones as part of the hydroxyl-apatite complex, but it is also a vital structural component of DNA and RNA. It is also found in the phospholipid bilayer that defines cell boundaries and adenosine triphosphate, the source of energy. From the total P supplied with the feed, only the part that is absorbed by intestinal epithelia can contribute to meeting the organism's requirements; this is called digestible P (Costa et al. 2023).

The amount of unavailable and available P varies depending on its source, i.e., plant, rock, and animal origins (Fig. 7.1). However, soil P loss from agricultural systems will limit food and feed production in the future. Thus, potential threats of global P limitation due to peak P have been discussed intensively in recent years (Alewell et al. 2020; Edixhoven et al. 2014).

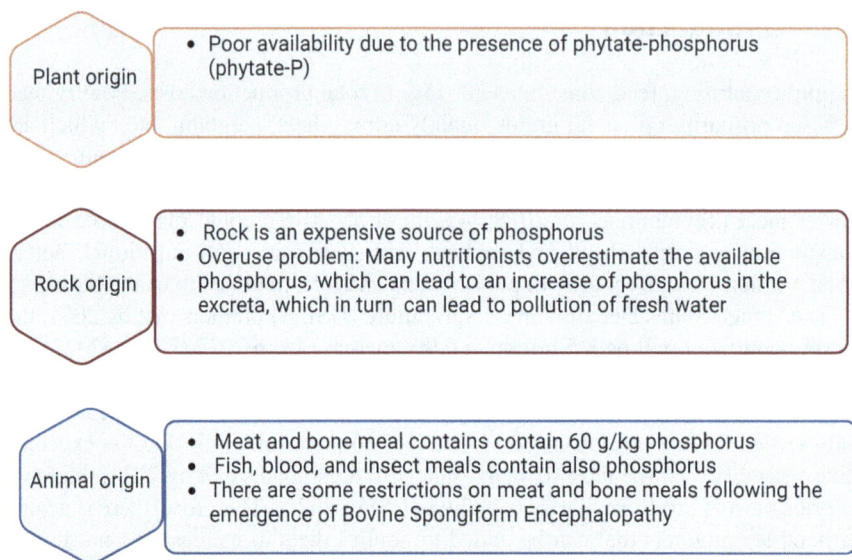

Fig. 7.1 Challenges related to the sources of phosphorus in poultry diets (Ma et al. 2019). The Figure was generated by BioRender

2.1 Phytic Acid and Phytate

Phytic acid (myo-inositol-1,2,3,4,5,6-hexakisdihydrogenphosphate, $C_6H_{18}P_6O_{24}$) possesses six phosphate groups, which are esterified with the hydroxyl groups of the cyclic sixfold alcohol myo-inositol (Pallauf and Rimbach 1997). Phytic acid is unstable in the free acid form (Reddy 1989) and occurs mainly as a complex with metal cations such as calcium (Ca^{++}), iron (Fe^{++}), zinc (Zn^{++}), magnesium (Mg^{++}), potassium (K^+), and manganese (Mn^{++}). These salts are called phytates (Cheryan and Rackis 1980; Morris 1986). Briefly, phytic acid is the primary P storage compound in seeds, typically contributing 50–80% of total P in plant seeds (Kumar et al. 2012). It helps control effective germination, allowing for a P-release boost when digested by seed phytase upon germination. The salt form of phytic acid is called phytate, and almost all phytic acid is present as a mixed salt (phytin). Phytate-P is poorly available to animals and can reduce the digestibility of other nutrients and the performance of animals owing to its anti-nutritional effect (Woyengo and Nyachoti 2013).

Plant-origin P contains phytate-phosphorus (phytate-P), which monogastric animals, including poultry, cannot completely degrade (Sommerfeld et al. 2018). Thus, with respect to livestock nutrition, there are four possible sources of phytase available for the animals for phytate degradation: endogenous mucosal phytase, gut microfloral phytase, plant phytase, and exogenous microbial phytase (Humer and Zebeli 2015). Nonetheless, P in the form of phytate is largely unavailable to monogastric animals (Evans and Irving 2022). There are differences in phytate concentration between different raw materials (Eeckhout and De Paepe 1994; Ravindran et al. 1994; Rodehutscord et al. 2016). For example, the concentration of phytate-P in cereal grains is approximately 0.17%–0.23%, whereas in oilseed meals and cereal-based bran, it can vary from 0.34% to 0.76% and 0.68% to 0.88%, respectively (Viveros et al. 2000). Half to two-thirds of the total P content is found in phytate in the majority of plant raw materials. This percentage decreases, though, if the material has undergone fermentation, as in the case of fermented feed, ensiled moist corn and corn cob mix, or distillers' grains. In summary, phytate is a component of plant feedstuffs that, if broken down in the digestive system or during feed processing, significantly increases the amount of digestible P available to animals (Woyengo and Nyachoti 2013).

Environmental pollution is a huge concern in the poultry industry (Alewell et al. 2020). Phytate has a P content of 282 g/kg (phytate-P), but chickens do not effectively utilize the P component of phytate, which leads to P excretion (Selle et al. 2023). Phosphorus pollution of freshwater reserves causes algal blooms and eutrophication, which poses a risk to the environment (Correll 1998; Sharpley 1999). Indeed, the amount of P used as fertilizer (82%) is far beyond the amount used in animal feeds (7%) (Prud'Homme 2010).

2.2   Factors Affecting Phytate-P Hydrolysis

It has been suggested that the deficiency of endogenous phosphatase and phytase activity causes poor phytate-P digestion in chickens and other monogastrics (Humer and Zebeli 2015). Numerous negative charges found in phytic acid, as previously mentioned, will bind to cations to form insoluble phytate complexes.

2.2.1   Effect of Calcium

Since the development of insoluble complexes is the source of the limited availability of phytate-P, it makes sense that a change in the concentration of cations that cause these complexes to form will have an impact on the hydrolysis of phytate-P (Selle et al. 2009). Since Ca is the most common cation in the gastrointestinal system lumen due to the bird's high need, it is also the most significant cation when talking about the creation of phytate complexes. A soluble Ca source is more likely to form a complex with phytate and thus reduce the efficacy of the exogenous phytase (Zanu et al. 2020). Another factor that is crucial to Ca-phytate formation is gut pH. The acidic pH in the foregut of the chicken increases the solubility and susceptibility to degradation of phytate. However, Ca, especially as limestone, has a high acid-binding capacity and tends to increase digesta pH (Angel et al. 2002; Selle et al. 2009). This may present challenges for laying hens since they need adequate Ca levels to generate high-quality eggshells and to maintain strong bones. A broiler's feed usually contains 10 g/kg of Ca, whereas laying hens need about 40 g/kg of Ca from their diet. Ileal phytate hydrolysis increased from 9% to 33% when Ca levels were lowered from 40 g/kg to 30 g/kg, according to (Van der Klis et al. 1997). This demonstrates that the laying hen's digestive tract is capable of breaking down phytic acid. However, as this meal would not satisfy the bird's Ca requirement, a Ca level of 30 g/kg should not be suggested.

The link between Ca and P homeostasis may confound the benefits of lowering Ca. Greater absorption of Ca and P from the intestines may result from a diet low in Ca compared to P (Tamim et al. 2004). When examining the hydrolysis of phytic acid, it is critical to take into account both the overall values and the ratio of Ca to P (Lei et al. 1994).

2.2.2 Effect of Other Nutrients

Phytic acid hydrolysis can be inhibited by other cations such as magnesium (Mg), manganese (Mn), iron (Fe), and zinc (Zn), according to research conducted in vitro (Maenz et al. 1999). For example, according to (Banks et al. 2004), the combination of 250 ppm Cu and 600 FTU of phytase/kg diet resulted in a decrease in apparent P retention, BW gain, feed consumption, feed conversion, and tibia ash and toe ash weights as compared with birds that were fed 600 FTU of phytase/kg only. However, the addition of 600 FTU of phytase/kg to the diet containing 250 ppm copper (Cu) increased apparent P retention as a percentage of total P and as a percentage of the diet.

Vitamin D has been demonstrated in numerous studies to enhance phytic acid hydrolysis and phytate-P utilization (Carlos and Edwards Jr 1998; DeLuca 1979; Edwards Jr 1993; Mitchell and Edwards Jr 1996). This could be due to vitamin D from food can improve the body's ability to absorb Ca (Bauman 1983). The creation of Ca–phytate complexes will decrease in parallel with the concentration of Ca cations in the gastrointestinal tract lumen. An increase in vitamin D will, therefore, increase the solubility of phytic acid and so increase is susceptibility to phytase activity. Vitamin D can also increase the transport of P within the small intestine (Tanaka and DeLuca 1974). This will increase the retention of any P hydrolyzed from phytic acid and so increase the retention of phytate-P will be greater.

The hydrolysis of phytic acid produces orthophosphate by catalyzing the stepwise removal of the phosphate groups, which has been shown to reduce the activity of phytase (Konietzny and Greiner 2002; Lei and Stahl 2000). The level of dietary inorganic non-phytate P will strongly effect the concentration of orthophosphate present in the gastrointestinal tract, and so logically, the hydrolysis of phytate-P is dependent upon the amount of non-phytate P present in the diet.

3   Anti-nutritional Factors

In feeds for poultry, especially in Europe and outside the USA, wheat and barley are the most widely used energy sources after maize (Amerah et al. 2015; Ravn et al. 2017). The non-starch polysaccharides (NSPs) can be classified chemically into three groups: cellulose, non-cellulosic polymers (such as arabinoxylans (AX) and mixed β-glucans), and pectic polysaccharides (Bedford and Apajalahti 2018).

Cereal grains are mostly composed of arabinoxylans (AX), along with cellulose and mixed β-glucans, which are non-cellulosic polymers; pectins are present in very small amounts (Knudsen 2014). While maize, sorghum, and rice have less soluble NSPs and are classified as non-viscous, the NSPs in wheat, barley, rye, triticale, and oats frequently generate viscous digesta because a significant amount of them are soluble, high-molecular-weight NSPs (Choct 2015). In wheat, AX makes up the majority of soluble NSPs; however, in barley, it is β-glucan. Arabinoxylans are frequently referred to as pentosans since they are both pentose sugars, xylose, and arabinose. While the unbound AX is soluble or water-extractable (WE-AX) and can produce extremely viscous solutions, the insoluble or water-unextractable AX (WU-AX) is tethered to the cell walls or other AX (Choct 1997; Moers et al. 2005). Briefly, the cell wall of most cereal grains is composed of 60–70% AX located in the aleurone and endosperm (Bach Knudsen et al. 2017). (Jaworski et al. 2015) investigated the carbohydrate composition of major feedstuffs and found that 48.7% of the total NSP content of corn consists of AX. Sorghum, wheat, and wheat bran had AX levels of 44.3%, 63%, and 64.3% of the total NSP composition, respectively. The amount of insoluble AX, as well as the arabinose to xylose (A:X) ratio, determines how complex the AX structure is, the higher the degree of arabinose substitution, the higher the crosslinking between AX and the lower the enzymatic degradation (Tiwari and Jha 2017).

Non-starch polysaccharides have been reported to increase the viscosity of digesta in the small intestine, leading to reduced absorption and digestibility of nutrients in monogastric animals (Baker et al. 2021). Studies showed that arabinoxylans were not digested in the small intestine in broilers, which subsequently created a viscous "chyme" in the intestines. This viscous chyme led to a proliferation of pathogenic bacteria, intestinal inflammation, and impairment of barrier function in the intestine (Bedford and Cowieson 2012; Morgan et al. 2018). This correlation between an increase in digesta viscosity and an increase in pathogenic load within the small intestine can lead to increased oxidative stress and inflammation (Moore 2016). These negative effects can be alleviated by xylanase's ability to decrease digesta viscosity. It has been reported innumerable times that xylanase inclusion, especially in high wheat-containing diets, reduces the viscosity of both jejunal and ileal digesta (Engberg et al. 2004; Taylor et al. 2018). Additionally, xylanases increase digesta passage rates and nutrient digestion in the small intestine, which restricts the growth of fermentative microorganisms. By decreasing host–enteric microbiota competition, nutrient utilization is enhanced (Choct et al. 2004; Raza et al. 2019).

4 Enzymes in Poultry Diets: Key to Unlocking Hidden Nutrients

The first commercial use of feed enzymes dates back to 1984 in Finland, where opportunities existed to significantly improve the nutritional quality of barley-based rations by the inclusion of enzymes derived from the brewing industry (Bedford 2002). The enzymes were first created for the brewing sector and then applied to animal feed. The application of feed enzymes has several advantages:

(i) Decreasing anti-nutritional effects and enhancing feed efficiency by enabling the animals to obtain more nutritional value from the feed ingredients through their mode of action, which targets specific anti-nutrients found in essential raw materials in the feed (Evans and Irving 2022). For example, the elimination of the anti-nutritive properties of certain dietary NSPs by their enzymatic hydrolysis to prebiotic type components, which, in turn, may facilitate gut development and health in young chickens (Slominski 2011).

(ii) Modifying the feed formulas and reducing the usage of costly substances (such as fat, synthetic amino acids, and inorganic phosphorus) while still satisfying the animals' nutritional needs, producers have been able to realize direct and immediate economic gains. For example, phytases aid in the release of P from phytate hydrolysis (Slominski 2011). Also, enzymes help in the hydrolysis of certain types of carbohydrate–protein linkages and, therefore, improve the availability of amino acids.

(iii) Reducing excretion of undigested feed components in animal production systems. This is particularly crucial in markets where disposing of waste on land is a concern or where environmental regulations are stringent (Evans and Irving 2022).

(iv) Improving animal performance and using less feed to reach a market body weight for livestock faster by increasing feed efficiency. Since feed accounts for the majority of production expenses (about 70%), every action that lowers the price per ton and the feed cost per animal in the system would increase producer profitability (Barletta 2011). For example, solubilization of cell wall NSP resulted in more effective hindgut fermentation and improved overall energy utilization (Slominski 2011). Enzyme activities that target the primary potential anti-nutrients present in animal diets, such as phytate, lectins, and proteinaceous substances, have dominated the market to date. Additionally, starch digestion may be hampered, in which case adding exogenous amylase may be beneficial (Evans and Irving 2022).

The most widely used enzymes (Fig. 7.2) in the poultry industry are phytases that catalyze the hydrolysis of phytate (Adeola and Cowieson 2011). The second most widely used enzymes in the poultry industry are carbohydrases, which cleave the NSPs in viscous cereals (Adeola and Cowieson 2011; Pirgozliev et al. 2019). These two types of enzymes account for approximately 90% of the enzyme market (Adeola and Cowieson 2011). Proteases have become more relevant in the poultry industry for their ability to improve protein and amino acid digestibility. They have become particularly relevant in young birds where there is a low relative activity of protease (Bedford and Apajalahti 2022; Walk et al. 2018).

Enzymes can be produced commercially from microorganisms, plants, and animals, although enzymes from plants and animals are limited in production (Alabi et al. 2019). The total concentration of enzymes in plants is low, so producing enough plants to meet the enzyme productivity requirements is unfeasible. Enzymes produced from animals are mostly by-products of the meat industry, which already has other competition (Alabi et al. 2019). However, microorganisms can produce

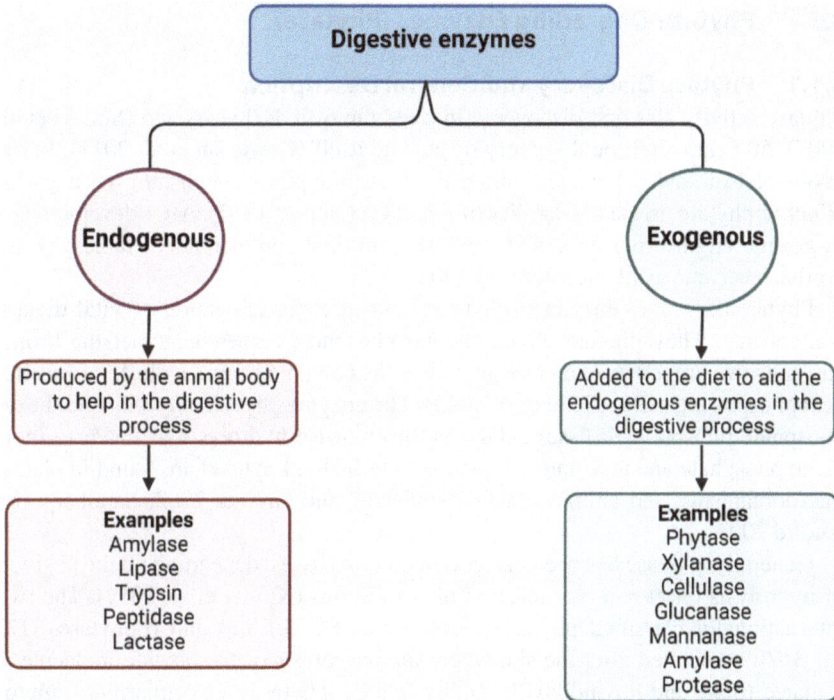

Fig. 7.2 Classification of digestive enzymes (Bedford and Apajalahti 2022). The Figure was generated by BioRender

large amounts of enzymes and can be sufficient to meet market demands (Adeola and Cowieson 2011). These microbes are not affected by seasonality, and it is possible to manipulate their genetics and environment to increase yields and improve the enzymes (Alabi et al. 2019). Phytases, carbohydrases, and proteases are naturally excreted by a range of different bacteria and fungi, who use them to meet their own metabolic needs (Adeola and Cowieson 2011). They also produce a wide array of properties that make them suitable for specific applications. The most common strains for commercial production include *Aspergillus niger, Bacillus subtilis, Trichoderma longibrachiatum, Trichoderma viride*, and *Humicola insolens* (Alabi et al. 2019). The genetically modified strains of these microbes are cultured on sugar and starch hydrolysis substrates using fermentation technology (Liu and Kokare 2023). Most commercially available exogenous enzymes are obtained from these fermentation systems, which rely on genetically modified fungi or bacteria (Adeola and Cowieson 2011; Shah et al. 2017; Tanruean et al. 2021). In the following section, we will shed light on the main enzymes used in poultry.

4.1 Phytate-Degrading Enzymes (Phytase)

4.1.1 Phytase Discovery and General Description

Phytase activity was first discovered in rice bran over 100 years ago (Suzuki et al. 1907), 50 years after the discovery of phytate itself (Cowieson et al. 2011). In the 1960s Nelson and colleagues started the first in-depth research into the negative effect of phytate on the availability of P and Ca (Nelson 1967). After developments in genetic engineering, it was in 1991 that the first commercially viable phytase products became available (Bedford 2002).

Phytate-degrading enzymes (phytases) enhance the utilization of vital dietary components. They liberate phytate-bound phosphorus and counteract the broad, anti-nutritive properties of dietary phytate or the polyanionic molecule, myo-inositol hexaphosphate (InsP6) (Selle et al. 2023). The enzyme phytase (myo-inositol hexaphosphate phosphohydrolase) catalyzes the stepwise hydrolysis of InsP6 to inorganic phosphate and myo-inositol via InsP5 to InsP1. Phytases are found in plants, microorganisms, and animal tissues (Konietzny and Greiner 2002; Sandberg and Andlid 2002).

Generally, phytase has been categorized on two bases, depending on the position of hydrolysis onset and on preferred pH conditions (Kumar et al. 2012). The two internationally classified phytases, 3-phytases (EC 3.1.3.8) and 6-phytases (EC 3.1.3.26) are named after the site where the hydrolysis of the phytate molecule is initiated (Selle and Ravindran 2007). While the 3-phytases are primarily of microbial origin, the 6-phytases are mainly isolated from plant sources (Cosgrove et al. 1970). However, there are exceptions, as soybean phytases are 3-phytases and E. coli phytases are 6-phytases (Sandberg and Andlid 2002).

Several factors affect phytase efficiency: pH stability (acidic pH in the gastric and neutral in the small intestine), proteolytic stability as well as temperature stability (Konietzny and Greiner 2002). As the temperature optimum of most phytases ranges from 50 to 60 °C (Konietzny and Greiner 2002), the animal's body temperature beyond this optimum is a further reason for a P digestibility as low as 65%, despite phytase supplementation (Weremko et al. 1997). As phytase is a protein, it is sensitive to high temperatures, which causes denaturation at excessive heat. Spraying liquid phytase onto cooled pellets could be one opportunity to maintain its activity when diets are pelleted. Phytase is also sensitive to humidity. Therefore, heat stability, proper storage, and handling procedures should be considered to maintain efficacy (Jacela et al. 2010).

There are four possible sources of phytate-degrading enzyme activity in respect of poultry nutrition. These include endogenous phytase generated by the small intestinal mucosa or by microfloral phytase activity mainly present in the large intestine, intrinsic plant phytase activity derived from certain feedstuff, and the dietary inclusion of exogenous enzymes (Selle et al. 2009; Selle and Ravindran 2007).

4.1.2 Intrinsic Plant Phytase

It has to be mentioned that the activity of phytase is commonly expressed as FTU, which is defined as the amount of phytase that liberates 1 mmol of inorganic phosphate per minute from 0.0051 mol/L sodium phytate at pH 5.5 and at a temperature of 37 °C (Aoac 2000). Of the cereal grains, rye has the highest intrinsic phytase activity, followed by triticale, wheat, and barley (Rodehutscord et al. 2016). Feed that goes through procedures like pelleting or extrusion that need temperatures higher than 65 °C has virtually no intrinsic phytase activity since plant intrinsic phytase is not very thermotolerant (Rodehutscord et al. 2022). However, when unprocessed feed is provided, plant intrinsic phytase can be active in the animals' digestive tract and is known to increase phosphorus digestibility. The presence of intrinsic phytases 1, 2, 3, 4, and 5 were considered the predominant plant-based intrinsic feed phytases, and they were responsible for phytate degradation in the stomach. The minimal phytate degradation in the stomach and proximal duodenum with a maize and soybean meal-based diet supplemented with low intrinsic phytase activity is 35 FTU/kg dry matter (DM) (Kemme et al. 2006). Rapp and others found that diets with high intrinsic plant phytase exhibited greater phytate hydrolysis in the stomach compared to very low intrinsic phytase activity (43 FTU/kg DM) (Rapp et al. 2001).

4.1.3 Exogenous Phytase

Phytase was originally developed to reduce phosphorus pollution by increasing the digestibility of phytate-bound P leading to a reduction in the level of inorganic phosphorus that is added to the diet (Ahmad et al. 2023). The use of phytase eventually became popular for reducing the anti-nutritional effects of phytate and phytic acid. The first attempt at commercializing fungal phytase for animal feed nutrition was in 1962, although there were no significant results (Lei et al. 2013). The first successful attempt at commercialization of fungal phytase for animal feed was in 1991 by a German chemical company (Dersjant-Li et al. 2015). However, in 1999 bacterial phytases were found to be superior compared to fungal phytases, consequently leading to the popularity of bacterial phytases for animal feed nutrition (Lei et al. 2013). Nonetheless, some studies (Attia et al. 2020; Saleh et al. 2021) showed that fungal phytase could improve the overall performance of broiler chickens.

Few phytases on the market are able to completely dephosphorylate phytic acid into myo-inositol due to the presence of the phosphate at position two on the ring. However, mucosal phosphatases and a few microbial esterases are able to cleave this phosphate, thus generating free myo-inositol in the gastrointestinal tract of the animal (Adeola and Cowieson 2011). Phytase is now the most widely used exogenous enzyme for non-ruminants, accounting for approximately 60% of the enzyme market (Corrêa et al. 2015).

About two-thirds of the total P in poultry feed is present in the form of phytate, and poultry has a low capacity to utilize phytate-bound P. Poultry produces some phytase, but it is insufficient for hydrolyzing phytates (Mullaney and Ullah 2003). Phytase hydrolyzes phytate into myo-inositol and P groups, releasing a usable form of inorganic P (Mullaney and Ullah 2003). Myo-inositol concentrations were increased in gizzard digesta when phytase was added to laying hen diets (Taylor et al. 2018). The addition of phytase reduced phytate but increased inositol

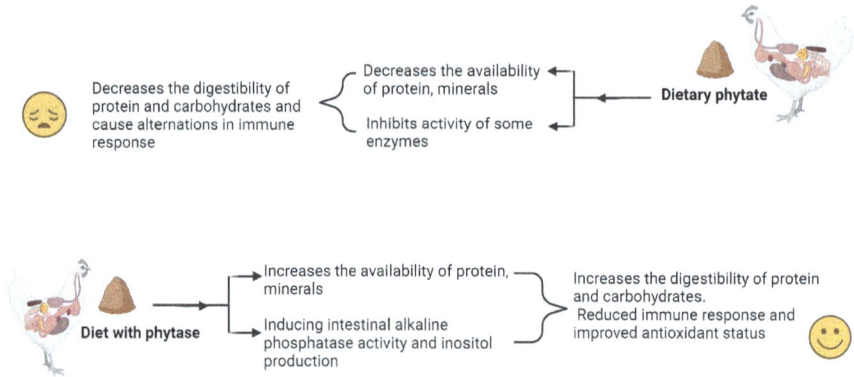

Fig. 7.3 Problems associated with undigested dietary phytate and the beneficial effects of phytase supplementation in poultry (Valente Junior et al. 2024). The Figure was generated by BioRender

concentrations in the gizzard and ileal digesta (Kriseldi et al. 2021). Inositol concentrations were also increased in broilers fed with added phytase (Gautier et al. 2018). Phosphorus digestibility and utilization were improved when phytase was added to both broiler and layer diets (Gautier et al. 2018; Taylor et al. 2018). Phytase allows nutritionists to make P more available in poultry and swine diets.

Phytase has been shown to improve more than just P availability (Valente Junior et al. 2024) (Fig. 7.3). When phytase was added to laying hen diets, it increased hen-day production and daily egg mass, resulting in increased overall layer production (Taylor et al. 2018). When phytase was supplemented to broilers, it improved growth performance (BW gain: 763 g vs 759 g) and bone ash content (52.4% vs 51.7% of dry defatted tibia ash at d 23) (Gautier et al. 2018). Microbial phytase supplementation improved mineral retention in broilers (Gallardo et al. 2018; Ptak et al. 2013) as well as increased apparent ileal digestibility of some amino acids, specifically alanine, aspartic acid, glycine, threonine, isoleucine, leucine, serine, and tyrosine, in addition to increasing apparent metabolizable energy (AME) (Selle et al. 2012).

4.1.4 Effect of Phytase on Phosphorus

Numerous studies have shown that the addition of phytase to a poultry diet can improve the availability of phytate-P and increase P retention through the hydrolysis of phytic acid (Augspurger and Baker 2004; Cowieson et al. 2011; Denbow et al. 1995; Onyango et al. 2005; Ravindran et al. 2000).

The improvement in phytate-P utilization allows nutritionists to decrease the amount of non-phytate P within the diet. Phytase has been shown to reverse negative effects caused by low P diets in pigs, broilers, and laying hens (Denbow et al. 1995; Gordon and Roland Sr 1998; Jalal and Scheideler 2001). The decrease of non-phytate P within diets and improvement of phytate-P absorption can be used to reduce excreted P and therefore P pollution.

4.1.5 Effect of Phytase on Protein

Increasing phytase inclusion increased the retention of nitrogen (Shirley and Edwards Jr 2003). Meanwhile, multiple studies have shown that phytase can improve amino acid digestibility (Selle et al. 2000). Phytase benefitted the true ileal digestibility coefficient of 16 amino acids in maize, wheat, rapeseed meal, and soybean meal by 3.90, 12.94, 9.31, and 6.39%, respectively, when fed to broilers (Rutherfurd et al. 1997). Diluting a diet to reduce the level of crude protein (17 vs 15%), metabolizable energy (2900 vs 2800 kcal ME/kg), Ca (4.2 vs. 3.8%), and P (0.375 vs 0.28%) had no negative effect on performance if supplemented with phytase at 600 FTU/kg (Lima et al. 2010). Adding 300 FTU/kg phytase to three different laying hen diets (Snow et al. 2003). The control was a maize–soybean diet, the treatment diets had either 75 g/kg meat and bone meal or 100 g/kg wheat middling added. Although numerical increases between 2 and 3% were seen, there was no significant phytase effect on the digestibility of amino acids. The addition of phytase to a Ca and P deficient diet improved the nitrogen and amino acid digestibility in comparison with a negative control diet (Liu et al. 2007). Phytase could significantly improve the utilization of protein, energy, and amino acids in diets with low levels of Ca (Agbede et al. 2009). This effect was reduced somewhat when the Ca concentration was higher. In their second trial, phytase failed to significantly affect the digestibility of nitrogen or amino acids across a range of dietary P levels. Effects of phytase supplementation on pre-cecal amino acid digestibility in broiler chickens and on pc InsP6 degradation can differ among oilseed meals (Krieg et al. 2020). This suggests that phytase dosage might be optimized based on the composition of the diet.

4.1.6 Effect of Phytase on Energy

In broiler diets, the inclusion of phytase consistently increases the apparent metabolizable energy of a diet (Selle and Ravindran 2007) including phytase at 600 FTU/kg in a wheat-based diet increased the AME from 14.22 MJ/kg to 14.56 MJ/kg (Selle et al. 2005). Adding 1000 FTU/kg phytase in layer diets improved the AME in both maize- and wheat-based diets by 3.5 and 3.6%, respectively (Scott et al. 2001). However, Liebert et al. (2005) found that 300 FTU/kg phytase did not affect AME in layer diets. The difference seen between these studies is probably due to the difference in phytase dose. A reduction in metabolizable energy from 12.14 to 11.72 MJ ME/kg had no negative effect with the inclusion of 600 FTU/kg (Lima et al. 2010). Adding phytase improved the AME of a broiler diet by 1.1%, and this was due to improved ileal digestibility of fat, protein, and starch by 3.5, 2.6, and 1.4% (Camden et al. 2001). To conclude, phytase can affect the AME of a diet, but it appears that higher doses are required for the improvements to become significant.

4.1.7 Effect of Phytase on Minerals and Bone Mineralization

As phytase breaks down the phytic acid it will also reduce the amount of Ca and other minerals that become bound within the insoluble phytate complexes. Phytase is not as effective at hydrolyzing phytic acid when it is already bound to Ca

(Plumstead et al. 2008). That said, exogenous phytase acts in the foregut and so can hydrolyze phytate before mineral phytate complexes can form within the small intestines unless the pH of the crop rises. Laying hens have a high Ca demand for eggshell production. This will increase the amount of Ca available for Ca–phytate complex formation and can increase the pH of the digestive tract, more specifically the crop, toward an environment more suitable for complex formation; this may inhibit the effect exogenous phytase can have on breaking down phytate. However, 1000 FTU/kg did hydrolyze phytate to lower esters in layer hen diets with high Ca concentrations of about 45 g/kg DM (Agbede et al. 2009). Increasing phytase inclusion increased the retention of dietary Ca in broiler diets (Shirley and Edwards Jr 2003). Using a basal diet deficient in Ca and added either phytase at 500 FTU/kg or one of five graded doses of Ca (0–4 g/kg) to calculate equivalence values. They found that their phytase released the equivalence of 0.72 g Ca/kg (Augspurger and Baker 2004).

According to Cardoso Júnior et al. (2010), bone mineralization reflects adequate bone quality, which is associated with beneficial effects on broiler performance and is important in supporting muscular development. Furthermore, bone mineralization data is commonly used to estimate and validate inorganic P release by phytase and is an efficient parameter to quantify phytate-P released in corn and soybean meal-based diets (Dersjant-Li et al. 2020). It is worth noting that Ca and P are closely related in bone mineralization results because they are both stored together in bone, and Ca is stored almost entirely as hydroxyapatite crystals of Ca phosphate in bone (Bougouin et al. 2014).

In a study by Hernandez et al. (2022), the three treatments with phytase supplementation (negative control (NC) + 1000 FTU coated phytase, NC + 1000 FTU uncoated phytase, and NC + 500 FTU coated +500 FTU uncoated phytase) had improved results for all bone parameters in comparison to the NC. Similarly, Chung et al. (2013) reported improvements in bone mineral content and bone mineral density in broilers fed diets with two inclusion levels (500 and 1000 FTU/kg) of a coated phytase compared to broilers fed low-P diets. In addition, several other authors have observed that tibia ash and other bone parameters improve with phytase addition (Babatunde et al. 2022; Babatunde et al. 2019; Leyva-Jimenez et al. 2019). Moreover, these results suggest that phytase supplementation is beneficial for bone mineralization and is in accordance with the data obtained for growth performance measurements. However, no clear effects were observed for the combination of different phytase forms.

4.1.8 High Phytase Dosing

Phytase is typically added to chicken diets at about a concentration of 500 FTU/kg of diet, resulting in increased digestibility of P and Ca and improvement in growth performance (Cowieson 2005; Viveros et al. 2000). However, there is renewed interest in the use of higher levels of phytase (>1500 FTU/kg) because these supplementation levels have been associated with further improvements in performance that cannot be solely attributed to increased release of P (Józefiak et al. 2010).

A phytase dose is considered to be above the standard inclusion level when the phytase activity is at or greater than 2500 FTU/kg. A number of studies have now shown that the inclusion of high levels of phytase within diets can benefit animal performance (Cowieson et al. 2011; Walker et al. 2024). Adding phytase improves broiler performance based on regression analyses. A high dose of 2973 FTU/kg had the best weight gain from 1 to 21 days of age. From 21 to 42 days, 2051 FTU/kg and 2101 FTU/kg showed the best weight gain and feed conversion ratio (FCR), respectively (Broch et al. 2018).

Using low doses of phytase (500–1000 FTU/kg) is adequate for growth and bone development of the broilers fed with Ca- and P-balanced diets (0.50% Ca, 0.25% non-phytate-P, and Ca-to- non-phytate-P ratio = 2.0) (Shi et al. 2022). However, high doses of phytase (2500–10,000 FTU/kg) are needed to alleviate the negative effects of Ca and P imbalance (1.00% Ca, 0.25% non-phytate-P, and Ca-to-non-phytate-P ratio = 4.0) on growth performance and bone mineralization of broiler chickens from 1 to 42 days of age (Shi et al. 2022).

The aim of the high inclusion levels is to hydrolyze all phytic acid in the foregut down to the lowest esters. Within the small intestines, a rise in pH will strengthen the association between the higher phytic acid ester and charged nutrients, resulting in precipitation. Hydrolyzing phytic acid into the lower esters before it enters the duodenum reduces the formation of phytate complexes, as the lower esters are more soluble and have less ability to bind cations (Persson et al. 1998). This means they are more available to endogenous phytase, which can further hydrolyze the phytic acid to myo-inositol, InsP1, and P. These can then be absorbed and utilized by the bird. A number of mechanisms have been suggested to explain how a high phytase diet can improve performance (Cowieson et al. 2011); the first is through the release of additional P. The complete hydrolysis of phytic acid will release almost all the phytate-P (Nelson et al. 1971). This is a substantial amount of P. However, this will only improve the performance if P is a limiting nutrient, and so is unlikely to be the main mechanism seen in the trials with adequate P. Excessive P will just be excreted in the manure (Angel et al. 2002). Although this is probably not the mechanism that improves performance, the release of all of the phytate-P will allow further reduction in non-phytate-P and therefore could reduce the cost and environmental impact of monogastric diets (Cowieson et al. 2011).

Phytase targets the higher esters of phytic acid (Wyss et al. 1999), IP6 and IP5; this is important because these phytic acid molecules have a greater Ca binding capacity than the lesser esters, InsP4 and InsP3. Therefore, initial phytase activity will release disproportionately high amounts of Ca compared to phosphorus. The use of high phytase doses can deliver a more balanced release of P and Ca, closer to the 1.1 Ca:P ratio normally used for phytase matrices. It may be that the benefits seen in the higher inclusion trials are due to a restoration of the Ca:P ratio (Cowieson et al. 2011).

The complete hydrolysis of phytic acid within the gastrointestinal tract will also have a benefit through the removal of the extra-phosphoric effects of phytate. As previously mentioned, phytate complex formation hinders mineral, protein, and energy nutrition. Dephosphorylation of phytic acid earlier within the tract has the potential to remove these negative extra-phosphoric effects by preventing the formation of phytate complexes (Shi et al. 2022).

Finally, there is the benefit of increased myo-inositol availability. After phytate has been dephosphorylated to the lower esters, it can be absorbed and further broken down into free P and myo-inositol. Myo-inositol is known to be involved with the transport of fats and fat-soluble nutrients, and it has been shown to increase growth in rats and decrease fatty liver syndrome (Katayama 1997). It is also worth noting that myo-inositol is the precursor of multiple compounds that are important for maintaining normal cellular function. For example, InsP3 can be used as a secondary messenger during cell signalling or in the control of intracellular Ca ion concentration (Berridge 1993).

5   Carbohydrases

Non-starch polysaccharides (NSP; soluble and insoluble), one of several ANFs, which are present in large amounts in feed ingredients such as wheat, barley, sunflower meal, and canola meal are indigestible by poultry, as they lack the endogenous enzymes necessary to digest the beta type of linkages present in these ingredients (Raza et al. 2019). Owing to the lack of endogenous enzymes that degrade dietary fibers (structural carbohydrates or non-digestible components that make up the plant cell wall), including high-molecular weight-soluble NSP, intestinal viscosity increases, which slows down the migration and absorption of nutrients (Raza et al. 2019). Consequently, this affects birds' health and increases production costs. Another concern for poultry producers is subclinical pathogenic challenges owing to the use of such low-quality ingredients. These subclinical pathogens also lead to other pathogenic infections, such as necrotic enteritis (Kaldhusdal et al. 2016).

The poultry industry has become very receptive to the use of enzymes, including carbohydrases due to their ability to improve nutrient utilization. They help the absorption of essential nutrients and improve feed utilization (Alabi et al. 2019). Carbohydrases became commercialized for poultry diets in the late 1980s (Aftab and Bedford 2018). They were originally used for their ability to minimize wet excreta, digestion, and metabolizable energy issues due to high fiber in the diets. Years ago, over 80% of the carbohydrases used in the global market were accounted for by xylanase and glucanase (Adeola and Cowieson 2011). Bacterial and fungal enzymes have been proven effective in degrading AX and β-glucans present in viscous cereals (Lei and Porres 2003; Silva and Smithard 2002). Carbohydrase enzymes have been known to produce varying effects. This is likely due to the variation in NSP chemical structure, as the NSP profile varies from feedstuff to feedstuff (Knudsen 2014). Thus, an enzyme may achieve beneficial results in one feedstuff but may not be able to produce those results in another. This means that there is a need for a broad range of carbohydrase enzymes that can act on varying substrates (Adeola and Cowieson 2011; Raza et al. 2019).

5.1   Non-starch Polysaccharide-Degrading Enzymes (NSPases)

According to Paloheimo et al. (2010), NSPases account for approximately half or more of the feed enzyme market's value, with xylanases having a much greater share than β-glucanases and mannanases, which are employed in a few minor niche applications.

5.1.1   NSPases Mode of Action

The birds' own enzymes cannot breakdown the cell walls of the cereal endosperm, and the small intestine still contains intact cells, indicating that the gizzard cannot completely breakdown all of the cells (Bedford and Autio 1996; Parkkonen et al. 1997). Therefore, it was proposed that adding NSPases to the diet could cause endosperm cells to penetrate more completely. This would help breakdown the cell wall matrix, making it easier for digestive enzymes to enter the body and increase the diet's digestibility. The use of NSPases decreased the amount of intact cell walls observed in the ileum and jejunum, according to a microscopic analysis of the intestinal contents (Bedford and Autio 1996). Additionally, the presence of exogenous NSPase increased the release of starch and protein from ground cereal samples undergoing simulated gastrointestinal digestion processes (Mirzaie et al. 2012).

5.1.2   Advantages of NSPases

The good impact of NSPases on feed's nutritional value has been attributed to their ability (Fig. 7.4):

- Reducing viscosity and breaking down cell walls.
- Prebiotic properties. In recent times, there has been a growing body of evidence indicating that the prebiotic effect of oligosaccharides produced from feed hemicelluloses by NSPases may be one of the reasons feed additives containing

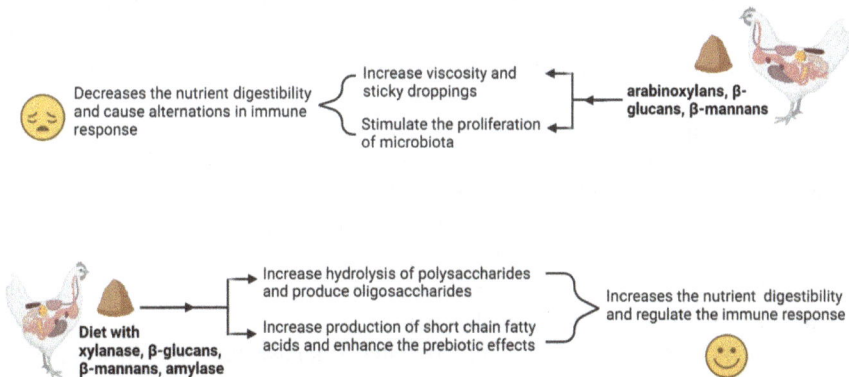

Fig. 7.4 Problems associated with undigested dietary non-starch polysaccharides and the beneficial effects of exogenous carbohydrases supplementation in poultry (Valente Junior et al. 2024). The Figure was generated by BioRender

xylanases and β-glucanases have been beneficial (Bedford and Apajalahti 2018). These oligosaccharides, which range in length from 3 to 10 sugar residues, would encourage the synthesis of short-chain fatty acids (SCFAs), such as butyric acid, by the gut microbiota. These SCFAs can function as various signals within the animal, enhancing its overall performance (Craig et al. 2020).

- Improving nutrient absorption and animal performance.
- Maintaining gut health.

The market's offers can be broadly categorized into three groups, which correspond to the various stages of product development:

(i) First-generation multi-enzyme preparations made from traditional (nonGMO) strains with both xylanase and cellulase activity.
(ii) Second-generation monocomponent preparations are made from a GMO with a chosen main activity, frequently from a thermostable source.
(iii) Monocomponent of a protein-engineered thermostable molecule.

5.1.3 Xylanase

Xylanases are produced naturally by fungi, bacteria, yeast, marine algae, etc., but are not produced by mammals and must be supplemented. Xylanolytic organisms have been reported in extreme environments such as thermal springs, marines, Antarctic environments, and soda lakes (Chakdar et al. 2016). Xylanase hydrolyzes the β-1,4 glycosidic bonds of xylan into xylose. Xylans are the second most common plant cell polysaccharide in the world and account for one-third of renewable organic carbon sources. Xylans differ between plant species and are essential for cell wall integrity as well as plant growth and development (Faik 2013).

Xylanase breaks down NSP, specifically arabinoxylans, in cereal grains such as barley, rye, wheat, and oat. These viscous cereals contain higher amounts of soluble NSP (26.1%, 25.6%, 21.7%, and 13.3% on DM basis, respectively) compared to non-viscous cereals such as corn (11.8%) (Knudsen 2014). Arabinoxylans reduce the digestibility of starch, fat, and proteins in poultry diets. Xylanase has been shown to increase the concentrations of arabinose, galactose, and glucuronic acid in the digesta when added to wheat diets (Craig et al. 2020).

Xylanase supplementation has been shown to improve feed conversion ratio when added to both layer and broiler diets (Arczewska-Wlosek et al. 2019; Gonzalez-Ortiz et al. 2017; Olukosi et al. 2020). There is some variability in xylanase's performance in young, immature birds versus that of older, more mature birds (Arczewska-Wlosek et al. 2019). Although xylanase improved the FCR in broilers from 1 to 21 days of age, it did not influence the feed conversion ratio in older birds (Arczewska-Wlosek et al. 2019).

Moreover, Morgan et al. (2022) reported that xylanase improved growth performance across wheat-, corn-, and barley-based diets. Xylanase also had a positive effect on NSP degradability and free oligosaccharide digestibility across all three diets. Supplementation of laying hen diets with xylanase improved feed efficiency and increased Ca and DM digestibility (Taylor et al. 2018).

Jejunal digesta viscosity was decreased in broilers fed both corn and wheat diets; however, wheat diets had a greater digesta viscosity decrease (-31%) when compared to that of corn diets (-10%) (Munyaka et al. 2016). Xylanase has shown promising results in other diets that contain larger amounts of NSPs, like rye. Amerah and others found that xylanase significantly decreased the relative weight and length of all gut components as well as reduced the viscosity of small intestine digesta in diets consisting of high concentrations of rye (20%) (Amerah et al. 2008). Research on the effects of xylanase in primarily corn diets is conflicting; this is most likely due to the lower amount of arabinoxylans in corn.

5.1.4 β-Glucanase

β-Glucanase is a type of cellulase enzyme produced commonly by various bacteria, fungi, and actinomycetes (Esteve-Garcia et al. 1997). Monogastric animals cannot produce naturally the β-glucanase but can be found in the rumen bacteria of ruminants. β-Glucanase hydrolyzes glycosidic bonds in β-glucans and has been reported to reduce the anti-nutritional effects of β-glucans. β-Glucans are NSP from cell walls of endosperm cells and are present in large amounts in barley and are also found in small amounts in soybean and canola meals. Barley diets supplemented with β-glucanase have the potential to significantly increase the nutritive value of barley (Sun et al. 2018).

The β-glucanase has been shown to improve growth performance in broilers fed barley-based diets (Esteve-Garcia et al. 1997; Karunaratne et al. 2021; Munyaka et al. 2016; Sun et al. 2018). The β-glucanase increased gross energy and crude protein digestibility in broilers fed barley-based diets (75.5% vs. 74.6% and 77.2% vs 75.5%, respectively) (Sun et al. 2018). The β-glucanase supplementation decreased intestinal viscosity in broiler chickens (Esteve-Garcia et al. 1997; Karunaratne et al. 2021). Noted similar results, stating that dietary β-glucanase supplementation (0.1%) decreased ileal viscosity (3.78 centipoise vs 5.39 centipoise) in broilers. Supplementation of β-glucanase improved egg production in laying ducks and increased the activity of amylase and chymotrypsin in the duodenal digesta in addition to reducing digesta glucan content (Chen et al. 2021). Thus, β-glucanase has proven to be beneficial in alleviating the anti-nutritional effects of β-glucans in barley.

5.1.5 Mannanases

β-Mannan is a polymer of plant sugar made up of a linear chain of β-1,4-mannose that can have units of galactose, glucose, or glucuronic acid partially substituted for it. Legumes like soybeans and palm kernels, which are relatively uncommon feed components, both contain mannans in comparatively large quantities (Jackson et al. 2004). Fungal and yeast cell walls also include unique mannans, such as α-1,6-mannan with α-1,2- and α-1,3-linked branches. β-Mannans are a component of wood hemicellulose, and bacteria and fungi that break down lignocellulosic materials frequently feature β-mannanase in their cellulase–hemicellulase enzyme profile (Li et al. 2014). Although β-mannanases and their thermostable or alkaline forms have found application in other industries, such as oil and detergents, their usage in feed has been restricted (Chauhan et al. 2012; Soni and Kango 2013). There are commercial fungal mannanases from *T. reesei* and *A. niger* that belong to the glycoside hydrolase family GH5, and these have been utilized as feed additives on occasion (Ademark et al. 1998; Dhawan and Kaur 2007; Van Zyl et al. 2010). It has been proposed that β-mannanases can lower metabolizable energy loss by preventing or reducing a feed-induced immune response in chicken-fed diets containing β-mannan or β-galactomannan (Arsenault et al. 2017).

5.1.6 Amylases

Starch is a heterogeneous structure that varies significantly in amylose and amylopectin composition. Both amylose and amylopectin are polymers made up of D-glucose molecules. They differ based solely on the type of chain that is formed. Amylose forms a linear chain polymer of D-glucose, whereas amylopectin forms a branch-chain polymer. Amylose consists of only α 1–4 glycosidic bonds, while amylopectin has both α 1–4 glycosidic and α 1–6 glycosidic bonds. Due to these α 1–6 glycosidic bonds, amylopectin is more soluble, amorphous, and readily digestible than amylose (Cowieson et al. 2019b).

Corn is the most widely used starch source for poultry on a global scale and it contributes the majority of dietary energy. About 86% of the corn starch is found in the endosperm. Starch is classified as either normal, waxy, or amylo. Waxy corn has an amylopectin content of around 99%, with the other 1% being amylose. Normal corn is around 75% amylopectin and 25% amylose, and high amylose corn is 25% amylopectin and 75% amylose (Cowieson et al. 2019b).

Amylase is a starch-digesting enzyme used to provide more available energy from increased starch digestion. Starch digestion in swine begins with salivary amylase in the mouth, but in poultry, starch digestion does not occur until pancreatic amylase encounters ingested starch polymers (Cowieson et al. 2019b). The α-amylase hydrolyzes amylose into maltose, which is then degraded further into glucose by maltase. There is a limited amount of starch digestion that occurs in the crop and proventriculus due to α-amylase. Poultry can almost completely digest starch by the time it reaches the end of the ileum, however, this can vary based on starch structure and solubility, age of the bird, and various other factors. As birds mature, they develop an increased capacity to digest starches (Cowieson et al. 2019b).

During the first 7-day period, the growth performance of birds fed amylase improved by 9.4% and feed conversion by 4.2% (Gracia et al. 2003). It was suggested that in the first week, post-hatch, birds have a higher sensitivity to amylase as they require assistance from the enzyme to offset their immature pancreatic amylase production due to their intestinal tracts being less developed (Cowieson et al. 2019b). Evidence supports that, when compared to their juvenile counterparts, older birds produce more pancreatic amylase (Krogdahl and Sell 1989). However, this does not mean that older birds do not benefit from amylase supplementation. The interactions between age and amylase effect over a 42-day period in broilers fed corn/soy diets have been reported (Gracia et al. 2003). Weight gain increases were noted from 0 to 7 day of 9.4%, from 0 to 21 day of 3.6%, and from 21 to 42 day of 5.5%. The FCR ratio decreases were also reported from 0 to 7 day of 5 points (1.13

vs. 1.18), from 0 to 21 day of 0 points (1.41 vs 1.41), and from 21 to 42 day of 5 points (1.62 vs. 1.67). This indicates that birds may not only have an amylase sensitivity post-hatch but also in the grower/finisher phase when there is a high starch intake compared to metabolic weight (Cowieson et al. 2019b). This data agrees with Vieira et al. (2015), who saw similar quadratic interactions between age and amylase supplementation in broilers fed corn/soy diets over a 40-day period. They reported that the FCR of the broilers decreased by 4.5, 2.6, 1.7, 0, 2, and 2.7 over 6 weeks, respectively.

Amylase had a more substantial effect on FCR and egg quality in all corn diets compared to all wheat diets fed to laying hens (Olgun et al. 2018). Amylase supplementation in corn diets had a more profound effect on increasing AME when compared to a complete corn–soybean meal diet (Schramm et al. 2021). Amylase supplementation has also been shown to improve AME in the diet of broilers at d 7 old age (2976 vs. 2927 kcal/kg; $P \leq 0.05$) (Gracia et al. 2003). Aderibigbe et al. (2020) reported that α-amylase decreased jejunal viscosity while increasing total tract digestibility of starch and energy. Similarly, amylase improved energy utilization when added to corn–soybean meal diets. Evidence also suggests that increasing levels of amylase linearly increases the ileal digestibility of resistant starch (Schramm et al. 2021).

6 Proteases

All animals need proteins as a necessary nutrient in their diet. Protein is a significant cost factor in animal feed intended for production. About half of the protein in a typical broiler diet comes from protein meals, including soybean and rapeseed meals, which have high-protein concentrations (35–50%). The remaining half comes from cereal grains, like wheat and maize, which have lower protein concentrations (8–15%) (Beski et al. 2015).

Proteases are enzymes that may break down proteins by hydrolyzing the peptide bonds that connect amino acid residues, thereby slicing lengthy polypeptide chains into smaller pieces. Proteases are a large family of enzymes with substantial structural and sequence variation. As a result, the protease space exhibits an amazing range of functionality, including temperature, pH profiles, specificity, stability, and stability (Barrett et al. 2012).

Proteases are further divided into exo- and endopeptidases. The main endogenous proteases that are involved in the digestion of proteins such as pepsin, trypsin, chymotrypsin, elastase, and carboxypeptidase A, B. Exogenous proteases have the ability to complement endogenous proteases or work together with them to provide functions that the animals themselves are unable to provide (Yegani and Korver 2013).

Exogenous proteases have been used more frequently in recent years to decrease endogenous amino acid loss and to lessen variations in amino acid digestibility (Cho et al. 2020). Lewis et al. (1955) and Baker et al. (1956) were the first to augment production animal diets with exogenous protease. At first, their research concentrated on the impact of additional pepsin and pancreatin on feed conversion.

Many more reports on the application of exogenous proteases in animal nutrition have been published since this initial groundbreaking work, and they indicate a significant potential to improve performance in a variety of diet types (Angel et al. 2011; Cowieson et al. 2019a; Hessing 1996; Rooke et al. 1998; Thorpe and Beal 2001; Zuo et al. 2015).

6.1 Advantages of Proteases

- Increasing protein digestibility and utilization.
- Improvement of gut health by reducing the proteins in the hindgut, which mitigates harmful bacteria.
- Reduction of the incidence of footpad lesions due to enhanced protein digestion, which also results in less nitrogen in the excreta.
- Environmental impact by reduction of nitrogen emissions (Lemme et al. 2019).

It is not surprising that the response to exogenous enzymes can vary, and this can be attributed to a number of factors, including the substrate concentrations, duration of studies, limiting nutrients, and the intrinsic quality of the diets provided (Bedford 2002). Similar to this, there may be some variation in the response to exogenous protease. This variation may be related to the type of protease fed (pH optimum, source organism, substrate specificity), the kind of protein in the diet, the age and species of the animal, and a number of other factors (Cowieson and Roos 2016).

The addition of protease to poultry diets is of interest to improve protein and amino acid digestibility and reduce anti-nutritional factors such as trypsin inhibitors, β-conglycinin, and glycinin. This may be particularly important for young birds where protease activity may be suboptimal (Walk et al. 2018). Trypsin inhibitors are found naturally in various legumes, including soybeans. Trypsin inhibitors are a type of serine protease inhibitor that reduces the effectiveness of trypsin. Trypsin is an enzyme involved in the breakdown of various proteins during digestion (Cohen et al. 2019). The β-conglycinin and glycinin are the two most important soybean proteins. These proteins are not digested but are then absorbed into the intestinal mucosa. Protease supplementation could help alleviate these anti-nutritional factors in broilers (Wang et al. 2022).

Proteases can improve growth performance, especially post-hatch, due to the pancreatic protease activity being low from hatch to around 21 days (Jin et al. 1998). The digestive protease could be limiting protein digestion due to the immaturity of the digestive systems of chicks. The addition of protease may aid pancreatic enzymes and improve the rate of protein breakdown in the intestines. Protease may also be able to reduce the need for additional amino acid supplementation (Lourenco et al. 2020).

Research into protease has been mostly focused on supplementation with other enzymes, while very little research has been done on protease supplementation alone. Cowieson and Roos found that protease supplementation improved the gain-to-feed ratio (Cowieson and Roos 2016). The addition of protease improved growth performance and apparent metabolizable energy in broilers (Jabbar et al. 2021). When protease was added to lower crude protein diets in broilers, growth performance, intestinal health, and carcass traits improved (Maqsood et al. 2022). Other studies have found that protease supplementation did not improve growth performance above that of a nutritionally adequate diet (Walk et al. 2019).

Environmental pollution is a huge issue in the agricultural production industry. Low-protein feeds are being fed to poultry to attempt to reduce the amount of excess nitrogen that is excreted into the environment (Lemme et al. 2019). Protease supplementation may be used to offset the effects of low-protein diets on broiler growth. Liu et al. (2024) have shown that the direct addition of this new protease can effectively improve the performance of poultry, increase the protein digestibility, and greatly increase the economic benefits of broiler breeding (Liu et al. 2024). Also, when protease was added to low-protein diets it improved the growth performance of broilers (Wang et al. 2022).

7   Strategies of Enzymes Applications

Enzymes can be included in the diet in the form of liquid or dry products. Since liquid product forms are typically not thermostable, a post-pelleting liquid application system is required to add them to the feed. Post-pelleting liquid application systems have several benefits despite their relative complexity. Enzyme application to feed produced under challenging processing circumstances is made possible by post-pelleting liquid application. These processing conditions can be employed for several purposes, including reducing the risk of *Salmonella* spp. contamination. The post-pelleting liquid application provides dosing flexibility, allowing the enzyme dose, particularly phytase, to be adjusted from batch to batch. This is a benefit over the alternative of including the enzyme in the premix. Like liquid amino acids, liquid enzymes should ideally be sufficiently thermostable to be dosed straight into the feed mixer before pelleting (Evans and Irving 2022). Compared to post-pelleting liquid application, this would provide a considerably simpler and less expensive dosing method along with automation and dosing flexibility.

Pelleting is widely used to improve feed properties and increase the nutritional value of the feed (Aftab and Bedford 2018). The stability, for example, of phytase during the pelleting process is a major concern because phytase activity is negatively affected by pressure and high temperature (Abdollahi et al. 2013). De Jong et al. (2017) incurred when corn and soybean meal diets supplemented with various microbial phytases are conditioned at increasing conditioning temperatures. Authors reported a 1.9% decrease in phytase activity with every 1 °C increase in conditioning temperature. The forms of dry products might be non-thermostable or thermostable. Compared to post-pelleting liquid application, the dry thermostable product has the benefit of being easier to add to the feed mixer or incorporate into the premix, therefore, simplifying operations. Nowadays, most feed enzymes are dry thermostable compounds because most monogastric feed is pelleted. Usually, non-pelletized dry, non-thermostable products are utilized in mash feed (Evans and Irving 2022). Lastly, handling convenience and safety should be considered.

8   Regulations of Nutritional Enzymes

Regulations controlling the introduction of feed enzymes into the market have been enforced in the majority of global markets (EFSA 2019). Those manufacturing must demonstrate the safety and effectiveness of the enzyme for the intended species feed enzymes or releasing them onto the market. Additionally, evidence of the product's quality must be provided, including stability and uniformity. Market-specific approval times can vary from 3 months to more than 2 years, contingent upon the degree of information required in the regulatory dossier. The European Union (EU), the USA, and Brazil have the strictest regulations, although there is a global trend toward more regulations overall (Evans and Irving 2022).

Enzymes must obtain clearance under Regulation EC 1831/2003 in order to be sold in the EU. The European Food Safety Authority must receive a comprehensive dossier describing the enzyme's identity, characteristics, and usage guidelines (EFSA). Since microorganisms continue to be the primary source of feed enzymes, comprehensive information regarding the manufacturing organism's characteristics is also necessary. A comprehensive toxicity test must be used to prove the enzyme's safety, and studies on tolerance that take into account the target species must also be shown. Since enzyme products are proteinaceous, it is assumed that they are respiratory sensitizers. Particle size information is utilized to evaluate the users' risk and likelihood of exposure (Evans and Irving 2022).

Proof of efficacy for the intended use is also required by the EU dossier, in addition to safety assessments. Research must show measurable effects at the lowest dose advised in three studies for each main target species, carried out in accordance with standard agricultural, animal husbandry, and feed production procedures in the EU. After the dossier has been developed and submitted, EFSA takes about 2 years to approve it (EFSA 2019). Any product intended to be used as an ingredient in animal food is regarded as food in the USA, and the Center for Veterinary Medicine of the Food and Drug Administration is in charge of regulating animal food products. Producers have the option to self-certify as Generally Recognized as Safe.

9   Conclusions and Future Trends

The animal production sector is currently going through a transitional phase. Ensuring that the animal receives all the nutrients it needs to grow to its maximum potential is crucial, but it is also critical to restrict the amount or type of undigested nutrients that enter the terminal ileum or hindgut to avoid giving non-beneficial bacterial populations a substrate on which to grow. As feed enzymes are known to increase digestibility and reduce the quantity of undigested nutrients in the lower gastrointestinal tract, they have a significant impact on this site. All major enzyme producers are expanding their research disciplines to understand how enzymes can continue to improve via (i) nutrition (where enzymes are traditionally important), (ii) the microbiome, and (iii) gut/immune function. The ultimate aim is to help animal and feed producers navigate the options that they have available and to demonstrate the importance of enzyme technology in the fast-changing world of animal production.

- Even though phytase has been used commercially for more than 25 years, further studies are needed to fill the knowledge gaps concerning the enzyme's functioning and the ideal dosage for animal feed. Even though pre-cecal InsP6 disappearance can currently be increased to 90% with supplemental phytase, some InsP6 and lower inositol esters remain undegraded, which presents a challenge for additional study and enzyme development to optimize poultry phytate-P usage. The literature shows contradictory results about the effects of enzymes on the digestibility of minerals and amino acids other than P. Accurate measurements of the phytate content of the feed raw materials that are really being used are, therefore, necessary for exact phytase application.
- The information substantiates some of the widely held beliefs in the field today and the ways in which these beliefs impact the choice of an NSP enzyme for use in commercial in-feed applications. The degree of substrate hydrolysis in vitro or in vivo, the relative availability of the substrate, or the amount of substrates provided by a particular diet might not be the only factors that need to be considered when choosing an NSP enzyme. While measures based on alternative responses, such as gut morphology, energy or nutrient digestibility, gut flora, and its metabolites, or fermentation profiles, are helpful in gaining a deeper comprehension of the phenomenon, they should not be interpreted as the exclusive yardstick for determining an NSP enzyme's utility.
- Exogenous proteases and amylases may improve the digestibility of starch, energy, and amino acids in poultry diets. To completely understand the mechanism of action and the complementarity of effect with nearby feed enzymes, more investigations are necessary. Optimizing the utilization of these enzymes and enabling their strategic incorporation into enzyme admixtures will be made possible by a deeper understanding of the variables that influence the variation in the retention of starch and protein in farm animals. Future work should focus on a better understanding of starch and protein structure and how exogenous enzymes may augment the endogenous enzyme array. In the future, exogenous amylases and proteases will likely play a more significant role in the feed enzyme marketplace and complement incumbent enzymes further to enhance the sustainability of the protein food chain.

References

Abdollahi M, Ravindran V, Svihus B (2013) Pelleting of broiler diets: an overview with emphasis on pellet quality and nutritional value. Anim Feed Sci Technol 179(1–4):1–23

Ademark P, Varga A, Medve J, Harjunpää V, Drakenberg T, Tjerneld F, Stålbrand H (1998) Softwood hemicellulose-degrading enzymes from aspergillus Niger: purification and properties of a β-mannanase. J Biotechnol 63(3):199–210

Adeola O, Cowieson A (2011) Board-invited review: opportunities and challenges in using exogenous enzymes to improve nonruminant animal production. J Anim Sci 89(10):3189–3218

Aderibigbe A, Cowieson A, Sorbara J, Adeola O (2020) Intestinal starch and energy digestibility in broiler chickens fed diets supplemented with α-amylase. Poult Sci 99(11):5907–5914

Aftab U, Bedford M (2018) The use of NSP enzymes in poultry nutrition: myths and realities. Worlds Poult Sci J 74(2):277–286

Agbede JO, Kluth H, Rodehutscord M (2009) Studies on the effects of microbial phytase on amino acid digestibility and energy metabolisability in caecectomised laying hens and the interaction with the dietary phosphorus level. Br Poult Sci 50(5):583–591

Ahmad B, Tahir M, Naz S, Alhidary IA, Gul S (2023) Bacterial or fungal origin phytase enzyme affects the performance and mineralization of calcium and phosphorus differently in broiler chickens fed deficient calcium and phosphorous diets. J Appl Anim Res 51(1):669–676

Alabi O, Shoyombo A, Akpor O, Oluba O, Adeyonu A (2019) Exogenous enzymes and the digestibility of nutrients by broilers: a mini review. Int J Poult Sci 18(9):404–409

Alewell C, Ringeval B, Ballabio C, Robinson DA, Panagos P, Borrelli P (2020) Global phosphorus shortage will be aggravated by soil erosion. Nat Commun 11(1):4546

Amerah A, Ravindran V, Lentle R, Thomas D (2008) Influence of particle size and xylanase supplementation on the performance, energy utilisation, digestive tract parameters and digesta viscosity of broiler starters. Br Poult Sci 49(4):455–462

Amerah A, Van de Belt K, van Der Klis J (2015) Effect of different levels of rapeseed meal and sunflower meal and enzyme combination on the performance, digesta viscosity and carcass traits of broiler chickens fed wheat-based diets. Animal 9(7):1131–1137

Angel C, Saylor W, Vieira S, Ward N (2011) Effects of a monocomponent protease on performance and protein utilization in 7-to 22-day-old broiler chickens. Poult Sci 90(10):2281–2286

Angel R, Tamim N, Applegate T, Dhandu A, Ellestad L (2002) Phytic acid chemistry: influence on phytin-phosphorus availability and phytase efficacy. J Appl Poult Res 11(4):471–480

Aoac HW (2000) International a: official methods of analysis of the AOAC international. The Association, Arlington County, VA, USA

Arczewska-Wlosek A, Swiatkiewicz S, Bederska-Lojewska D, Orczewska-Dudek S, Szczurek W, Boros D, Fras A, Tomaszewska E, Dobrowolski P, Muszynski S (2019) The efficiency of xylanase in broiler chickens fed with increasing dietary levels of rye. Animals 9(2):46

Arsenault R, Lee J, Latham R, Carter B, Kogut M (2017) Changes in immune and metabolic gut response in broilers fed β-mannanase in β-mannan-containing diets. Poult Sci 96(12):4307–4316

Attia YA, Bovera F, Iannaccone F, Al-Harthi MA, Alaqil AA, Zeweil HS, Mansour AE (2020) Microbial and fungal phytases can affect growth performance, nutrient digestibility and blood profile of broilers fed different levels of non-phytic phosphorous. Animals 10(4):580

Augspurger N, Baker D (2004) High dietary phytase levels maximize phytate-phosphorus utilization but do not affect protein utilization in chicks fed phosphorus-or amino acid-deficient diets. J Anim Sci 82(4):1100–1107

Babatunde O, Bello A, Dersjant-Li Y, Adeola O (2022) Evaluation of the responses of broiler chickens to varying concentrations of phytate phosphorus and phytase. II. Grower phase (day 12–23 post hatching). Poult Sci 101(3):101616

Babatunde O, Cowieson A, Wilson J, Adeola O (2019) Influence of age and duration of feeding low-phosphorus diet on phytase efficacy in broiler chickens during the starter phase. Poult Sci 98(6):2588–2597

Bach Knudsen KE, Nørskov NP, Bolvig AK, Hedemann MS, Laerke HN (2017) Dietary fibers and associated phytochemicals in cereals. Mol Nutr Food Res 61(7):1600518

Baker JT, Duarte ME, Holanda DM, Kim SW (2021) Friend or foe? Impacts of dietary xylans, xylooligosaccharides, and xylanases on intestinal health and growth performance of monogastric animals. Animals 11(3):609

Baker R, Lewis C, Wilbur R, Hartman P, Speer V, Ashton G, Catron D (1956) Supplementation of baby pig diets with enzymes. J Anim Sci

Banks K, Thompson K, Rush J, Applegate T (2004) Effects of copper source on phosphorus retention in broiler chicks and laying hens. Poult Sci 83(6):990–996

Barletta A (2011) Introduction: current market and expected developments. In: Enzymes in farm animal nutrition, pp 1–11

Barrett AJ, Woessner JF, Rawlings ND (2012) Handbook of proteolytic enzymes, Volume 1. Elsevier

Bauman V (1983) Vitamin D, calcium-binding protein and calcium absorption in the intestines. Prikl Biokhim Mikrobiol 19(1):11–19

Bedford M (2002) The foundation of conducting feed enzyme research and the challenge of explaining the results. J Appl Poult Res 11(4):464–470

Bedford M, Apajalahti J (2018) Exposure of a broiler to a xylanase for 35d increases the capacity of cecal microbiome to ferment soluble xylan. Poult Sci 97(1):3263–3274

Bedford M, Autio K (1996) Microscopic examination of feed and digesta from wheat-fed broiler chickens and its relation to bird performance. Poult Sci 75(1):1–14

Bedford M, Cowieson A (2012) Exogenous enzymes and their effects on intestinal microbiology. Anim Feed Sci Technol 173(1–2):76–85

Bedford MR, Apajalahti JH (2022) The role of feed enzymes in maintaining poultry intestinal health. J Sci Food Agric 102(5):1759–1770

Berridge MJ (1993) Inositol trisphosphate and calcium signalling. Nature 361(6410):315–325

Beski SS, Swick RA, Iji PA (2015) Specialized protein products in broiler chicken nutrition: a review. Anim Nutr 1(2):47–53

Bougouin A, Appuhamy J, Kebreab E, Dijkstra J, Kwakkel R, France J (2014) Effects of phytase supplementation on phosphorus retention in broilers and layers: a meta-analysis. Poult Sci 93(8):1981–1992

Broch J, Nunes RV, Eyng C, Pesti GM, de Souza C, Sangalli GG, Fascina V, Teixeira L (2018) High levels of dietary phytase improves broiler performance. Anim Feed Sci Technol 244:56–65

Camden B, Morel P, Thomas D, Ravindran V, Bedford M (2001) Effectiveness of exogenous microbial phytase in improving the bioavailabilities of phosphorus and other nutrients in maize-soya-bean meal diets for broilers. Anim Sci 73(2):289–297

Cardoso Júnior A, Rodrigues PB, Bertechini AG, Freitas RTFD, Lima RRD, Lima GFR (2010) Levels of available phosphorus and calcium for broilers from 8 to 35 days of age fed rations containing phytase. Rev Bras Zootec 39:1237–1245

Carlos A, Edwards H Jr (1998) The effects of 1, 25-dihydroxycholecalciferol and phytase on the natural phytate phosphorus utilization by laying hens. Poult Sci 77(6):850–858

Chakdar H, Kumar M, Pandiyan K, Singh A, Nanjappan K, Kashyap PL, Srivastava AK (2016) Bacterial xylanases: biology to biotechnology. 3 Biotech 6:1–15

Chauhan PS, Puri N, Sharma P, Gupta N (2012) Mannanases: microbial sources, production, properties and potential biotechnological applications. Appl Microbiol Biotechnol 93:1817–1830

Chen W, Wang S, Xu R, Xia W, Ruan D, Zhang Y, Mohammed KA, Azzam MM, Fouad AM, Li K (2021) Effects of dietary barley inclusion and glucanase supplementation on the production performance, egg quality and digestive functions in laying ducks. Anim Nutr 7(1):176–184

Cheryan M, Rackis JJ (1980) Phytic acid interactions in food systems. Crit Rev Food Sci Nutr 13(4):297–335

Cho HM, Hong JS, Kim YB, Nawarathne SR, Choi I, Yi Y-J, Wu D, Lee H, Han SE, Nam KT (2020) Responses in growth performance and nutrient digestibility to a multi-protease supplementation in amino acid-deficient broiler diets. J Anim Sci Technol 62(6):840

Choct M (1997) Feed non-starch polysaccharides: chemical structures and nutritional significance. Feed Mill Int 191(1):13–26

Choct M (2015) Feed non-starch polysaccharides for monogastric animals: classification and function. Anim Prod Sci 55(12):1360–1366

Choct M, Kocher A, Waters D, Pettersson D, Ross G (2004) A comparison of three xylanases on the nutritive value of two wheats for broiler chickens. Br J Nutr 92(1):53–61

Chung T, Rutherfurd S, Thomas D, Moughan P (2013) Effect of two microbial phytases on mineral availability and retention and bone mineral density in low-phosphorus diets for broilers. Br Poult Sci 54(3):362–373

Cohen M, Davydov O, Fluhr R (2019) Plant serpin protease inhibitors: specificity and duality of function. J Exp Bot 70(7):2077–2085

Corrêa TLR, de Queiroz MV, de Araújo EF (2015) Cloning, recombinant expression and characterization of a new phytase from Penicillium chrysogenum. Microbiol Res 170:205–212

Correll DL (1998) The role of phosphorus in the eutrophication of receiving waters: a review. J Environ Qual 27(2):261–266

Cosgrove D, Irving G, Bromfield S (1970) Inositol phosphate phosphatases of microbiological origin. The isolation of soil bacteria having inositol phosphate phosphatase activity. Aust J Biol Sci 23(2):339–344

Costa G, Dilelis F, Vasconcellos T, Reis T, Souza C, Lima C (2023) True ileal digestibility of phosphorus from dicalcium phosphate in diets for broilers. Anim Feed Sci Technol 298:115601

Cowieson A, Smith A, Sorbara J, Pappenberger G, Olukosi O (2019a) Efficacy of a mono-component exogenous protease in the presence of a high concentration of exogenous phytase on growth performance of broiler chickens. J Appl Poult Res 28(3):638–646

Cowieson A, Vieira S, Stefanello C (2019b) Exogenous microbial amylase in the diets of poultry: what do we know? J Appl Poult Res 28(3):556–565

Cowieson A, Wilcock P, Bedford M (2011) Super-dosing effects of phytase in poultry and other monogastrics. Worlds Poult Sci J 67(2):225–236

Cowieson AJ (2005) Factors that affect the nutritional value of maize for broilers. Anim Feed Sci Technol 119(3–4):293–305

Cowieson AJ, Roos FF (2016) Toward optimal value creation through the application of exogenous mono-component protease in the diets of non-ruminants. Anim Feed Sci Technol 221:331–340

Craig A, Khattak F, Hastie P, Bedford MR, Olukosi O (2020) Xylanase and xylo-oligosaccharide prebiotic improve the growth performance and concentration of potentially prebiotic oligosaccharides in the ileum of broiler chickens. Br Poult Sci 61(1):70–78

De Jong J, Woodworth J, DeRouchey J, Goodband R, Tokach M, Dritz S, Stark C, Jones C (2017) Stability of four commercial phytase products under increasing thermal conditioning temperatures. Trans Anim Sci 1(3):255–260

DeLuca HF (1979) The vitamin D system in the regulation of calcium and phosphorus metabolism. Nutr Rev (USA) 37(6):161

Denbow D, Ravindran V, Kornegay E, Yi Z, Hulet R (1995) Improving phosphorus availability in soybean meal for broilers by supplemental phytase. Poult Sci 74(11):1831–1842

Dersjant-Li Y, Villca B, Sewalt V, de Kreij A, Marchal L, Velayudhan DE, Sorg RA, Christensen T, Mejldal R, Nikolaev I (2020) Functionality of a next generation biosynthetic bacterial 6-phytase in enhancing phosphorus availability to weaned piglets fed a corn-soybean meal-based diet without added inorganic phosphate. Anim Nutr 6(1):24–30

Dersjant-Li Y, Awati A, Schulze H, Partridge G (2015) Phytase in non-ruminant animal nutrition: a critical review on phytase activities in the gastrointestinal tract and influencing factors. J Sci Food Agric 95(5):878–896

Dhawan S, Kaur J (2007) Microbial mannanases: an overview of production and applications. Crit Rev Biotechnol 27(4):197–216

Edixhoven J, Gupta J, Savenije H (2014) Recent revisions of phosphate rock reserves and resources: a critique. Earth Syst Dynam 5(2):491–507

Edwards HM Jr (1993) Dietary 1, 25-dihydroxycholecalciferol supplementation increases natural phytate phosphorus utilization in chickens. J Nutr 123(3):567–577

Eeckhout W, De Paepe M (1994) Total phosphorus, phytate-phosphorus and phytase activity in plant feedstuffs. Anim Feed Sci Technol 47(1–2):19–29

EFSA (2019) Safety and efficacy of APSA PHYTAFEED® 20,000 GR/L (6-phytase) as a feed additive for chickens for fattening, chickens reared for laying and minor growing poultry species. EFSA J 17(5):e05692

Engberg RM, Hedemann MS, Steenfeldt S, Jensen BB (2004) Influence of whole wheat and xylanase on broiler performance and microbial composition and activity in the digestive tract. Poult Sci 83(6):925–938

Esteve-Garcia E, Brufau J, Perez-Vendrell A, Miquel A, Duven K (1997) Bioefficacy of enzyme preparations containing beta-glucanase and xylanase activities in broiler diets based on barley or wheat, in combination with flavomycin. Poult Sci 76(12):1728–1737

Evans C, Irving H (2022) The feed enzyme market in 2020 and beyond. In: Enzymes in farm animal nutrition. CABI GB, pp 1–9

Faik A (2013) "Plant cell wall structure-pretreatment" the critical relationship in biomass conversion to fermentable sugars. Green biomass pretreatment for biofuels production, 1–30

Gallardo C, Dadalt JC, Trindade Neto MA (2018) Nitrogen retention, energy, and amino acid digestibility of wheat bran, without or with multicarbohydrase and phytase supplementation, fed to broiler chickens. J Anim Sci 96(6):2371–2379

Gautier A, Walk C, Dilger R (2018) Effects of a high level of phytase on broiler performance, bone ash, phosphorus utilization, and phytate dephosphorylation to inositol. Poult Sci 97(1):211–218

Gonzalez-Ortiz G, Sola-Oriol D, Martinez-Mora M, Perez J, Bedford M (2017) Response of broiler chickens fed wheat-based diets to xylanase supplementation. Poult Sci 96(8):2776–2785

Gordon R, Roland D Sr (1998) Influence of supplemental phytase on calcium and phosphorus utilization in laying hens. Poult Sci 77(2):290–294

Gracia M, Aranibar MJ, Lazaro R, Medel P, Mateos G (2003) Alpha-amylase supplementation of broiler diets based on corn. Poult Sci 82(3):436–442

Hernandez JR, Gulizia JP, Adkins JB, Rueda MS, Haruna SI, Pacheco WJ, Downs KM (2022) Effect of Phytase level and form on broiler performance, tibia characteristics, and residual fecal Phytate phosphorus in broilers from 1 to 21 days of age. Animals 12(15):1952

Hessing M (1996) Quality of soybean meals (SBM) and effect of microbial enzymes in degrading soya antinutritional compounds (ANC). 2nd International Soybean Processing and Utilization Conference, Bangkok, Thailand, 1996

Humer E, Zebeli Q (2015) Phytate in feed ingredients and potentials for improving the utilization of phosphorus in ruminant nutrition. Anim Feed Sci Technol 209:1–15

Jabbar A, Tahir M, Khan RU, Ahmad N (2021) Interactive effect of exogenous protease enzyme and dietary crude protein levels on growth and digestibility indices in broiler chickens during the starter phase. Trop Anim Health Prod 53:1–5

Jacela JY, DeRouchey JM, Tokach MD, Goodband RD, Nelssen JL, Renter DG, Dritz SS (2010) Feed additives for swine: fact sheets–high dietary levels of copper and zinc for young pigs, and phytase. J Swine Health Prod 18(10):87–91

Jackson M, Geronian K, Knox A, McNab J, McCartney E (2004) A dose-response study with the feed enzyme beta-mannanase in broilers provided with corn-soybean meal based diets in the absence of antibiotic growth promoters. Poult Sci 83(12):1992–1996

Jalal M, Scheideler S (2001) Effect of supplementation of two different sources of phytase on egg production parameters in laying hens and nutrient digestiblity. Poult Sci 80(10):1463–1471

Jaworski N, Lærke HN, Bach Knudsen K, Stein H-H (2015) Carbohydrate composition and in vitro digestibility of dry matter and nonstarch polysaccharides in corn, sorghum, and wheat and coproducts from these grains. J Anim Sci 93(3):1103–1113

Jin S-H, Corless A, Sell J (1998) Digestive system development in post-hatch poultry. Worlds Poult Sci J 54(4):335–345

Józefiak D, Ptak A, Kaczmarek S, Maćkowiak P, Sassek M, Slominski B (2010) Multi-carbohydrase and phytase supplementation improves growth performance and liver insulin receptor sensitivity in broiler chickens fed diets containing full-fat rapeseed. Poult Sci 89(9):1939–1946

Kaldhusdal M, Benestad SL, Løvland A (2016) Epidemiologic aspects of necrotic enteritis in broiler chickens–disease occurrence and production performance. Avian Pathol 45(3):271–274

Karunaratne ND, Classen HL, Ames NP, Bedford MR, Newkirk RW (2021) Effects of hulless barley and exogenous beta-glucanase levels on ileal digesta soluble beta-glucan molecular weight, digestive tract characteristics, and performance of broiler chickens. Poult Sci 100(3):100967

Katayama T (1997) Effects of dietary myo-inositol or phytic acid on hepatic concentrations of lipids and hepatic activities of lipogenic enzymes in rats fed on corn starch or sucrose. Nutr Res 17(4):721–728

Kemme PA, Schlemmer U, Mroz Z, Jongbloed AW (2006) Monitoring the stepwise phytate degradation in the upper gastrointestinal tract of pigs. J Sci Food Agric 86(4):612–622

Knudsen KEB (2014) Fiber and nonstarch polysaccharide content and variation in common crops used in broiler diets. Poult Sci 93(9):2380–2393

Konietzny U, Greiner R (2002) Molecular and catalytic properties of phytate-degrading enzymes (phytases). Int J Food Sci Technol 37(7):791–812

Krieg J, Siegert W, Berghaus D, Bock J, Feuerstein D, Rodehutscord M (2020) Phytase supplementation effects on amino acid digestibility depend on the protein source in the diet but are not related to InsP6 degradation in broiler chickens. Poult Sci 99(6):3251–3265

Kriseldi R, Walk C, Bedford M, Dozier W III (2021) Inositol and gradient phytase supplementation in broiler diets during a 6-week production period: 2. Effects on phytate degradation and inositol liberation in gizzard and ileal digesta contents. Poult Sci 100(3):100899

Krogdahl Å, Sell JL (1989) Influence of age on lipase, amylase, and protease activities in pancreatic tissue and intestinal contents of young turkeys. Poult Sci 68(11):1561–1568

Kumar A, Bold R, Plumstead P (2012) Comparative efficacy of Buttiauxella and E. Coli phytase on growth performance in broilers. Poult Sci 91(Suppl 1):90

Lei X, Ku P, Miller E, Yokoyama M, Ullrey D (1994) Calcium level affects the efficacy of supplemental microbial phytase in corn-soybean meal diets of weanling pigs. J Anim Sci 72(1):139–143

Lei X, Stahl C (2000) Nutritional benefits of phytase and dietary determinants of its efficacy. J Appl Anim Res 17(1):97–112

Lei XG, Porres JM (2003) Phytase enzymology, applications, and biotechnology. Biotechnol Lett 25:1787–1794

Lei XG, Weaver JD, Mullaney E, Ullah AH, Azain MJ (2013) Phytase, a new life for an "old" enzyme. Annu Rev Anim Biosci 1(1):283–309

Lemme A, Hiller P, Klahsen M, Taube V, Stegemann J, Simon I (2019) Reduction of dietary protein in broiler diets not only reduces n-emissions but is also accompanied by several further benefits. J Appl Poult Res 28(4):867–880

Lewis C, Catron D, Liu C, Speer V, Ashton G (1955) Swine nutrition, enzyme supplementation of baby pig diets. J Agric Food Chem 3(12):1047–1050

Leyva-Jimenez H, Alsadwi AM, Gardner K, Voltura E, Bailey CA (2019) Evaluation of high dietary phytase supplementation on performance, bone mineralization, and apparent ileal digestible energy of growing broilers. Poult Sci 98(2):811–819

Li Y-F, Calley JN, Ebert PJ, Helmes EB (2014) Paenibacillus lentus sp. nov., a β-mannanolytic bacterium isolated from mixed soil samples in a selective enrichment using guar gum as the sole carbon source. Int J Syst Evol Microbiol 64(Pt_4):1166–1172

Liebert F, Htoo J, Sünder A (2005) Performance and nutrient utilization of laying hens fed low-phosphorus corn-soybean and wheat-soybean diets supplemented with microbial phytase. Poult Sci 84(10):1576–1583

Lima MR, Costa FGP, Givisiez PEN, Silva JHV, Sakomura NK, Lima DFF (2010) Reduction of the nutritional values of diets for hens through supplementation with phytase. Rev Bras Zootec 39:2207–2213

Liu N, Liu G, Li F, Sands J, Zhang S, Zheng A, Ru Y (2007) Efficacy of phytases on egg production and nutrient digestibility in layers fed reduced phosphorus diets. Poult Sci 86(11):2337–2342

Liu T, Ma W, Wang J, Wei Y, Wang Y, Luo Z, Zhang Y, Zeng X, Guan W, Shao D (2024) Dietary protease supplementation improved growth performance and nutrients digestion via modulating intestine barrier, immunological response, and microbiota composition in weaned piglets. Antioxidants 13(7):816

Liu X, Kokare C (2023) Microbial enzymes of use in industry. In: Biotechnology of microbial enzymes. Elsevier, pp 405–444

Lourenco JM, Nunn SC, Lee EJ, Dove CR, Callaway TR, Azain MJ (2020) Effect of supplemental protease on growth performance and excreta microbiome of broiler chicks. Microorganisms 8(4):475

Ma Q, Rodehutscord M, Novotny M, Li L, Yang L (2019) Phytate and phosphorus utilization by broiler chickens and laying hens fed maize-based diets. Front Agric Sci Eng 6:380–387

Maenz DD, Engele-Schaan CM, Newkirk RW, Classen HL (1999) The effect of minerals and mineral chelators on the formation of phytase-resistant and phytase-susceptible forms of phytic acid in solution and in a slurry of canola meal. Anim Feed Sci Technol 81(3–4):177–192

Maqsood MA, Khan EU, Qaisrani SN, Rashid MA, Shaheen MS, Nazir A, Talib H, Ahmad S (2022) Interactive effect of amino acids balanced at ideal lysine ratio and exogenous protease supplemented to low CP diet on growth performance, carcass traits, gut morphology, and serum metabolites in broiler chicken. Trop Anim Health Prod 54(3):186

Mirzaie S, Zaghari M, Aminzadeh S, Shivazad M, Mateos G (2012) Effects of wheat inclusion and xylanase supplementation of the diet on productive performance, nutrient retention, and endogenous intestinal enzyme activity of laying hens. Poult Sci 91(2):413–425

Mitchell R, Edwards H Jr (1996) Additive effects of 1, 25-dihydroxycholecalciferol and phytase on phytate phosphorus utilization and related parameters in broiler chickens. Poult Sci 75(1):111–119

Moers K, Celus I, Brijs K, Courtin CM, Delcour JA (2005) Endoxylanase substrate selectivity determines degradation of wheat water-extractable and water-unextractable arabinoxylan. Carbohydr Res 340(7):1319–1327

Moore RJ (2016) Necrotic enteritis predisposing factors in broiler chickens. Avian Pathol 45(3):275–281

Morgan N, Bhuiyan M, Hopcroft R (2022) Non-starch polysaccharide degradation in the gastrointestinal tract of broiler chickens fed commercial-type diets supplemented with either a single dose of xylanase, a double dose of xylanase, or a cocktail of non-starch polysaccharide-degrading enzymes. Poult Sci 101(6):101846

Morgan N, Choct M, Toghyani M, Wu S (2018) Effects of dietary insoluble and soluble non-starch polysaccharides on performance and ileal and excreta moisture. 29th Annual Australian Poultry Science Symposium,

Morris ER (1986) Phytate and dietary mineral bioavailability

Mullaney EJ, Ullah AH (2003) The term phytase comprises several different classes of enzymes, vol 312, p 179

Munyaka P, Nandha N, Kiarie E, Nyachoti C, Khafipour E (2016) Impact of combined β-glucanase and xylanase enzymes on growth performance, nutrients utilization and gut microbiota in broiler chickens fed corn or wheat-based diets. Poult Sci 95(3):528–540

Nelson T (1967) The utilization of phytate phosphorus by poultry—a review. Poult Sci 46(4):862–871

Nelson T, Shieh T, Wodzinski R, Ware J (1971) Effect of supplemental phytase on the utilization of phytate phosphorus by chicks. J Nutr 101(10):1289–1293

Olgun O, Altay Y, Yildiz AO (2018) Effects of carbohydrase enzyme supplementation on performance, eggshell quality, and bone parameters of laying hens fed on maize- and wheat-based diets. Br Poult Sci 59(2):211–217

Olukosi O, González-Ortiz G, Whitfield H, Bedford M (2020) Comparative aspects of phytase and xylanase effects on performance, mineral digestibility, and ileal phytate degradation in broilers and turkeys. Poult Sci 99(3):1528–1539

Onyango E, Bedford M, Adeola O (2005) Efficacy of an evolved Escherichia coli phytase in diets of broiler chicks. Poult Sci 84(2):248–255

Pallauf J, Rimbach G (1997) Nutritional significance of phytic acid and phytase. Arch Anim Nutr 50(4):301–319

Paloheimo M, Piironen J, Vehmaanperä J (2010) Xylanases and cellulases as feed additives. In: Enzymes in farm animal nutrition, pp 12–53

Parkkonen T, Tervilä-Wilo A, Hopeakoski-Nurminen M, Morgan A, Poutanen K, Autio K (1997) Changes in wheat micro structure following in vitro digestion. Acta Agric Scand–B Soil Plant Sci 47(1):43–47

Persson H, Türk M, Nyman M, Sandberg A-S (1998) Binding of Cu2+, Zn2+, and Cd2+ to inositol tri-, tetra-, penta-, and hexaphosphates. J Agric Food Chem 46(8):3194–3200

Pirgozliev V, Brearley C, Rose S, Mansbridge S (2019) Manipulation of plasma myo-inositol in broiler chickens: effect on growth performance, dietary energy, nutrient availability, and hepatic function. Poult Sci 98(1):260–268

Plumstead P, Leytem A, Maguire R, Spears J, Kwanyuen P, Brake J (2008) Interaction of calcium and phytate in broiler diets. 1. Effects on apparent prececal digestibility and retention of phosphorus. Poult Sci 87(3):449–458

Prud'Homme M (2010) World phosphate rock flows, losses and uses. In: International Fertilizer Industry Association, Phosphates 2010 International Conference, Brussels

Ptak A, Józefiak D, Kierończyk B, Rawski M, Żyła K, Świątkiewicz S (2013) Effect of different phytases on the performance, nutrient retention and tibia composition in broiler chickens. Arch Anim Breed 56(1):1028–1038

Rapp C, Lantzsch HJ, Drochner W (2001) Hydrolysis of phytic acid by intrinsic plant and supplemented microbial phytase (aspergillus Niger) in the stomach and small intestine of minipigs fitted with re-entrant cannulas: 3. Hydrolysis of phytic acid (IP6) and occurrence of hydrolysis products (IP5, IP4, IP3 and IP2). J Anim Physiol Anim Nutr 85(11–12):420–430

Ravindran V, Cabahug S, Ravindran G, Selle P, Bryden W (2000) Response of broiler chickens to microbial phytase supplementation as influenced by dietary phytic acid and non-phytate phosphorous levels. II. Effects on apparent metabolisable energy, nutrient digestibility and nutrient retention. Br Poult Sci 41(2):193–200

Ravindran V, Ravindran G, Sivalogan S (1994) Total and phytate phosphorus contents of various foods and feedstuffs of plant origin. Food Chem 50(2):133–136

Ravn JL, Thøgersen JC, Eklöf J, Pettersson D, Ducatelle R, Van Immerseel F, Pedersen NR (2017) GH11 xylanase increases prebiotic oligosaccharides from wheat bran favouring butyrate-producing bacteria in vitro. Anim Feed Sci Technol 226:113–123

Raza A, Bashir S, Tabassum R (2019) An update on carbohydrases: growth performance and intestinal health of poultry. Heliyon 5(4):e01437

Reddy N (1989) Interactions of phytate with proteins and minerals. Phytates Cereals Legumes:57–70

Rodehutscord M, Rückert C, Maurer HP, Schenkel H, Schipprack W, Bach Knudsen KE, Schollenberger M, Laux M, Eklund M, Siegert W (2016) Variation in chemical composition and physical characteristics of cereal grains from different genotypes. Arch Anim Nutr 70(2):87–107

Rodehutscord M, Sommerfeld V, Kühn I, Bedford MR (2022) Phytases: potential and limits of Phytate destruction in the digestive tract of pigs and poultry. In: Enzymes in farm animal nutrition. CABI GB, pp 1–29

Rooke J, Slessor M, Fraser H, Thomson J (1998) Growth performance and gut function of piglets weaned at four weeks of age and fed protease-treated soya-bean meal. Anim Feed Sci Technol 70(3):175–190

Rutherfurd S, Edwards A, Selle P (1997) Effect of phytase on lysine-rice pollard complexes. Manipulating Pig Production VI. Australasian Pig Science Association, 248

Saleh AA, Elsawee M, Soliman MM, Elkon RY, Alzawqari MH, Shukry M, Abdel-Moneim A-ME, Eltahan H (2021) Effect of bacterial or fungal phytase supplementation on the performance, egg quality, plasma biochemical parameters, and reproductive morphology of laying hens. Animals 11(2):540

Sandberg AS, Andlid T (2002) Phytogenic and microbial phytases in human nutrition. Int J Food Sci Technol 37(7):823–833

Schramm V, Massuquetto A, Bassi L, Zavelinski V, Sorbara J, Cowieson A, Félix A, Maiorka A (2021) Exogenous α-amylase improves the digestibility of corn and corn–soybean meal diets for broilers. Poult Sci 100(4):101019

Scott T, Kampen R, Silversides F (2001) The effect of adding exogenous phytase to nutrient-reduced corn- and wheat-based diets on performance and egg quality of two strains of laying hens. Can J Anim Sci 81(3):393–401

Selle P, Ravindran V, Caldwell A, Bryden W (2000) Phytate and phytase: consequences for protein utilisation. Nutr Res Rev 13(2):255–278

Selle, P., Ravindran, V., Ravindran, G., & Bryden, W. (2005). Amino acid digestibility and growth performance interactions to phytase and lysine supplementation of lysine-deficient broiler diets

Selle PH, Cowieson AJ, Cowieson NP, Ravindran V (2012) Protein–phytate interactions in pig and poultry nutrition: a reappraisal. Nutr Res Rev 25(1):1–17

Selle PH, Cowieson AJ, Ravindran V (2009) Consequences of calcium interactions with phytate and phytase for poultry and pigs. Livest Sci 124(1–3):126–141

Selle PH, Macelline SP, Chrystal PV, Liu SY (2023) The contribution of phytate-degrading enzymes to chicken-meat production. Animals 13(4):603

Selle PH, Ravindran V (2007) Microbial phytase in poultry nutrition. Anim Feed Sci Technol 135(1–2):1–41

Shah PC, Kumar VR, Dastager SG, Khire JM (2017) Phytase production by aspergillus Niger NCIM 563 for a novel application to degrade organophosphorus pesticides. AMB Express 7:1–11

Sharpley A (1999) Agricultural phosphorus, water quality, and poultry production: are they compatible? Poult Sci 78(5):660–673

Shi C, Lv X, Wu L, Liu M, He L, Zhang T, Qiao Y, Hao J, Wang G, Cui Y (2022) High doses of Phytase alleviate the negative effects of calcium and phosphorus imbalance on growth performance and bone mineralization in broiler chickens. Br J Poult Sci 24(4):eRBCA-2021-1568

Shirley R, Edwards H Jr (2003) Graded levels of phytase past industry standards improves broiler performance. Poult Sci 82(4):671–680

Silva S, Smithard R (2002) Effect of enzyme supplementation of a rye-based diet on xylanase activity in the small intestine of broilers, on intestinal crypt cell proliferation and on nutrient digestibility and growth performance of the birds. Br Poult Sci 43(2):274–282

Slominski BA (2011) Recent advances in research on enzymes for poultry diets. Poult Sci 90(9):2013–2023

Snow J, Douglas M, Parsons C (2003) Phytase effects on amino acid digestibility in molted laying hens. Poult Sci 82(3):474–477

Sommerfeld V, Schollenberger M, Kühn I, Rodehutscord M (2018) Interactive effects of phosphorus, calcium, and phytase supplements on products of phytate degradation in the digestive tract of broiler chickens. Poult Sci 97(4):1177–1188

Soni H, Kango N (2013) Microbial mannanases: properties and applications. Advances in enzyme biotechnology, 41–56

Sun HY, Ingale SL, Rathi P, Kim IH (2018) Influence of β-glucanase supplementation on growth performance, nutrient digestibility, blood parameters, and meat quality in broilers fed wheat–barley–soybean diet. Can J Anim Sci 99(2):384–391

Suzuki U, Yoshimura K, Takaishi M (1907) Ueber ein Enzym "Phytase" das "Anhydro-oxy-methylen diphosphorsaure" Spaltet. Bull Coll Agric Tokyo Imp Univ 7:503–512

Tamim N, Angel R, Christman M (2004) Influence of dietary calcium and phytase on phytate phosphorus hydrolysis in broiler chickens. Poult Sci 83(8):1358–1367

Tanaka Y, DeLuca H (1974) Role of 1, 25-dihydroxyvitamin D3 in maintaining serum phosphorus and curing rickets. Proc Natl Acad Sci 71(4):1040–1044

Tanruean K, Penkhrue W, Kumla J, Suwannarach N, Lumyong S (2021) Valorization of lignocellulosic wastes to produce phytase and cellulolytic enzymes from a thermophilic fungus, Thermoascus aurantiacus SL16W, under semi-solid state fermentation. J Fungi 7(4):286

Taylor A, Bedford M, Pace S, Miller H (2018) The effects of phytase and xylanase supplementation on performance and egg quality in laying hens. Br Poult Sci 59(5):554–561

Teymouri H, Zarghi H, Golian A (2018) Evaluation of hull-less barley with or without enzyme cocktail in the finisher diets of broiler chickens. J Agric Sci Technol 20(3):469–483

Thorpe J, Beal J (2001) Vegetable protein meals and the effects of enzymes. Enzymes in farm animal nutrition, 125–143

Tiwari UP, Jha R (2017) Nutrients, amino acid, fatty acid and non-starch polysaccharide profile and in vitro digestibility of macadamia nut cake in swine. Anim Sci J 88(8):1093–1099

USDA (2022) Census of agriculture. https://www.nass.usda.gov/AgCensus/. Accessed 20 Dec 2022

Valente Junior DT, Genova JL, Kim SW, Saraiva A, Rocha GC (2024) Carbohydrases and Phytase in poultry and pig nutrition: a review beyond the nutrients and energy matrix. Animals 14(2):226

Van der Klis J, Versteegh H, Simons P, Kies A (1997) The efficacy of phytase in corn-soybean meal-based diets for laying hens. Poult Sci 76(11):1535–1542

Van Zyl WH, Rose SH, Trollope K, Görgens JF (2010) Fungal β-mannanases: mannan hydrolysis, heterologous production and biotechnological applications. Process Biochem 45(8):1203–1213

Vieira SL, Stefanello C, Rios HV, Serafini N, Hermes R, Sorbara J (2015) Efficacy and metabolizable energy equivalence of an α-amylase-β-glucanase complex for broilers. Br J Poult Sci 17:227–235

Viveros A, Centeno C, Brenes A, Canales R, Lozano A (2000) Phytase and acid phosphatase activities in plant feedstuffs. J Agric Food Chem 48(9):4009–4013

Waititu S, Sanjayan N, Hossain M, Leterme P, Nyachoti C (2018) Improvement of the nutritional value of high-protein sunflower meal for broiler chickens using multi-enzyme mixtures. Poult Sci 97(4):1245–1252

Walk C, Juntunen K, Paloheimo M, Ledoux D (2019) Evaluation of novel protease enzymes on growth performance and nutrient digestibility of poultry: enzyme dose response. Poult Sci 98(11):5525–5532

Walk C, Pirgozliev V, Juntunen K, Paloheimo M, Ledoux D (2018) Evaluation of novel protease enzymes on growth performance and apparent ileal digestibility of amino acids in poultry: enzyme screening. Poult Sci 97(6):2123–2138

Walker H, Vartiainen S, Apajalahti J, Taylor-Pickard J, Nikodinoska I, Moran CA (2024) The effect of including a mixed-enzyme product in broiler diets on performance, Metabolizable energy, phosphorus and calcium retention. *Animals* 14(2):328

Wang T, Ling H, Zhang W, Zhou Y, Li Y, Hu Y, Peng N, Zhao S (2022) Protease or clostridium butyricum addition to a low-protein diet improves broiler growth performance. Appl Microbiol Biotechnol 106(23):7917–7931

Weremko D, Fandrejewski H, Zebrowska T, Han IK, Kim J, Cho W (1997) Bioavailability of phosphorus in feeds of plant origin for pigs-review. Asian Australas J Anim Sci 10(6):551–566

Woyengo T, Nyachoti C (2013) Anti-nutritional effects of phytic acid in diets for pigs and poultry–current knowledge and directions for future research. Can J Anim Sci 93(1):9–21

Wyss M, Brugger R, Kronenberger A, Rémy R, Fimbel R, Oesterhelt G, Lehmann M, Van Loon AP (1999) Biochemical characterization of fungal phytases (myo-inositol hexakisphosphate phosphohydrolases): catalytic properties. Appl Environ Microbiol 65(2):367–373

Yegani M, Korver D (2013) Effects of corn source and exogenous enzymes on growth performance and nutrient digestibility in broiler chickens. Poult Sci 92(5):1208–1220

Zanu H, Keerqin C, Kheravii S, Morgan N, Wu S, Bedford M, Swick R (2020) Influence of meat and bone meal, phytase, and antibiotics on broiler chickens challenged with subclinical necrotic enteritis: 1. Growth performance, intestinal pH, apparent ileal digestibility, cecal microbiota, and tibial mineralization. Poult Sci 99(3):1540–1550

Zuo J, Ling B, Long L, Li T, Lahaye L, Yang C, Feng D (2015) Effect of dietary supplementation with protease on growth performance, nutrient digestibility, intestinal morphology, digestive enzymes and gene expression of weaned piglets. Anim Nutr 1(4):276–282

The Use of Phytogenic Substances Against Chronic Stress: Opportunities and Challenges

8

Shereen Basiouni, Hesham El-Saedi, Guillermo Tellez-Isaias, Wolfgang Eisenreich, and Awad A. Shehata

Contents

S. Basiouni
Cilia Cell Biology, Institute of Molecular Physiology, Johannes-Gutenberg University, Mainz, Germany

H. El-Saedi
Pharmacognosy Group, Department of Pharmaceutical Biosciences, Uppsala University, SE, Uppsala, Sweden
e-mail: hesham.el-seedi@farmbio.uu.se

G. Tellez-Isaias
Division of Agriculture, Department of Poultry Science, University of Arkansas, Fayetteville, AR, USA

W. Eisenreich · A. A. Shehata (✉)
Structural Membrane Biochemistry, Bavarian NMR Center, Technical University of Munich (TUM), Garching, Bayern, Germany
e-mail: Awad.shehata@tum.de

© The Author(s), under exclusive license to Springer Nature Switzerland AG 2024
A. A. Shehata et al. (eds.), *Alternatives to Antibiotics against Pathogens in Poultry*, https://doi.org/10.1007/978-3-031-70480-2_8

Abstract

The demand for biotics as alternatives to antibiotics that promote growth and lessen the use of antimicrobials in poultry farms is fueled by recent worldwide regulations and consumer expectations. Phytogenic substances are becoming increasingly valuable options because many of these natural compounds possess anti-inflammatory and antioxidant properties that are only less commercialized. The usage of phytogenic substances can also help maintain a balance of healthy microorganisms in the gut, which is beneficial for the digestive system to withstand various chronic stressors, both infectious and non-infectious. Although some phytogenic substances are commercially available, these are typically hampered by inconsistencies in their effectiveness, low bioavailability, and low stability. This chapter discusses the current and potential phytogenic compounds that can exert antioxidant and anti-inflammatory properties to improve poultry production and restore intestinal microbiota. Additionally, the main challenges still related to phytogenic products will be discussed.

Keywords

Anti-inflammatory substances · Antioxidants · Poultry · Phytogenic substances

1   Introduction

Phytogenic feed additives (PFAs) are substances that are used to improve the sensory qualities and taste of animal feed (Silvestro et al. 2021). They can also improve nutrient digestibility (Farahat et al. 2017), boost intestinal health (Gao et al. 2018), and improve growth performance (Gao et al. 2018) in chickens. PFAs are currently often utilized in poultry feeding programs. PFAs can also exhibit biological activities, including anti-inflammatory and antioxidant activities, improving performance and reducing environmental emissions. Additionally, PFAs might exhibit pharmacological effects and modulate the gut flora. Chap. 2 discussed how oxidative stress (OxS) and inflammation can be derived from the chronic stress conditions typically found in commercial poultry farming. In this chapter, the possible applications of botanical substances as anti-inflammatory and antioxidants to serve against chronic stress in poultry will be discussed. Moreover, the main challenges related to phytotherapy will be highlighted.

2   General Mechanisms of Phytogenic Substances

2.1   Antioxidants and Anti-inflammatory Effects

Generally, the antioxidant mechanisms in animals are based on endogenous and exogenous antioxidants. These mechanisms of living cells depend upon three levels. (i) prevention level by avoiding free radicals formation via removing the precursors of OxS; (ii) protection level by removal of hydrogen peroxide (H_2O_2). Vitamin E, ubiquinol, polyphenols, carotenoids, vitamin A, and ascorbic acid can do this; and (iii) elimination level by repairing the damaged molecules with lipases, proteases, and other enzymes (Basiouni et al. 2022).

Polyphenols are phytochemical compounds classified into flavonoids, non-flavonoids, and tannins (Serra et al. 2021) (Fig. 8.1). They are found in several parts of many plants, including leaves, bark, stems, roots, fruits, and flowers. Polyphenols, especially flavonoids, have potent antioxidant properties that help protect the body against cellular damage.

Several studies described the antioxidant effect of polyphenols. Polyphenols can trigger reactive oxygen species (ROS) scavenger enzymes. The mechanism of action of polyphenols can be summarized in Fig. 8.2 as follows (Basiouni et al. 2022):

Transfer of H-atoms from the OH group(s) of polyphenols to free radical(s).

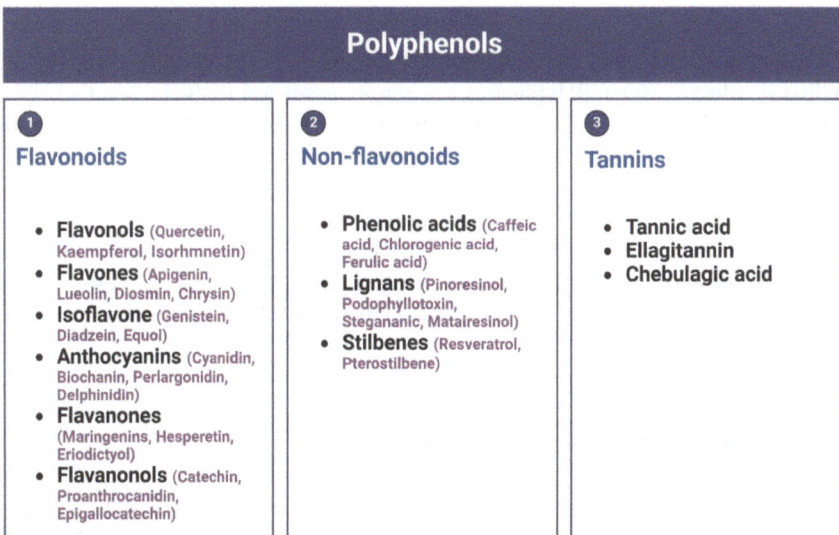

Polyphenols

1   Flavonoids

- **Flavonols** (Quercetin, Kaempferol, Isorhmnetin)
- **Flavones** (Apigenin, Lueolin, Diosmin, Chrysin)
- **Isoflavone** (Genistein, Diadzein, Equol)
- **Anthocyanins** (Cyanidin, Biochanin, Perlargonidin, Delphinidin)
- **Flavanones** (Maringenins, Hesperetin, Eriodictyol)
- **Flavanonols** (Catechin, Proanthrocanidin, Epigallocatechin)

2   Non-flavonoids

- **Phenolic acids** (Caffeic acid, Chlorogenic acid, Ferulic acid)
- **Lignans** (Pinoresinol, Podophyllotoxin, Stegananic, Matairesinol)
- **Stilbenes** (Resveratrol, Pterostilbene)

3   Tannins

- **Tannic acid**
- **Ellagitannin**
- **Chebulagic acid**

Fig. 8.1 Classification of polyphenols. The figure was generated using BioRender

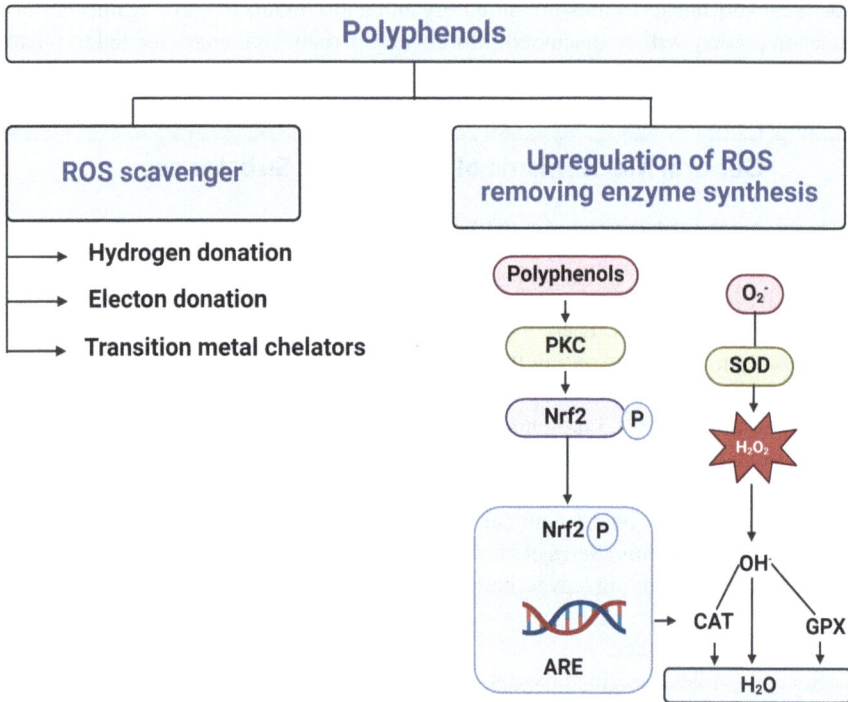

Fig. 8.2 Antioxidant effect of polyphenols as natural antioxidants (Basiouni et al. 2022). *ROS* reactive oxygen species, *PKC* Protein kinase C, *Nrf* NF-E2 p45-related factor 2, *ARE=*, *SOD* superoxide dismutase, H_2O_2 hydrogen peroxide, *CAT* catalase, *GPX* glutathione peroxidase

Transfer of a single electron to free radicals.

Chelating metal ions, particularly Fe^{2+} and Cu^{2+}, consequently mitigate the hydroxyl radicals (HO•). Polyphenols can eliminate several ROS and reactive nitrogen species (RNS) by these mechanisms, such as HO•, lipid peroxyl radicals (ROO•), superoxide anion radicals (O2•−), and peroxynitrite (ONOO−). In copper/hydrogen peroxide systems, polyphenols have been found to exhibit pro-oxidant properties, which prevent the generation of HO•.

Several plants, including Boswellia, Cannabis, Capsaicin, Cinnamaldehyde, Curcumin, Ginger, Piperamides, and Thyme, exhibit anti-inflammatory effects. The mechanism of action is reviewed by Basiouni et al. (2022). A summary is shown in Table 8.1.

2.2   Immunomodulatory Effects

Phytogenic substances enhance the activities of immune cells, which in turn modulate the immune system through several mechanisms (Mahima et al. 2012; Kiczorowska et al. 2017; Wang et al. 2024), including:

Table 8.1 Selected plants and phytogenic substances which possess antiinflammtory effects. Adapted from Basiouni et al. (2022)

Bioactive substance	Mechanism of action
Boswellia	Suppression of 5-lipoxygenase, lowering cytokines levels (interleukins and tumor necrosis factor α) and decreasing reactive oxygen species production
Cannabis	Upregulation of T-regulatory cells and suppression of cytokines
Capsaicin	Diminishing cyclooxygenase-2 mRNA expression
Cinnamaldehyde	Downregulation of interleukin-1 β, and 6, and tumor necrosis factor α as well as inducible nitric oxide synthase and cyclooxygenase-2 expression
Curcumin	Decreasing mRNA expression patterns of interleukin-1β, mucin 2, cyclooxygenase-2, and mitogen-activated protein kinase P38 Enhancing the expression of interleukins 1, 6, 12, and 18
Ginger	Inhibition of cyclooxygenase-1 and 2 Suppression of leukotrienes synthesis by suppression 5-lipoxygenase
Piperamides	Suppression of interleukin-6 and tumor necrosis factor α by downregulation of nuclear factor-κB and extracellular signal-regulated kinase
Thyme	Reducing tumor necrosis factor α, interleukins1B, and 6 Inhibiting the phosphorylation of nuclear factor-κB and mitogen-activated protein kinases Downregulation of interleukin-6, tumor necrosis factor α, and inducible nitric oxide synthase

Enhancement of phagocytosis
Modulation of cytokine secretion
Modulation of histamine release
Activation of Igs secretion

Several phytogenic substances, including carvacrol, cinnamaldehyde, and capsicum (Pirgozliev et al. 2019), turmeric oleoresin (Lee et al. 2013a, b), *Allium hookeri, Artemisia annua* (Song et al. 2018), thyme powder (Hassan and Awad 2017), *Echinacea purpurea* L. (Landy et al. 2011; Böhmer et al. 2009; Rahimi et al. 2011; Maass et al. 2005) have immunomodulatory properties in chickens (Table 8.2).

2.3   Antimicrobial Activity

Extensive studies have been conducted on the antibacterial properties of phytogenic bioactive compounds. For example, see O'Bryan et al. (2015); Mahfuz et al. (2021). However, the exact method and the specific targets by which these drugs exert their effects are not fully understood. It is believed that phytogenic extracts may work in various ways to produce their antimicrobial effects. (i) One way is that most phytogenic extracts are hydrophobic, which means that they can interfere with the structures of mitochondria and cell membranes, thereby benefitting the mammalian cells and/or damaging the bacterial cell walls (O'Bryan et al. 2015; Rossi et al. 2020). For pathogens, this may lead to the disruption of cellular functions, which in turn

Table 8.2 Selected immunogenic compounds in poultry

Phytogenic compound	Application	Immunomodulatory effects	References
Cinnamon	Broilers 4 and 8 g powder/kg diet	Increasing lymphocyte count	(Najafi and Taherpour 2014)
	Broilers 5 g powder /L Drinking water	Increasing of antibodies against NDV	(Sadeghi et al. 2012)
Curcuma longa	Broilers 2.5, 5, and 7.5 g rhizome powder/kg diet	Increasing IgA, IgG, and IgM levels, and a decreased ratio of monocytes	(Emadi and Kermanshahi 2007)
	Broilers 2 g rhizome powder /kg diet	Increasing the total antibody titer, and decreased H/L ratio	(Akhavan-Salamat and Ghasemi 2016)
	Broilers 35 mg rhizome powder /kg diet	Increasing antibodies against Eimeria microneme protein	(Kim et al. 2013)
Echinacea purpurea L.	Broilers 5 and 10 g aerial part powder/kg diet	Increasing antibodies against SRBC and NDV	(Landy et al. 2011)
	Laying hens 0.25 mL pressed juice/kg BW	Increasing lymphocytes and NDV antibodies in the blood	(Böhmer et al. 2009)
	Broilers 1 mL aqueous extract drinking/L water	Increasing antibodies against SRBC	(Rahimi et al. 2011)

causes bacterial death. (ii) Some bioactive substances harm the activities of bacterial RNA polymerase and DNA topoisomerase, which are important in the production of DNA and RNA during replication and transcription (Mahfuz et al. 2021). (iii) Some bioactive chemicals can affect the synthesis of proteins, lipids, and polysaccharides and interfere with membrane proteins and intracellular targets (Rossi et al. 2020). The antimicrobial effects of selected phytogenic substances on avian pathogens are shown in Table 8.3.

2.4 Modulation of Intestinal Microbiota

Several bioactive substances modulate the intestinal microbiota (Fig. 8.3) and thereby provide robustness and better performances in poultry, reviewed by Basiouni et al. (2022).

Table 8.3 Antimicrobial effects of some selected phytogenic substances

Study	Phytogenic substances	Antimicrobial effect	Reference
In vitro	Tannin-rich extracts	*Campylobacter jejuni*	(Anderson et al. 2012)
	Ethanolic cinnamon	*Salmonella*	(Bonilla and Sobral 2017)
	Cinnamaldehyde	*Brachyspira hyodysenteriae*	(Vande Maele et al. 2016)
	Curcumin (diferuloylmethane)	*Eimeria tenella*	(Khalafalla et al. 2011)
In vivo	Cinnamaldehyde	*Salmonella*	(Upadhyaya et al. 2015)
	Quercetin	*Salmonella enterica* serotype Typhimurium, *Escherichia coli*, *Staphylococcus aureus*, and *Pseudomonas aeruginosa*	(Iqbal et al. 2020)
	Acetone and cold water extracts of *Alchornea laxiflora*, *Ficus exasperata*, *Morinda lucida*, *Jatropha gossypiifolia*, *Ocimum gratissimum*, and *Acalypha wilkesiana*	Antifungal activity	(Olawuwo et al. 2022)

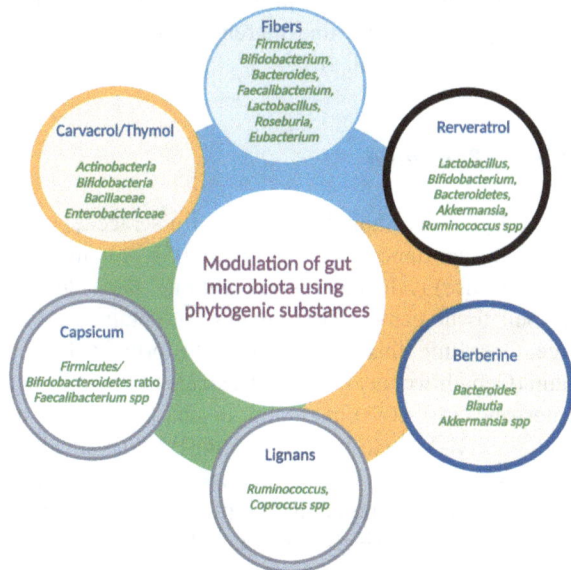

Fig. 8.3 Modulatory effects of selected phytogenic substances. The figure was generated using BioRender (adapted from Basiouni et al. (2022))

3   Common Bioactive Substances Used in Poultry

3.1   Berberine

Berberine is a bioactive substance of several plants, including *Coptis chinensis*, *Cortex phellodendri*, and *Berberis asiatica* (Xu et al. 2021).

Berberine is converted by enzymes such as bacterial nitroreductases that transform berberine into dihydroberberine and active metabolites that can exhibit several biological activities.

Modulation of gut microbiota despite its low oral bioavailability. In mice, it enhanced beneficial bacteria, including *Bacteroides* spp., *Lactobacillus* spp., and *Bifidobacterium* spp. (Habtemariam 2020; Lyu et al. 2021). Administering berberine at a dose of 2 g/kg feed to broilers significantly modulated the intestinal microbiota (Dehau et al. 2023).

Providing anti-inflammatory, anti-diabetic, anti-atherosclerotic, and cardioprotective benefits (Zeng et al. 2015; Takahara et al. 2019; Ilyas et al. 2020). Berberine could modulate several proinflammatory mediators, including downregulating TNF-α, IL-1β, and 6 (Gu et al. 2009, 2011; Li et al. 2010; Gong et al. 2017). The anti-inflammatory effect of berberine was also reported in broiler chickens by downregulation of cyclooxygenase-2, TNF-α, IL-1β, NF-κB, and nitrite synthase (Yang et al. 2019).

Enhancing the secretion of short-chain fatty acids (SCFAs), which could exhibit positive effects against several enteric pathogens and diseases, including *Clostridium perfringens* and *Eimeria* infection (Malik et al. 2016; Xiang et al. 2017) (Habtemariam 2020).

Enhancing drug absorption and bioavailability. In broilers, it suppresses P-glycoprotein (P-gp) and thereby improves the pharmacodynamics of the P-gp substrate (Zhang et al. 2019).

3.2   Boswellia

Boswellia (order Sapindales, family Burseraceae) includes several species that have medical importance (Weeks et al. 2005). Among these species, *B. serrata* (Indian frankincense) is the most prominent one. *Boswellia* trees produce resins and balms in their tissue canals. These resins are obtained by making 6–8 incisions on each tree. Over time, this resin solidifies to form frankincense, also referred to as olibanum (Gebrehiwot et al. 2003). The resin of the tree contains volatile oils (3–8%) and triterpenes (30–60%), including α- and β-boswellic acids, 11-keto-boswellic acid, and 3-acetyl-11-keto-boswellic acid (BAs) (Almeida-da-Silva et al. 2022).

BAs act as anti-inflammatory agents by several mechanisms:

Inhibition/downregulation of proinflammatory mediators, including 5-lipoxygenase (Krieglstein et al. 2001), interleukins, and tumor necrosis factor-α (Ammon 2006), prostaglandin E2 synthase-1 (Siemoneit et al. 2011), human leukocyte elastase (Safayhi et al. 1997), cathepsin G (Tausch et al. 2009), NF-κB (Cuaz-Pérolin et al. 2008), COX-1, and 2 (Siemoneit et al. 2008).

Increasing lymphocyte proliferation at low doses and downregulating NF-κB, TNF-α, IL-1, IL-2, IL-4, IL-6, and γ-interferon (Ammon 2010).

The combination of frankincense and myrrh multiplied their anti-inflammatory and analgesic effects, making this combination more therapeutically useful than using individual frankincense extracts (Su et al. 2012). Frankincense and myrrh are potent antioxidants in mouse macrophages by suppression of nitric oxide (Cao et al. 2019).

The use of *B. serrata* in poultry can provide various benefits. Specifically, it can improve the birds' antioxidant status, raise globulin levels and superoxide dismutases (SOD), and encourage the production of digestive enzymes such as amylase and lipase. Moreover, it reduces total cholesterol, low-density lipoprotein, and malondialdehyde (Zeng et al. 2021). Adding *Boswellia* (3% and 4%) to broiler chicken diets improved body weight, digestion efficiency, and carcass traits in a study (Ilyas et al. 2020). When 0.3% of *B. serrata* and *Salix alba* were added to leghorn hens, a significant improvement in immune response toward infectious viruses causing bronchitis was reported (Ke et al. 2020).

3.3 Calendula officinalis

Calendula officinalis, also known as marigold, has several beneficial effects, including antibacterial, antioxidant, and anti-inflammatory activities (Zitterl-Eglseer et al. 1997; Yoshikawa et al. 2001; Ukiya et al. 2006; Muley et al. 2009). *C. officinalis* is rich in triterpenoids, which were found to reduce H_2O_2, INF-γ, and TNF-α (Dall'Acqua et al. 2016). It has been shown that broilers fed with *C. officinalis* flower extract have improved growth performance and immune response (Rajput et al. 2012). However, Foroutankhah et al. (2019) found that 0.5% and 1% of *C. officinalis* dried powder did not positively impact the growth performance of broiler chickens (Foroutankhah et al. 2019). Further investigations are needed to assess the potential impact of *Calendula* on the digestive health of broiler chickens.

3.4 Cannabis

Cannabis spp., family Cannabaceae, are rich in cannabinoids. The main bioactive substances of *Cannabis* spp. are the psycho-active tetrahydrocannabinol (THC) and the non-psychoactive cannabidiol (CBD), which have several biological activities, including anti-inflammatory effects. Additionally, phytocannaboids can regulate intestinal motility ion transport and help maintain gut health (Klein and Cabral 2006; Izzo and Sharkey 2010).

CBD has been used as a medication for several years due to its anti-inflammatory, antispasmodic, and analgesic properties (Konieczka et al. 2020). There are two types of receptors, CB1 and CB2, where phytocannabinoids can bind. CB1 receptors are mostly located in the brain, while CB2 receptors are mostly present in immune cells. CBD has immunological properties that promote T-regulatory cell activation, and it also has an anti-inflammatory effect by reducing the production of cytokines and chemokines (Nagarkatti et al. 2009). A study conducted by Vispute et al. in 2019 demonstrated that adding 3 g of hemp seeds per kg of feed to the diet of commercial broiler chickens can improve serum quality and gut health when taken alone or in conjunction with other supplements (Vispute et al. 2019). This supplementation also caused a significant reduction in the levels of aspartate aminotransferase (AST) and alanine aminotransferase (ALT), but not for alkaline phosphatase. Another study by Konieczka et al. in 2020 found that the performance of chickens was enhanced by the addition of 15 g/kg of CBD (Konieczka et al. 2020). In addition, mixing *C. sativa* extract and nano-selenium strengthened the intestinal barrier and increased protection against *Clostridium* infections.

3.5 Capsaicin

Capsaicin is the natural ingredient responsible for chili peppers' hot and spicy taste. It has been proven to have strong antioxidant properties (Cho et al. 2020; Zamljen et al. 2021). Capsaicin also contains a phenolic hydroxyl group that facilitates the effective transfer of hydrogen and reduces the activity of free radicals, ultimately preventing DNA damage (Pérez-González et al. 2020; Cheng et al. 2022; Liu et al. 2021). Additionally, capsaicin has strong anti-inflammatory effects (Cho et al. 2020; Zamljen et al. 2021). In a study conducted on rats with induced gastritis, capsaicin downregulated TNF-α, IL-1β, and IL-6 (Mendivil et al. 2019). Additionally, capsaicin has anti-inflammatory properties by downregulating COX-2 mRNA expression (Liu et al. 2021).

Liu et al. (2021) found that adding 150 mg/kg capsaicin to the diet of laying ducks increased their appetite and egg production performance (Liu et al. 2021). In broilers, capsaicin (80 mg/kg feed) enhanced the animals' nutritional digestibility and immune system response (Liu et al. 2021).

3.6 Cinnamaldehyde

Cinnamon, family Lauraceae, and its primary bioactive compound is cinnamaldehyde. *Cinnamomum zeylanicum*, *C. cassia*, *C. burmanni*, and *C. loureiori* are the worldwide economically important species (Rao and Gan 2014; Ali et al. 2021).

Cinnamon contains many phytocompounds, such as volatile oils, flavonoids, curcuminoids, coumarins, tannins, alkaloids, xanthones, terpenoids, phenolics, and others. Extraction of essential oil from the bark and leaves revealed that cinnamaldehyde and eugenol are the main compounds, respectively. These compounds exhibit antioxidative, antibacterial, and anti-inflammatory properties (Ali et al. 2021). Cinnamaldehyde has an anti-inflammatory effect by reducing several cytokines in vitro, including iNOS, COX-2, IL-1β and 6, and TNF-α (Pannee et al. 2014). Additionally, in lipopolysaccharide-activated murine macrophage-like cells, enhanced IL-10 release has been reported. Cinnamaldehyde is an immunostimulant bioactive substance (Pannee et al. 2014). Nitric oxide production by macrophages was enhanced by cinnamaldehyde when it was given at dosages between 1.2 and 5.0 g/mL. Cinnamaldehyde also demonstrated anticoccidial activity (Lee et al. 2011). Cinnamon can be used in poultry for several purposes, including improving digestion, boosting immune response, maintaining intestinal health, regulating the balance of water and electrolytes, and promoting feed intake. Therefore, it can also be used to mitigate the harmful effects of heat stress (Ali et al. 2021).

3.7 Curcumin

Curcuma longa is a member of the Zingiberaceae family. *Curcuma* contains curcumin (diferuloylmethane), demethoxycurcumin, and bis-dimethoxycurmarin (Bober et al. 2018).

Curcumin is a powerful natural antioxidant, anti-inflammatory, antiviral, and antimicrobial bioactive substance (Leyva-Diaz et al. 2021; Petrone-Garcia et al. 2021). It modifies the mitogen-activated protein kinase p38 pathway, which inhibits cytokines. It upregulates IL-1, IL-6, IL-12, IL-18, and TNF-α (Lee et al. 2010).

Numerous investigations have indicated that curcumin in poultry has health advantages. It can lessen the severity of *C. perfringens* (Solis-Cruz et al. 2019), salmonellosis, aflatoxicosis (Solis-Cruz et al. 2019), and coccidiosis (Petrone-Garcia et al. 2021). Curcumin also has a hepatoprotective effect by modulating lipogenesis- and lipolysis-related genes (Bober et al. 2018; Xie et al. 2019). Additionally, curcumin enhances the secretion of digestive enzymes (Bober et al. 2018).

3.8 Ginger Extracts

Ginger, a member of the family Zingiberaceae, is thought to be found in India and Southern Asia. It has numerous therapeutic and nutritional benefits due to several active compounds, including phenolics (paradols, shogaols, and gingerols) and terpenes (Nile and Park 2015).

It has antioxidant (Stoner 2013), anti-inflammatory (Zhang et al. 2016), antibacterial (Kumar et al. 2014), and anticancer (Karangiya et al. 2016) effects. The antioxidant effect is attributed to gingerols, shogaols, gingerdiols, and gingerdiones (Fuhrman et al. 2000; Zhao et al. 2011). The anti-inflammatory properties of ginger could be explained by the reduction of prostaglandin (An et al. 2019), resulting from inhibiting COX-1 and 2 (Nile and Park 2015). It also suppresses 5-LOX through decreasing leukotriene synthesis.

Ginger has been shown to have several advantageous effects on poultry. Supplementing chickens with ginger (0.1 or 0.2%) in feed increased the expression SOD, glutathione peroxidase (GSH-Px), and total antioxidant capacity. Additionally, it increased serum malondialdehyde (Zhao et al. 2011).

3.9 Isoflavones (ISF) and Isoquinoline Alkaloids

Isoflavones (ISFs), including genistein, glycitein, and daidzein, have numerous health advantages, including reducing cholesterol, strengthening immunity, and exhibiting antioxidant properties (Ajdžanović et al. 2014; Messina et al. 2004).

It was found that administering 10–20 mg/kg of ISFs to commercial broiler chickens reduced the expression of infectious bursal disease and minimized the pathological lesions in the Bursa of Fabricius (Azzam et al. 2019). Administration of either 40 or 80 mg ISF/kg body weight exhibited antioxidant properties in terms of enhancement of the antioxidant capacity and SOD activity (Jiang et al. 2007).

ISFs supplements are becoming more popular for layers, especially in the later phases of egg production (Jiang et al. 2007). The addition of ISFs to the diet significantly increased egg production in laying hens (Shi et al. 2013). Indeed, ISFs affect the estrogen levels. However, this activity depends on the dose of phytoestrogens and the supplementation period (Cassidy 2003). Additionally, supplementing chickens with ISFs enhanced the fertility of eggs (Zhao et al. 2005, 2013; Ni et al. 2012). Similarly, quails given ISFs resulted in improved bones and higher-quality eggs (Sahin et al. 2007).

Isoquinoline alkaloids (IQAs) are found in *Macleaya cordata* or plume poppies extract. IQAs improved the gut health of pigs that were exposed to heat stress (Le et al. 2020), which might be by enhancing the Zonula occludens-1 (ZO-1) and claudin-1 (Liu et al. 2016a). However, studies on the impact of IQAs on poultry have not been done yet.

3.10   Piperamides

Black pepper (*Piper nigrum*), family Piperaceae, contains glutathione and glucose-6-phosphate dehydrogenase (Karthikeyan and Rani 2003). Black pepper is used in Eastern medicine for pain and infections. Piperlongumine, an analog of piperine, has antioxidant effects and enhances the utilization of nutrients such as selenium, vitamin B complex, beta-carotene, and curcumin (Khalaf et al. 2008; Yadav et al. 2020). Black pepper helps to combat free radicals and to influence benzopyrene metabolism through cytochrome P450; therefore, black pepper has a significant effect on the metabolism and transportation of xenobiotics (Reen et al. 1996).

Piperine, the active phytogenic substance found in black pepper, promotes the thermogenesis of lipids (Malini et al. 1999). In addition to stimulation of the flow of digestive juice (Moorthy et al. 2009), it protects against DNA damage, and increases gut motility and intestinal microvilli, which all contribute to enhancing nutrient absorption.

Piperlongumine was found to improve gut integrity (Lee et al. 2013a, b), through its anti-inflammatory effect by suppressing the secretion of IL-6 and TNF-α. Additionally, *P. nigrum* and *Capsicum annum* black pepper enhanced the broiler performance and lowered the blood cholesterol levels (Al-Kassie et al. 2012). In commercial broilers, it improved the body weight and feed conversion ratio (Ghazalah et al. 2007; Tollba et al. 2007; Mansoub 2011). In contrast, supplementing black pepper did not show a significant improvement in the body weight of broilers (Akbarian 2012).

3.11 Phenolic Derivatives

The olive tree, designated *Olea europaea* L., has been shown to exhibit a number of biological properties, including hypocholesterolemic and anti-inflammatory properties (El and Karakaya 2009), due to the presence of polyphenols, phenolics, and their derivatives (Talhaoui et al. 2016). These phytochemicals mitigate intestinal inflammation (Deiana et al. 2018; Farràs et al. 2020) and reduce pathogenic bacteria (Sarıca and Ürkmez 2016). In obese mice, it improved the gut integrity by enhancing the Zonula occludens-1 and occludin (Deiana et al. 2018; Farràs et al. 2020), as well as increasing mucin-2, Zonula occludens-1, and Trefoil factor (Vezza et al. 2017). In broilers, live pomace at a dose of 750 ppm increased the Firmicutes (Herrero-Encinas et al. 2020), suggesting its role in modulating intestinal microbiota.

3.12 Quercetin

Quercetin, a type of flavonoid (Hertog et al. 1992), is known for its potent antioxidant properties (Erlund 2004). It helps to restore gut health by enhancing Zonula occludens protein 2, occludin, claudin-1, and claudin-4 while blocking protein kinase Cδ (Suzuki and Hara 2009; Amasheh et al. 2009).

Carrasco-Pozo et al. tested the protection conferred by quercetin in indomethacin and rotenone-stimulated Caco-2 cells. The authors found that quercetin has a positive effect on Zonula occludens-1 and occludin and refrain from the reduction in their expression (Carrasco-Pozo et al. 2013). Quercetin also inhibits isoform-mixed protein kinase C and phosphoinositide-3-kinase (Agullo et al. 1997).

3.13 Resveratrol

Resveratrol, a polyphenolic substance that occurs naturally in several plants, including peanuts, *Polygonum cuspidatum*, and grapes (Zhang et al. 2017c), regulates energy metabolism. It also exerts a beneficial effect against inflammation and OxS (Manna et al. 2000; Madeo et al. 2019). Resveratrol also helps in gut integrity and lessens intestinal damage (Zhao et al. 2018).

In broilers, resveratrol exhibits antioxidant activity during heat stress. It stimulates the nuclear factor E2-related factor 2 pathways. At 400 mg/kg, it significantly increased vitamin E and decreased serum malondialdehyde. In stress-impaired broilers, it decreased the muscle malondialdehyde and lactate dehydrogenase (LDH) activity and increased total SOD and GSH-Px (Liu et al. 2014; Zhang et al. 2017a). It could also mitigate the negative impacts of heat-stress broilers at 400 mg/kg. Thus, it increased muscle total antioxidant capacity (T-AOC) and activity of antioxidant enzymes (CAT, GSH-PX) (Zhang et al. 2017b; Hu et al. 2019).

Resveratrol can potentially mitigate intestinal damage through the upregulation of Hsp70, Hsp90, and NF-kappaB mRNA and downregulate the mucosal production of epidermal growth factor (Liu et al. 2016b).

Resveratrol also exhibits a modulatory effect on intestinal microbiota. Specifically, it increases the *Lactobacillus* sp., *Bifidobacterium* sp., *Bacteroidetes*, *Akkermansia* sp., and *Ruminococcus* sp., while it decreases *Lactococcus* sp., *Clostridium* spp., *Oscillibacter* spp., and *Hydrogenoanaerobacterium* spp. (Pan et al. 2018).

Moreover, resveratrol maintains intestinal morphology (Feng et al. 2021; Wang et al. 2021) by enhancing the tight junctions and reducing its permeability to lipopolysaccharide (Alrafas et al. 2019; Cai et al. 2020; Wang et al. 2021).

3.14 Salix Extracts

The genus *Salix* (willows), belonging to the family Salicaceae, contains salicin and polyphenols. Salicin has several biological activities, including antioxidant, anti-inflammatory, analgesic, and antipyretic properties (Tawfeek et al. 2021). Indeed, Salix may be an effective alternative in alleviating the adverse impacts of heat stress (Al-fataftah and Abdelqader 2013).

In the gut, salicin is transformed into saligenin, which is then taken up and oxidized in the liver to produce salicylic acid. This compound exhibits analgesic and anti-inflammatory properties by inhibiting COX-1 and 2 (Shara and Stohs 2015; Saracila et al. 2021). Additionally, some *Salix* spp., including *S. daphnoides*, *S. purpurea*, and *S. fragilis*, exert anti-inflammatory effects in the human monocytes by inhibiting IFN-γ (Kelber et al. 2006).

Feeding broilers with 0.05% of *Salix* L. bark powder reduced the levels of malondialdehyde, GSH, and thiobarbituric acid reactive substances (liver tissue lipid peroxidation marker) (Panaite et al. 2020). Ishikado et al. explained the antioxidant effects of the ability of *Salix* to activate the nuclear factor erythroid-2-related factor 2 (Nrf2) pathway (Ishikado et al. 2013).

Indeed, *Salix* exhibited modulatory effects on intestinal microbiota by promoting Lactobacilli and decreasing *E. coli* and Staphylococci (Panaite et al. 2020). Additionally, *Salix* may induce hypocholesterolemia, but it did not show any significant improvement in productive performance in commercial broiler chickens (Saracila et al. 2018).

3.15 Thymol/Carvacrol

Thymol and carvacrol are present in plants belonging to the Lamiaceae family, such as thyme and oregano. The biological activities of thymol and carvacrol are shown in Fig. 8.4. They exhibit versatile pharmacological properties against OxS and inflammation (Yalçin et al. 2020). Thymol modulates proinflammatory mediators, including TNF-α, IL-1B, and IL-6 (Ocaña and Reglero 2012). In lipopolysaccharide-stimulated mouse mammary epithelial cells, thymol reduced the phosphorylation of NF-κB and mitogen-activated protein kinases, resulting in downregulation of IL-6, TNF-α, iNOS, and COX-2 (Liang et al. 2014).

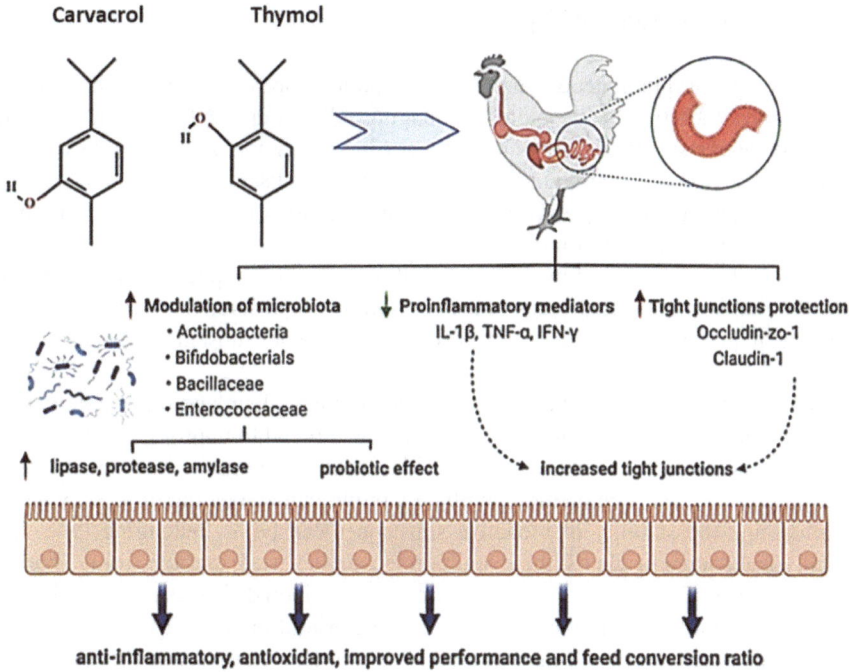

Fig. 8.4 Biological activities of thymol/carvacrol (Shehata et al. 2022a)

Furthermore, Hassan and Awad (2017) found that thymol at 10, 20, and 40 µg/mL inhibited the activation of mitogen-activated protein kinases, NF-B p65, extracellular signal-related kinase, C-Jun N-terminal kinase, and p38 in lipopolysaccharide-stimulated mouse mammary epithelial cells (Hassan and Awad 2017).

Thyme oil, 0.1 g/kg feed, improved digestion in chickens by promoting digestive enzymes (Lee et al. 2003). However, there are conflicting findings about how it affects the performance of animals. Thyme oil supplementation at a dose of 5 g/kg did not significantly affect broiler growth performance (Hassan and Awad 2017).

4   Challenges and Opportunities

The key challenges associated with the implementation of PFAs may be broadly summarized as follows (Shehata et al. 2022b):

Low bioavailability
Inconsistent efficiency
Risk of contamination with harmful substances
Public acceptance

4.1   Low Bioavailability

Determining the pharmacodynamics of compounds derived from plants, namely their bioavailability, can be quite challenging (Bhattaram et al. 2002). In some cases, bioactive substances show low efficacy in vivo as compared to in vitro. Only 2–15% of phytobiotic compounds can be absorbed (Kikusato 2021), which could be a reason for the low effectiveness of some phytobiotic compounds in vivo. Ineffectiveness can result, for example, from low concentrations of active principles in the target tissues caused by inadequate absorption of polyphenols. The substantial biotransformation that takes place in the colon and liver may also affect the biological activities and bioavailability of phytogenic substances (Paszkiewicz et al. 2012).

Moreover, bioactive substances may undergo metabolism and rapid elimination (Stevanović et al. 2018). Advanced techniques are crucial to improve the absorption and utilization of phytogenic chemicals. Moreover, studying the bioavailability of bioactive chemicals is challenging without a proper understanding of their synergy. Maintaining the stability of bioactive substances can be a challenging task. The quality and source of plants, the extraction techniques used, and the storage conditions are critical factors that must be carefully considered. For example, polyphenols and phenolics are highly susceptible to deterioration when exposed to light, heat, and oxygen during storage. Therefore, it is important to ensure that these factors are properly controlled to prevent the degradation of bioactive substances (Trucillo et al. 2018; Takano et al. 2021; Tolve et al. 2021). The condition mentioned above can have an impact on the stability of storage and convert active components into either inactive or harmful byproducts (Suganya and Anuradha 2017; Tolve et al. 2021).

4.2   Inconsistent Efficiency

It is crucial to focus more on addressing the inconsistent efficacy of botanical substances. There are numerous factors responsible for the discrepancies, such as variations in plant chemical profiles, different harvesting times, and plant origins. Other detrimental factors that could be responsible for the divergent effectiveness are the type of extraction techniques and the type of extracts (powder, extract, essential oil). Additionally, differences in the experimental evaluation and assessment protocols can lead to divergent results (Shehata et al. 2022b). There is a need for standardizing research procedures to ensure consistent and reliable outcomes. Thus, inconsistent effectiveness results from the variations due to lacking standardization (Vlaicu et al. 2021; Sugiharto and Ayasan 2022).

The efficacy of PFAs in promoting growth through lessening chronic stress has been described (Sugiharto and Ayasan 2022). If chickens are raised in an optimal environment, they may not benefit from these substances. It should also be noted that to evaluate the efficacy of PFAs, birds should be subjected to a certain level of stress (Sugiharto and Ayasan 2022) to determine significant differences between treated and untreated animals.

Indeed, the accurate dose of PFAs is challenging; inadequate doses (either too low or too high) can have undesired biological effects (Hafeez et al. 2020). There were no improvements observed in broiler performance after administering high doses of ajwain, fenugreek, and black cumin (Saleh et al. 2018). Some PFAs may negatively affect the performance of animals due to their flavor, leading to a reduction in feed consumption (Sugiharto 2021; Vlaicu et al. 2021).

4.3   Risk of Contamination with Harmful Substances

Certain PFAs may pose risks, with some exhibiting carcinogenic effects (Fennell et al. 2004). For instance, polygermander (*Teucrium polium*) can cause hepatotoxic effects. There is also a risk of contamination of phytogenic substances with mycotoxins and heavy metals. In light of these potential concerns, it becomes crucial to consider product purity.

4.4   Public Acceptance

The first paradigm of using alternatives to antimicrobials relies only on the PFAs that possess antimicrobial activities. However, bioactive compounds have several PFAs other than direct antimicrobial effects. Additionally, the oversimplification in which the industry can recommend PFAs for treating acute infections results in consumer dissatisfaction. Therefore, nowadays, the second generation of alternatives to antimicrobials focuses on host-mediated effects, not pathogens. This paradigm is interested in using biotics for several beneficial effects such as modulation of host immunity, modulation of and restoring intestinal microbiota, physiology, and metabolism, which subsequently increase the resistance to pathogens and improve performance. This paradigm provides a clear opportunity for next-generation PFAs as biotics.

5   Prospective and Recommendation

One approach to increase the effectiveness of PFAs is through encapsulation, a process that involves trapping bioactive compounds within a particle's core. Encapsulation offers several advantages, including: (i) enhancement of the stability of PFAs in the gut; (ii) masking unpleasant tastes and smells of PFAs leads to improving the palatability; (iii) prolonging the life span by protecting them from environmental stressors such as oxygen; and (iv) decreasing the dose and costs (Sugiharto and Ayasan 2022; Shehata et al. 2022b). Numerous substances are recognized as "generally recognized as safe" (GRAS) for encapsulation, including lipids (such as waxes, glycerides, and phospholipids), proteins, and carbohydrate polymers (Timilsena et al. 2020).

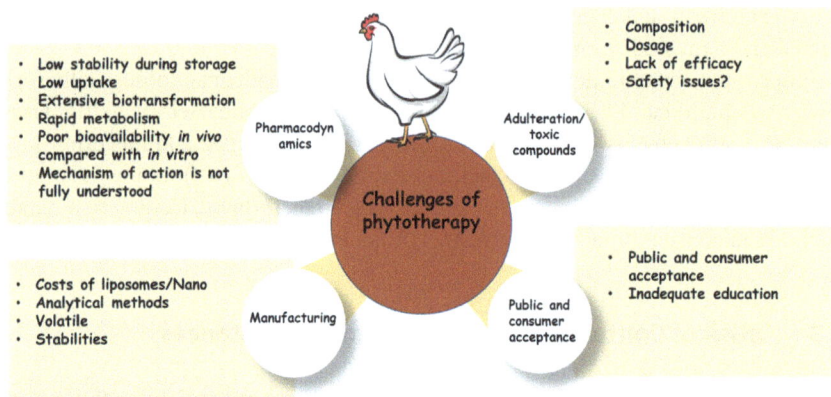

Fig. 8.5 Challenges related to the application of bioactive substances in poultry (Shehata et al. 2022b)

Encapsulation of PFAs has also been implemented for poultry, including essential oils and botanical extracts (Amiri et al. 2021; Lee et al. 2020; Meimandipour et al. 2017). However, this requires optimizing encapsulation methods of phytogenic compounds and gaining a comprehensive understanding of how microbial fermentation affects the degradation rate and kinetics of these compounds. Ongoing research can help to advance the implementation of encapsulated PFAs in poultry. To maintain ancestral stability and uniformity across plant populations, it is essential to expand the use of scientifically based procedures for the secure and trustworthy production of PFAs.

Traditional plant breeding techniques can improve the characteristics of plants, whether they are utilized for food or medicinal purposes. Nowadays, selective breeding and molecular markers are being employed more and more. There is an increasing need for raw plant materials since PFAs are becoming increasingly popular. Toxic components, pollution, variability and instability of extracts, and genetic and phenotypic variety must all be addressed through domestic cultivation under close supervision.

Recent advances in molecular techniques have enabled us to explore the constituents of gut flora more precisely and acquire a better comprehension of their dynamics. Moreover, growing global regulations and local consumer demands to minimize the application of antimicrobials that promote growth and limit antibiotic use have increased the need for alternatives, including PFAs (Fig. 8.5).

References

Agullo G, Gamet-Payrastre L, Manenti S, Viala C, Rémésy C, Chap H, Payrastre B (1997) Relationship between flavonoid structure and inhibition of phosphatidylinositol 3-kinase: a comparison with tyrosine kinase and protein kinase C inhibition. Biochem Pharmacol 53:1649–1657. https://doi.org/10.1016/s0006-2952(97)82453-7

Ajdžanović VZ, Medigović IM, Pantelić JB, Milošević VL (2014) Soy isoflavones and cellular mechanics. J Bioenerg Biomembr 46:99–107. https://doi.org/10.1007/s10863-013-9536-6

Akbarian A (2012) Influence of turmeric rhizome and black pepper on blood constituents and performance of broiler chickens. Afr J Biotechnol 11. https://doi.org/10.5897/AJB11.3318

Akhavan-Salamat H, Ghasemi HA (2016) Alleviation of chronic heat stress in broilers by dietary supplementation of betaine and turmeric rhizome powder: dynamics of performance, leukocyte profile, humoral immunity, and antioxidant status. Trop Anim Health Prod 48:181–188. https://doi.org/10.1007/s11250-015-0941-1

Al-fataftah A, Abdelqader A (2013) Effect of Salix babylonica, Populus nigra and Eucalyptus camaldulensis extracts in drinking water on performance and heat tolerance of broiler chickens during heat stress. Am-Eurasian J Agric Environ Sci:1309–1313

Ali A, Ponnampalam EN, Pushpakumara G, Cottrell JJ, Suleria HAR, Dunshea FR (2021) Cinnamon: a natural feed additive for poultry health and production—a review. Animals 11:2026. https://doi.org/10.3390/ani11072026

Al-Kassie GAM, Butris GY, Ajeena SJ (2012) The potency of feed supplemented mixture of hot red pepper and black pepper on the performance and some hematological blood traits on broiler diet. Int J Adv Biol Res:53–57

Almeida-da-Silva CLC, Sivakumar N, Asadi H, Chang-Chien A, Qoronfleh MW, Ojcius DM, Essa MM (2022) Effects of frankincense compounds on infection, inflammation, and Oral health. Molecules 27:4174. https://doi.org/10.3390/molecules27134174

Alrafas HR, Busbee PB, Nagarkatti M, Nagarkatti PS (2019) Resveratrol modulates the gut microbiota to prevent murine colitis development through induction of Tregs and suppression of Th17 cells. J Leukoc Biol 106:467–480. https://doi.org/10.1002/JLB.3A1218-476RR

Amasheh M, Grotjohann I, Amasheh S, Fromm A, Söderholm JD, Zeitz M, Fromm M, Schulzke J-D (2009) Regulation of mucosal structure and barrier function in rat colon exposed to tumor necrosis factor alpha and interferon gamma in vitro: a novel model for studying the pathomechanisms of inflammatory bowel disease cytokines. Scand J Gastroenterol 44:1226–1235. https://doi.org/10.1080/00365520903131973

Amiri N, Afsharmanesh M, Salarmoini M, Meimandipour A, Hosseini SA, Ebrahimnejad H (2021) Nanoencapsulation (in vitro and in vivo) as an efficient technology to boost the potential of garlic essential oil as alternatives for antibiotics in broiler nutrition. Animal 15:100022. https://doi.org/10.1016/j.animal.2020.100022

Ammon HPT (2006) Boswellic acids in chronic inflammatory diseases. Planta Med 72:1100–1116. https://doi.org/10.1055/s-2006-947227

Ammon HPT (2010) Modulation of the immune system by Boswellia serrata extracts and boswellic acids. Phytomedicine 17:862–867. https://doi.org/10.1016/j.phymed.2010.03.003

An S, Liu G, Guo X, An Y, Wang R (2019) Ginger extract enhances antioxidant ability and immunity of layers. Anim Nutr 5:407–409. https://doi.org/10.1016/j.aninu.2019.05.003

Anderson RC, Vodovnik M, Min BR, Pinchak WE, Krueger NA, Harvey RB, Nisbet DJ (2012) Bactericidal effect of hydrolysable and condensed tannin extracts on campylobacter jejuni in vitro. Folia Microbiol (Praha) 57:253–258. https://doi.org/10.1007/s12223-012-0119-4

Azzam MM, Jiang S, Chen J, Lin X, Gou Z, Fan Q, Wang Y, Li L, Jiang Z (2019) Effect of soybean Isoflavones on growth performance, immune function, and viral protein 5 mRNA expression in broiler chickens challenged with infectious bursal disease virus. Animals 9:247. https://doi.org/10.3390/ani9050247

Basiouni S, Tellez-Isaias G, Latorre DG, Graham DB, Petrone-Garcia MW, El-Sweedi H, Yalçın S, Wahab A, Visscher C, May-Simera LM, Huber C, Eisenreich W, Shehata AA (2022) Anti-inflammatory and Antioxidative phytogenic substances against secret killers in poultry: current status and prospects. Vet Res 10

Bhattaram VA, Graefe U, Kohlert C, Veit M, Derendorf H (2002) Pharmacokinetics and bioavailability of herbal medicinal products. Phytomedicine 9(Suppl 3):1–33. https://doi.org/10.1078/1433-187x-00210

Bober Z, Stępień A, Aebisher D, Ożog Ł, Bartusik-Aebisher D (2018) Medicinal benefits from the use of black pepper, curcuma and ginger. Eur J Clin Exp Med 16:133–145. https://doi.org/10.15584/ejcem.2018.2.9

Böhmer BM, Salisch H, Paulicks BR, Roth FX (2009) Echinacea purpurea as a potential immuno-stimulatory feed additive in laying hens and fattening pigs by intermittent application. Livest Sci 122:81–85. https://doi.org/10.1016/j.livsci.2008.07.013

Bonilla J, Sobral PJDA (2017) Antioxidant and antimicrobial properties of ethanolic extracts of guarana, boldo, rosemary and cinnamon. Braz J Food Technol 20. https://doi.org/10.1590/1981-6723.2416

Cai T-T, Ye X-L, Li R-R, Chen H, Wang Y-Y, Yong H-J, Pan M-L, Lu W, Tang Y, Miao H, Snijders AM, Mao J-H, Liu X-Y, Lu Y-B, Ding D-F (2020) Resveratrol modulates the gut microbiota and inflammation to protect against diabetic nephropathy in mice. Front Pharmacol 11:1249. https://doi.org/10.3389/fphar.2020.01249

Cao B, Wei X-C, Xu X-R, Zhang H-Z, Luo C-H, Feng B, Xu R-C, Zhao S-Y, Du X-J, Han L, Zhang D-K (2019) Seeing the unseen of the combination of two natural resins, frankincense and myrrh: changes in chemical constituents and pharmacological activities. Molecules 24:3076. https://doi.org/10.3390/molecules24173076

Carrasco-Pozo C, Morales P, Gotteland M (2013) Polyphenols protect the epithelial barrier function of Caco-2 cells exposed to indomethacin through the modulation of occludin and zonula occludens-1 expression. J Agric Food Chem 61:5291–5297. https://doi.org/10.1021/jf400150p

Cassidy A (2003) Potential risks and benefits of phytoestrogen-rich diets. Int J Vitam Nutr Res 73:120–126. https://doi.org/10.1024/0300-9831.73.2.120

Cheng J, Lin Y, Tang D, Yang H, Liu X (2022) Structural and gelation properties of five polyphenols-modified pork myofibrillar protein exposed to hydroxyl radicals. LWT 156:113073. https://doi.org/10.1016/j.lwt.2022.113073

Cho S-Y, Kim H-W, Lee M-K, Kim H-J, Kim J-B, Choe J-S, Lee Y-M, Jang H-H (2020) Antioxidant and anti-inflammatory activities in relation to the flavonoids composition of pepper (Capsicum annuum L.). Antioxidants (Basel) 9:E986. https://doi.org/10.3390/antiox9100986

Cuaz-Pérolin C, Billiet L, Baugé E, Copin C, Scott-Algara D, Genze F, Büchele B, Syrovets T, Simmet T, Rouis M (2008) Antiinflammatory and antiatherogenic effects of the NF-kappaB inhibitor acetyl-11-keto-beta-boswellic acid in LPS-challenged ApoE−/− mice. Arterioscler Thromb Vasc Biol 28:272–277. https://doi.org/10.1161/ATVBAHA.107.155606

Dall'Acqua S, Catanzaro D, Cocetta V, Igl N, Ragazzi E, Giron MC, Cecconello L, Montopoli M (2016) Protective effects of ψ taraxasterol 3-O-myristate and arnidiol 3-O-myristate isolated from Calendula officinalis on epithelial intestinal barrier. Fitoterapia 109:230–235. https://doi.org/10.1016/j.fitote.2016.01.007

Dehau T, Cherlet M, Croubels S, Van Immerseel F, Goossens E (2023) A high dose of dietary Berberine improves Gut Wall morphology, Despite an Expansion of Enterobacteriaceae and a Reduction in Beneficial Microbiota in Broiler Chickens. mSystems 8:e01239–e01222. https://doi.org/10.1128/msystems.01239-22

Deiana M, Serra G, Corona G (2018) Modulation of intestinal epithelium homeostasis by extra virgin olive oil phenolic compounds. Food Funct 9:4085–4099. https://doi.org/10.1039/c8fo00354h

El SN, Karakaya S (2009) Olive tree (Olea europaea) leaves: potential beneficial effects on human health. Nutr Rev 67:632–638. https://doi.org/10.1111/j.1753-4887.2009.00248.x

Emadi M, Kermanshahi H (2007) Effect of turmeric rhizome powder on immunity responses of broiler chickens. J Anim Vet Adv:833–836

Erlund I (2004) Review of the flavonoids quercetin, hesperetin, and naringenin. Dietary sources, bioactivities, bioavailability, and epidemiology. Nutr Res 24:851–874. https://doi.org/10.1016/j.nutres.2004.07.005

Farahat MH, Abdallah FM, Ali HA, Hernandez-Santana A (2017) Effect of dietary supplementation of grape seed extract on the growth performance, lipid profile, antioxidant status and immune response of broiler chickens. Animal 11:771–777. https://doi.org/10.1017/S1751731116002251

Farràs M, Martinez-Gili L, Portune K, Arranz S, Frost G, Tondo M, Blanco-Vaca F (2020) Modulation of the gut microbiota by olive oil phenolic compounds: implications for lipid metabolism, immune system, and obesity. Nutrients 12:2200. https://doi.org/10.3390/nu12082200

Feng J, Lu M, Wang J, Zhang H, Qiu K, Qi G, Wu S (2021) Dietary oregano essential oil supplementation improves intestinal functions and alters gut microbiota in late-phase laying hens. J Animal Sci Biotechnol 12:72. https://doi.org/10.1186/s40104-021-00600-3

Fennell CW, Lindsey KL, McGaw LJ, Sparg SG, Stafford GI, Elgorashi EE, Grace OM, van Staden J (2004) Assessing African medicinal plants for efficacy and safety: pharmacological screening and toxicology. J Ethnopharmacol 94:205–217. https://doi.org/10.1016/j.jep.2004.05.012

Foroutankhah M, Toghyani M, Landy N (2019) Evaluation of Calendula officinalis L. (marigold) flower as a natural growth promoter in comparison with an antibiotic growth promoter on growth performance, carcass traits and humoral immune responses of broilers. Anim Nutr 5:314–318. https://doi.org/10.1016/j.aninu.2019.04.002

Fuhrman B, Rosenblat M, Hayek T, Coleman R, Aviram M (2000) Ginger extract consumption reduces plasma cholesterol, inhibits LDL oxidation and attenuates development of atherosclerosis in atherosclerotic, apolipoprotein E-deficient mice. J Nutr 130:1124–1131. https://doi.org/10.1093/jn/130.5.1124

Gao X, Xiao Z-H, Liu M, Zhang N-Y, Khalil MM, Gu C-Q, Qi D-S, Sun L-H (2018) Dietary Silymarin supplementation alleviates Zearalenone-induced hepatotoxicity and reproductive toxicity in rats. J Nutr 148:1209–1216. https://doi.org/10.1093/jn/nxy114

Gebrehiwot K, Muys B, Haile M, Mitloehner R (2003) Introducing Boswellia papyrifera (Del.) Hochst and its non-timber forest product, frankincense. Int For Rev 5:348–353

Ghazalah AA, El-Hakim ASA, Refaie AM (2007) Response of broiler chicks to some dietary growth promoters throughout different growth period. Egypt Poult Sci J:53–57

Gong J, Hu M, Huang Z, Fang K, Wang D, Chen Q, Li J, Yang D, Zou X, Xu L, Wang K, Dong H, Lu F (2017) Berberine attenuates intestinal mucosal barrier dysfunction in type 2 diabetic rats. Front Pharmacol 8. https://doi.org/10.3389/fphar.2017.00042

Gu L, Li N, Gong J, Li Q, Zhu W, Li J (2011) Berberine ameliorates intestinal epithelial tight-junction damage and down-regulates myosin light chain kinase pathways in a mouse model of endotoxinemia. J Infect Dis 203:1602–1612. https://doi.org/10.1093/infdis/jir147

Gu L, Li N, Li Q, Zhang Q, Wang C, Zhu W, Li J (2009) The effect of berberine in vitro on tight junctions in human Caco-2 intestinal epithelial cells. Fitoterapia 80:241–248. https://doi.org/10.1016/j.fitote.2009.02.005

Habtemariam S (2020) Berberine pharmacology and the gut microbiota: a hidden therapeutic link. Pharmacol Res 155:104722. https://doi.org/10.1016/j.phrs.2020.104722

Hafeez A, Sohail M, Ahmad A, Shah M, Din S, Khan I, Shuiab M, Nasrullah, Shahzada W, Iqbal M, Khan RU (2020) Selected herbal plants showing enhanced growth performance, ileal digestibility, bone strength and blood metabolites in broilers. J Appl Anim Res 48:448–453. https://doi.org/10.1080/09712119.2020.1818569

Hassan FAM, Awad A (2017) Impact of thyme powder (Thymus vulgaris L.) supplementation on gene expression profiles of cytokines and economic efficiency of broiler diets. Environ Sci Pollut Res Int 24:15816–15826. https://doi.org/10.1007/s11356-017-9251-7

Herrero-Encinas J, Blanch M, Pastor JJ, Mereu A, Ipharraguerre IR, Menoyo D (2020) Effects of a bioactive olive pomace extract from Olea europaea on growth performance, gut function, and intestinal microbiota in broiler chickens. Poult Sci 99:2–10. https://doi.org/10.3382/ps/pez467

Hertog MGL, Hollman PCH, Katan MB (1992) Content of potentially anticarcinogenic flavonoids of 28 vegetables and 9 fruits commonly consumed in The Netherlands. J Agric Food Chem 40:2379–2383. https://doi.org/10.1021/jf00024a011

Hu R, He Y, Arowolo M, Wu S, He J (2019) Polyphenols as potential attenuators of heat stress in poultry production. Antioxidants 8:67. https://doi.org/10.3390/antiox8030067

Ilyas Z, Perna S, Al-thawadi S, Alalwan TA, Riva A, Petrangolini G, Gasparri C, Infantino V, Peroni G, Rondanelli M (2020) The effect of Berberine on weight loss in order to prevent obesity: a systematic review. Biomed Pharmacother 127:110137. https://doi.org/10.1016/j.biopha.2020.110137

Iqbal Y, Cottrell JJ, Suleria HAR, Dunshea FR (2020) Gut microbiota-polyphenol interactions in chicken: a review. Animals 10:1391. https://doi.org/10.3390/ani10081391

Ishikado A, Sono Y, Matsumoto M, Robida-Stubbs S, Okuno A, Goto M, King GL, Keith Blackwell T, Makino T (2013) Willow bark extract increases antioxidant enzymes and reduces oxidative stress through activation of Nrf2 in vascular endothelial cells and Caenorhabditis elegans. Free Radic Biol Med 65:1506–1515. https://doi.org/10.1016/j.freeradbiomed.2012.12.006

Izzo AA, Sharkey KA (2010) Cannabinoids and the gut: new developments and emerging concepts. Pharmacol Ther 126:21–38. https://doi.org/10.1016/j.pharmthera.2009.12.005

Jiang ZY, Jiang SQ, Lin YC, Xi PB, Yu DQ, Wu TX (2007) Effects of soybean isoflavone on growth performance, meat quality, and antioxidation in male broilers. Poult Sci 86:1356–1362. https://doi.org/10.1093/ps/86.7.1356

Karangiya VK, Savsani HH, Patil SS, Garg DD, Murthy KS, Ribadiya NK, Vekariya SJ (2016) Effect of dietary supplementation of garlic, ginger and their combination on feed intake, growth performance and economics in commercial broilers. Vet World 9:245–250. https://doi.org/10.14202/vetworld.2016.245-250

Karthikeyan J, Rani P (2003) Enzymatic and non-enzymatic antioxidants in selected piper species. Indian J Exp Biol 41:135–140

Ke X, Huang Y, Li L, Xin F, Xu L, Zhang Y, Zeng Z, Lin F, Song Y (2020) Berberine attenuates arterial plaque formation in atherosclerotic rats with damp-heat syndrome via regulating autophagy. DDDT 14:2449–2460. https://doi.org/10.2147/DDDT.S250524

Kelber O, Bonaterra GA, Kinscherf R, Weiser D, Metz J (2006) Inhibitorische Effekte von Weidenrindenextrakten auf proinflammatorische Prozesse in LPS-aktivierten Humanmonozyten. Z Rheumatol S31

Khalaf NA, Shakya A, Al-Othman A, Elagbar Z, Farah HS (2008) Antioxidant activity of some common plants. Turk J Biol:51–55

Khalafalla RE, Müller U, Shahiduzzaman M, Dyachenko V, Desouky AY, Alber G, Daugschies A (2011) Effects of curcumin (diferuloylmethane) on Eimeria tenella sporozoites in vitro. Parasitol Res 108:879–886. https://doi.org/10.1007/s00436-010-2129-y

Kiczorowska B, Samolińska W, Al-Yasiry ARM, Kiczorowski P, Winiarska-Mieczan A (2017) The natural feed additives as immunostimulants in monogastric animal nutrition—a review. Ann Anim Sci 17:605–625. https://doi.org/10.1515/aoas-2016-0076

Kikusato M (2021) Phytobiotics to improve health and production of broiler chickens: functions beyond the antioxidant activity. Anim Biosci 34:345–353. https://doi.org/10.5713/ab.20.0842

Kim DK, Lillehoj HS, Lee SH, Jang SI, Lillehoj EP, Bravo D (2013) Dietary Curcuma longa enhances resistance against Eimeria maxima and Eimeria tenella infections in chickens. Poult Sci 92:2635–2643. https://doi.org/10.3382/ps.2013-03095

Klein TW, Cabral GA (2006) Cannabinoid-induced immune suppression and modulation of antigen-presenting cells. J Neuroimmune Pharmacol 1:50–64. https://doi.org/10.1007/s11481-005-9007-x

Konieczka P, Szkopek D, Kinsner M, Fotschki B, Juśkiewicz J, Banach J (2020) Cannabis-derived cannabidiol and nanoselenium improve gut barrier function and affect bacterial enzyme activity in chickens subjected to C. Perfringens challenge. Vet Res 51:141. https://doi.org/10.1186/s13567-020-00863-0

Krieglstein CF, Anthoni C, Rijcken EJ, Laukötter M, Spiegel HU, Boden SE, Schweizer S, Safayhi H, Senninger N, Schürmann G (2001) Acetyl-11-keto-beta-boswellic acid, a constituent of a herbal medicine from Boswellia serrata resin, attenuates experimental ileitis. Int J Color Dis 16:88–95. https://doi.org/10.1007/s003840100292

Kumar NV, Murthy PS, Manjunatha JR, Bettadaiah BK (2014) Synthesis and quorum sensing inhibitory activity of key phenolic compounds of ginger and their derivatives. Food Chem 159:451–457. https://doi.org/10.1016/j.foodchem.2014.03.039

Landy N, Ghalamkari G, Toghyani M, Moattar F (2011) The effects of Echinacea purpurea L. (purple coneflower) as an antibiotic growth promoter substitution on performance, carcass characteristics and humoral immune response in broiler chickens. J Med Plants Res 5:2332–2338

Le HH, Shakeri M, Suleria HAR, Zhao W, McQuade RM, Phillips DJ, Vidacs E, Furness JB, Dunshea FR, Artuso-Ponte V, Cottrell JJ (2020) Betaine and Isoquinoline alkaloids protect against heat stress and colonic permeability in growing pigs. Antioxidants 9:1024. https://doi.org/10.3390/antiox9101024

Lee KW, Everts H, Kappert HJ, Frehner M, Losa R, Beynen AC (2003) Effects of dietary essential oil components on growth performance, digestive enzymes and lipid metabolism in female broiler chickens. Br Poult Sci 44:450–457. https://doi.org/10.1080/0007166031000085508

Lee J-W, Kim D-H, Kim Y-B, Jeong S-B, Oh S-T, Cho S-Y, Lee K-W (2020) Dietary encapsulated essential oils improve production performance of coccidiosis-vaccine-challenged broiler chickens. Animals (Basel) 10:481. https://doi.org/10.3390/ani10030481

Lee SH, Lillehoj HS, Hong YH, Jang SI, Lillehoj EP, Ionescu C, Mazuranok L, Bravo D (2010) In vitro effects of plant and mushroom extracts on immunological function of chicken lymphocytes and macrophages. Br Poult Sci 51:213–221. https://doi.org/10.1080/00071661003745844

Lee SH, Lillehoj HS, Jang SI, Lee KW, Park MS, Bravo D, Lillehoj EP (2011) Cinnamaldehyde enhances in vitro parameters of immunity and reduces in vivo infection against avian coccidiosis. Br J Nutr 106:862–869. https://doi.org/10.1017/S0007114511001073

Lee SH, Lillehoj HS, Jang SI, Lillehoj EP, Min W, Bravo DM (2013a) Dietary supplementation of young broiler chickens with capsicum and turmeric oleoresins increases resistance to necrotic enteritis. Br J Nutr 110:840–847. https://doi.org/10.1017/S0007114512006083

Lee W, Yoo H, Kim JA, Lee S, Jee J-G, Lee MY, Lee Y-M, Bae J-S (2013b) Barrier protective effects of piperlonguminine in LPS-induced inflammation in vitro and in vivo. Food Chem Toxicol 58:149–157. https://doi.org/10.1016/j.fct.2013.04.027

Leyva-Diaz AA, Hernandez-Patlan D, Solis-Cruz B, Adhikari B, Kwon YM, Latorre JD, Hernandez-Velasco X, Fuente-Martinez B, Hargis BM, Lopez-Arellano R, Tellez-Isaias G (2021) Evaluation of curcumin and copper acetate against salmonella typhimurium infection, intestinal permeability, and cecal microbiota composition in broiler chickens. J Animal Sci Biotechnol 12:23. https://doi.org/10.1186/s40104-021-00545-7

Li N, Gu L, Qu L, Gong J, Li Q, Zhu W, Li J (2010) Berberine attenuates pro-inflammatory cytokine-induced tight junction disruption in an in vitro model of intestinal epithelial cells. Eur J Pharm Sci 40:1–8. https://doi.org/10.1016/j.ejps.2010.02.001

Liang D, Li F, Fu Y, Cao Y, Song X, Wang T, Wang W, Guo M, Zhou E, Li D, Yang Z, Zhang N (2014) Thymol inhibits LPS-stimulated inflammatory response via down-regulation of NF-κB and MAPK signaling pathways in mouse mammary epithelial cells. Inflammation 37:214–222. https://doi.org/10.1007/s10753-013-9732-x

Liu L, Fu C, Yan M, Xie H, Li S, Yu Q, He S, He J (2016b) Resveratrol modulates intestinal morphology and HSP70/90, NF-κB and EGF expression in the jejunal mucosa of black-boned chickens on exposure to circular heat stress. Food Funct 7:1329–1338. https://doi.org/10.1039/c5fo01338k

Liu LL, He JH, Xie HB, Yang YS, Li JC, Zou Y (2014) Resveratrol induces antioxidant and heat shock protein mRNA expression in response to heat stress in black-boned chickens. Poult Sci 93:54–62. https://doi.org/10.3382/ps.2013-03423

Liu SJ, Wang J, He TF, Liu HS, Piao XS (2021) Effects of natural capsicum extract on growth performance, nutrient utilization, antioxidant status, immune function, and meat quality in broilers. Poult Sci 100:101301. https://doi.org/10.1016/j.psj.2021.101301

Liu W, Zhai Y, Heng X, Che FY, Chen W, Sun D, Zhai G (2016a) Oral bioavailability of curcumin: problems and advancements. J Drug Target 24:694–702. https://doi.org/10.3109/1061186X.2016.1157883

Lyu Y, Lin L, Xie Y, Li D, Xiao M, Zhang Y, Cheung SCK, Shaw PC, Yang X, Chan PKS, Kong APS, Zuo Z (2021) Blood-glucose-lowering effect of Coptidis Rhizoma extracts from different origins via gut microbiota modulation in db/db mice. Front Pharmacol 12:684358. https://doi.org/10.3389/fphar.2021.684358

Maass N, Bauer J, Paulicks BR, Böhmer BM, Roth-Maier DA (2005) Efficiency of Echinacea purpurea on performance and immune status in pigs. J Anim Physiol Anim Nutr (Berl) 89:244–252. https://doi.org/10.1111/j.1439-0396.2005.00501.x

Madeo F, Carmona-Gutierrez D, Hofer SJ, Kroemer G (2019) Caloric restriction mimetics against age-associated disease: targets, mechanisms, and therapeutic potential. Cell Metab 29:592–610. https://doi.org/10.1016/j.cmet.2019.01.018

Mahfuz S, Shang Q, Piao X (2021) Phenolic compounds as natural feed additives in poultry and swine diets: a review. J Animal Sci Biotechnol 12:48. https://doi.org/10.1186/s40104-021-00565-3

Mahima, Rahal A, Deb R, Latheef SK, Abdul Samad H, Tiwari R, Verma AK, Kumar A, Dhama K (2012) Immunomodulatory and therapeutic potentials of herbal, traditional/indigenous and ethnoveterinary medicines. Pak J Biol Sci 15:754–774. https://doi.org/10.3923/pjbs.2012.754.774

Malik TA, Kamili AN, Chishti MZ, Tanveer S, Ahad S, Johri RK (2016) Synergistic approach for treatment of chicken coccidiosis using berberine–A plant natural product. Microb Pathog 93:56–62. https://doi.org/10.1016/j.micpath.2016.01.012

Malini T, Arunakaran J, Aruldhas MM, Govindarajulu P (1999) Effects of piperine on the lipid composition and enzymes of the pyruvate-malate cycle in the testis of the rat in vivo. Biochem Mol Biol Int 47:537–545. https://doi.org/10.1080/15216549900201573

Manna SK, Mukhopadhyay A, Aggarwal BB (2000) Resveratrol suppresses TNF-induced activation of nuclear transcription factors NF-kappa B, activator protein-1, and apoptosis: potential role of reactive oxygen intermediates and lipid peroxidation. J Immunol 164:6509–6519. https://doi.org/10.4049/jimmunol.164.12.6509

Mansoub NH (2011) Comparison of using different level of black pepper with probiotic on performance and serum composition of broiler chickens. J Basic Appl Sci Res:2425–2428

Meimandipour A, Nouri Emamzadeh A, Soleimani A (2017) Effects of nanoencapsulated aloe vera, dill and nettle root extract as feed antibiotic substitutes in broiler chickens. Arch Anim Breed 60:1–7. https://doi.org/10.5194/aab-60-1-2017

Mendivil EJ, Sandoval-Rodriguez A, Meza-Ríos A, Zuñiga-Ramos L, Dominguez-Rosales A, Vazquez-Del Mercado M, Sanchez-Orozco L, Santos-Garcia A, Armendariz-Borunda J (2019) Capsaicin induces a protective effect on gastric mucosa along with decreased expression of inflammatory molecules in a gastritis model. J Funct Foods 59:345–351. https://doi.org/10.1016/j.jff.2019.06.002

Messina M, Ho S, Alekel DL (2004) Skeletal benefits of soy isoflavones: a review of the clinical trial and epidemiologic data. Curr Opin Clin Nutr Metab Care 7:649–658. https://doi.org/10.1097/00075197-200411000-00010

Moorthy M, Ravi S, Ravikumar M, Viswanathan K, Edwin SC (2009) Ginger, pepper and curry leaf powder as feed additives in broiler diet. Int J of Poult Sci 8:779–782

Muley B, Khadabadi S, Banarase N (2009) Phytochemical constituents and pharmacological activities of Calendula officinalis Linn (Asteraceae): a review. Trop J Pharm Res 8. https://doi.org/10.4314/tjpr.v8i5.48090

Nagarkatti P, Pandey R, Rieder SA, Hegde VL, Nagarkatti M (2009) Cannabinoids as novel anti-inflammatory drugs. Future Med Chem 1:1333–1349. https://doi.org/10.4155/fmc.09.93

Najafi S, Taherpour K (2014) Effects of dietary ginger (Zingiber Ofjicinale), cinnamon (Cinnamomum), Synbiotic and antibiotic supplementation on performance of broilers. J Anim Sci Adv:658–667

Ni YD, Wu J, Tong HY, Huang YB, Lu LZ, Grossmann R, Zhao RQ (2012) Effect of dietary daidzein supplementation on egg laying rate was associated with the change of hepatic VTG-II mRNA expression and higher antioxidant activities during the post-peak egg laying period of broiler breeders. Anim Feed Sci Technol 177:116–123. https://doi.org/10.1016/j.anifeedsci.2012.08.001

Nile SH, Park SW (2015) Chromatographic analysis, antioxidant, anti-inflammatory, and xanthine oxidase inhibitory activities of ginger extracts and its reference compounds. Ind Crop Prod 70:238–244. https://doi.org/10.1016/j.indcrop.2015.03.033

O'Bryan CA, Pendleton SJ, Crandall PG, Ricke SC (2015) Potential of plant essential oils and their components in animal agriculture - in vitro studies on antibacterial mode of action. Front Vet Sci 2:35. https://doi.org/10.3389/fvets.2015.00035

Ocaña A, Reglero G (2012) Effects of thyme extract oils (from Thymus vulgaris, thymus zygis, and Thymus hyemalis) on cytokine production and gene expression of oxLDL-stimulated THP-1-macrophages. J Obes 2012:104706. https://doi.org/10.1155/2012/104706

Olawuwo OS, Famuyide IM, McGaw LJ (2022) Antibacterial and Antibiofilm activity of selected medicinal plant leaf extracts against pathogens implicated in poultry diseases. Front Vet Sci 9:820304. https://doi.org/10.3389/fvets.2022.820304

Pan M-H, Wu J-C, Ho C-T, Lai C-S (2018) Antiobesity molecular mechanisms of action: resveratrol and pterostilbene. Biofactors 44:50–60. https://doi.org/10.1002/biof.1409

Panaite TD, Saracila M, Papuc CP, Predescu CN, Soica C (2020) Influence of dietary supplementation of Salix alba bark on performance, oxidative stress parameters in liver and gut microflora of broilers. Animals 10:958. https://doi.org/10.3390/ani10060958

Pannee C, Wacharee L, Chandhanee I (2014) Antiinflammatory effects of essential oil from the leaves of Cinnamomum cassia and cinnamaldehyde on lipopolysaccharide-stimulated J774A.1 cells. J Adv Pharm Technol Res 5:164. https://doi.org/10.4103/2231-4040.143034

Paszkiewicz M, Budzyńska A, Różalska B, Sadowska B (2012) The immunomodulatory role of plant polyphenols. Postepy Hig Med Dosw (Online) 66:637–646. https://doi.org/10.5604/17322693.1009908

Pérez-González A, Prejanò M, Russo N, Marino T, Galano A (2020) Capsaicin, a powerful •OH-inactivating ligand. Antioxidants (Basel) 9:1247. https://doi.org/10.3390/antiox9121247

Petrone-Garcia VM, Lopez-Arellano R, Patiño GR, Rodríguez MAC, Hernandez-Patlan D, Solis-Cruz B, Hernandez-Velasco X, Alba-Hurtado F, Vuong CN, Castellanos-Huerta I, Tellez-Isaias G (2021) Curcumin reduces enteric isoprostane 8-iso-PGF2α and prostaglandin GF2α in specific pathogen-free Leghorn chickens challenged with Eimeria maxima. Sci Rep 11:11609. https://doi.org/10.1038/s41598-021-90679-5

Pirgozliev V, Mansbridge SC, Rose SP, Lillehoj HS, Bravo D (2019) Immune modulation, growth performance, and nutrient retention in broiler chickens fed a blend of phytogenic feed additives. Poult Sci 98:3443–3449. https://doi.org/10.3382/ps/pey472

Rahimi S, Teymori Zadeh Z, Karimi Torshizi MA, Omidbaigi R, Rokni H (2011) Effect of the three herbal extracts on growth performance, immune system, blood factors and intestinal selected bacterial population in broiler chickens. JAST 13:527–539

Rajput N, Naeem M, Ali S, Rui Y, Tian W (2012) Effect of dietary supplementation of Marigold pigment on immunity, skin and meat color, and growth performance of broiler chickens. Braz J Poult Sci 14:233–304

Rao PV, Gan SH (2014) Cinnamon: a multifaceted medicinal plant. Evid Based Complement Alternat Med 2014:1–12. https://doi.org/10.1155/2014/642942

Reen RK, Roesch SF, Kiefer F, Wiebel FJ, Singh J (1996) Piperine impairs cytochrome P4501A1 activity by direct interaction with the enzyme and not by down regulation of CYP1A1 gene expression in the rat hepatoma 5L cell line. Biochem Biophys Res Commun 218:562–569. https://doi.org/10.1006/bbrc.1996.0100

Rossi B, Toschi A, Piva A, Grilli E (2020) Single components of botanicals and nature-identical compounds as a non-antibiotic strategy to ameliorate health status and improve performance in poultry and pigs. Nutr Res Rev 33:218–234. https://doi.org/10.1017/S0954422420000013

Sadeghi G, Karimi A, Padidar Jahromi S, Azizi T, Daneshmand A (2012) Effects of cinnamon, thyme and turmeric infusions on the performance and immune response in of 1- to 21-day-old male broilers. Rev Bras Cienc Avic 14:15–20. https://doi.org/10.1590/S1516-635X2012000100003

Safayhi H, Rall B, Sailer ER, Ammon HP (1997) Inhibition by boswellic acids of human leukocyte elastase. J Pharmacol Exp Ther 281:460–463

Sahin N, Onderci M, Balci TA, Cikim G, Sahin K, Kucuk O (2007) The effect of soy isoflavones on egg quality and bone mineralisation during the late laying period of quail. Br Poult Sci 48:363–369. https://doi.org/10.1080/00071660701341971

Saleh AA, Ebeid TA, Abudabos AM (2018) Effect of dietary phytogenics (herbal mixture) supplementation on growth performance, nutrient utilization, antioxidative properties, and immune response in broilers. Environ Sci Pollut Res Int 25:14606–14613. https://doi.org/10.1007/s11356-018-1685-z

Saracila M, Panaite TD, Papuc CP, Criste RD (2021) Heat stress in broiler chickens and the effect of dietary polyphenols, with special reference to willow (Salix spp.) bark supplements—a review. Antioxidants 10:686. https://doi.org/10.3390/antiox10050686

Saracila M, Tabuc C, Panaite TD, Papuc CP, Olteanu M, Criste RD (2018) Effect of the dietary willow bark extract (Salix Alba) on the Caecal microbial population of broilers (14-28 days) reared at 32°C. "Agriculture for Life, Life for Agriculture" Conference Proceedings 1:155–161. https://doi.org/10.2478/alife-2018-0023

Sarıca Ş, Ürkmez D (2016) The use of grape seed-, olive leaf- and pomegranate peel-extracts as alternative natural antimicrobial feed additives in broiler diets. Verlag Eugen Ulmer 80. https://doi.org/10.1399/eps.2016.121

Serra V, Salvatori G, Pastorelli G (2021) Dietary polyphenol supplementation in food producing animals: effects on the quality of derived products. Animals (Basel) 11:401. https://doi.org/10.3390/ani11020401

Shara M, Stohs SJ (2015) Efficacy and safety of white willow bark (Salix alba) extracts. Phytother Res 29:1112–1116. https://doi.org/10.1002/ptr.5377

Shehata AA, Attia Y, Khafaga AF, Farooq MZ, El-Seedi HR, Eisenreich W, Tellez-Isaias G (2022b) Restoring healthy gut microbiome in poultry using alternative feed additives with particular attention to phytogenic substances: challenges and prospects. Ger J Vet Res 2:32–42. https://doi.org/10.51585/gjvr.2022.3.0047

Shehata AA, Yalçın S, Latorre JD, Basiouni S, Attia YA, Abd El-Wahab A, Visscher C, El-Seedi HR, Huber C, Hafez HM, Eisenreich W, Tellez-Isaias G (2022a) Probiotics, prebiotics, and phytogenic substances for optimizing gut health in poultry. Microorganisms 10:395. https://doi.org/10.3390/microorganisms10020395

Shi SR, Gu H, Chang LL, Wang ZY, Tong HB, Zou JM (2013) Safety evaluation of daidzein in laying hens: part I. Effects on laying performance, clinical blood parameters, and organs development. Food Chem Toxicol 55:684–688. https://doi.org/10.1016/j.fct.2013.01.009

Siemoneit U, Hofmann B, Kather N, Lamkemeyer T, Madlung J, Franke L, Schneider G, Jauch J, Poeckel D, Werz O (2008) Identification and functional analysis of cyclooxygenase-1 as a molecular target of boswellic acids. Biochem Pharmacol 75:503–513. https://doi.org/10.1016/j.bcp.2007.09.010

Siemoneit U, Koeberle A, Rossi A, Dehm F, Verhoff M, Reckel S, Maier TJ, Jauch J, Northoff H, Bernhard F, Doetsch V, Sautebin L, Werz O (2011) Inhibition of microsomal prostaglandin E2 synthase-1 as a molecular basis for the anti-inflammatory actions of boswellic acids from frankincense. Br J Pharmacol 162:147–162. https://doi.org/10.1111/j.1476-5381.2010.01020.x

Silvestro S, Bramanti P, Mazzon E (2021) Role of quercetin in depressive-like behaviors: findings from animal models. Appl Sci 11:7116. https://doi.org/10.3390/app11157116

Solis-Cruz B, Hernandez-Patlan D, Petrone VM, Pontin KP, Latorre JD, Beyssac E, Hernandez-Velasco X, Merino-Guzman R, Arreguin MA, Hargis BM, Lopez-Arellano R, Tellez-Isaias G (2019) Evaluation of a bacillus-based direct-fed microbial on aflatoxin B1 toxic effects, performance, immunologic status, and serum biochemical parameters in broiler chickens. Avian Dis 63:659–669. https://doi.org/10.1637/aviandiseases-D-19-00100

Song ZH, Cheng K, Zheng XC, Ahmad H, Zhang LL, Wang T (2018) Effects of dietary supplementation with enzymatically treated Artemisia annua on growth performance, intestinal morphology, digestive enzyme activities, immunity, and antioxidant capacity of heat-stressed broilers. Poult Sci 97:430–437. https://doi.org/10.3382/ps/pex312

Stevanović ZD, Bošnjak-Neumüller J, Pajić-Lijaković I, Raj J, Vasiljević M (2018) Essential oils as feed additives-future perspectives. Molecules 23:E1717. https://doi.org/10.3390/molecules23071717

Stoner GD (2013) Ginger: is it ready for prime time? Cancer Prev Res (Phila) 6:257–262. https://doi.org/10.1158/1940-6207.CAPR-13-0055

Su S, Hua Y, Wang Y, Gu W, Zhou W, Duan J, Jiang H, Chen T, Tang Y (2012) Evaluation of the anti-inflammatory and analgesic properties of individual and combined extracts from Commiphora myrrha, and Boswellia carterii. J Ethnopharmacol 139:649–656. https://doi.org/10.1016/j.jep.2011.12.013

Suganya V, Anuradha V (2017) Microencapsulation and Nanoencapsulation: a review. IJPCR 9. https://doi.org/10.25258/ijpcr.v9i3.8324

Sugiharto S (2021) Herbal supplements for sustainable broiler production during post antibiotic era in Indonesia–an overview. Livest Res Rural Dev 8

Sugiharto S, Ayasan T (2022) Encapsulation as a way to improve the phytogenic effects of herbal additives in broilers—an overview. Ann Anim Sci 0:53. https://doi.org/10.2478/aoas-2022-0045

Suzuki T, Hara H (2009) Quercetin enhances intestinal barrier function through the assembly of zonula [corrected] occludens-2, occludin, and claudin-1 and the expression of claudin-4 in Caco-2 cells. J Nutr 139:965–974. https://doi.org/10.3945/jn.108.100867

Takahara M, Takaki A, Hiraoka S, Adachi T, Shimomura Y, Matsushita H, Nguyen TTT, Koike K, Ikeda A, Takashima S, Yamasaki Y, Inokuchi T, Kinugasa H, Sugihara Y, Harada K, Eikawa S, Morita H, Udono H, Okada H (2019) Berberine improved experimental chronic colitis by regulating interferon-γ- and IL-17A-producing lamina propria CD4+ T cells through AMPK activation. Sci Rep 9:11934. https://doi.org/10.1038/s41598-019-48331-w

Takano T, Satoh K, Doki T (2021) Possible antiviral activity of 5-Aminolevulinic acid in feline infectious peritonitis virus (feline coronavirus) infection. Front Vet Sci 8:647189. https://doi.org/10.3389/fvets.2021.647189

Talhaoui N, Vezza T, Gómez-Caravaca AM, Fernández-Gutiérrez A, Gálvez J, Segura-Carretero A (2016) Phenolic compounds and in vitro immunomodulatory properties of three Andalusian olive leaf extracts. J Funct Foods 22:270–277. https://doi.org/10.1016/j.jff.2016.01.037

Tausch L, Henkel A, Siemoneit U, Poeckel D, Kather N, Franke L, Hofmann B, Schneider G, Angioni C, Geisslinger G, Skarke C, Holtmeier W, Beckhaus T, Karas M, Jauch J, Werz O (2009) Identification of human cathepsin G as a functional target of boswellic acids from the anti-inflammatory remedy frankincense. J Immunol 183:3433–3442. https://doi.org/10.4049/jimmunol.0803574

Tawfeek N, Mahmoud MF, Hamdan DI, Sobeh M, Farrag N, Wink M, El-Shazly AM (2021) Phytochemistry, pharmacology and medicinal uses of plants of the genus Salix: An updated review. Front Pharmacol 12:593856. https://doi.org/10.3389/fphar.2021.593856

Timilsena YP, Haque MA, Adhikari B (2020) Encapsulation in the food industry: a brief historical overview to recent developments. FNS 11:481–508. https://doi.org/10.4236/fns.2020.116035

Tollba AAH, Azouz HMM, El-Samad AMH (2007) Antioxidants supplementation to diet of Egyptian chicken under different environmental condition: 2-the growth during cold winter stress. Egyptian Poult Sci J:727–748

Tolve R, Tchuenbou-Magaia F, Di Cairano M, Caruso MC, Scarpa T, Galgano F (2021) Encapsulation of bioactive compounds for the formulation of functional animal feeds: the bio-fortification of derivate foods. Anim Feed Sci Technol 279:115036. https://doi.org/10.1016/j.anifeedsci.2021.115036

Trucillo P, Campardelli R, Aliakbarian B, Perego P, Reverchon E (2018) Supercritical assisted process for the encapsulation of olive pomace extract into liposomes. J Supercrit Fluids 135:152–159. https://doi.org/10.1016/j.supflu.2018.01.018

Ukiya M, Akihisa T, Yasukawa K, Tokuda H, Suzuki T, Kimura Y (2006) Anti-inflammatory, anti-tumor-promoting, and cytotoxic activities of constituents of marigold (Calendula officinalis) flowers. J Nat Prod 69:1692–1696. https://doi.org/10.1021/np068016b

Upadhyaya I, Upadhyay A, Kollanoor-Johny A, Mooyottu S, Baskaran SA, Yin H-B, Schreiber DT, Khan MI, Darre MJ, Curtis PA, Venkitanarayanan K (2015) In-feed supplementation of trans-cinnamaldehyde reduces layer-chicken egg-borne transmission of salmonella enterica serovar enteritidis. Appl Environ Microbiol 81:2985–2994. https://doi.org/10.1128/AEM.03809-14

Vande Maele L, Heyndrickx M, Maes D, De Pauw N, Mahu M, Verlinden M, Haesebrouck F, Martel A, Pasmans F, Boyen F (2016) In vitro susceptibility of Brachyspira hyodysenteriae to organic acids and essential oil components. J Vet Med Sci 78:325–328. https://doi.org/10.1292/jvms.15-0341

Vezza T, Algieri F, Rodríguez-Nogales A, Garrido-Mesa J, Utrilla MP, Talhaoui N, Gómez-Caravaca AM, Segura-Carretero A, Rodríguez-Cabezas ME, Monteleone G, Gálvez J (2017)

Immunomodulatory properties of Olea europaea leaf extract in intestinal inflammation. Mol Nutr Food Res 61:1601066. https://doi.org/10.1002/mnfr.201601066

Vispute MM, Sharma D, Mandal AB, Rokade JJ, Tyagi PK, Yadav AS (2019) Effect of dietary supplementation of hemp (Cannabis sativa) and dill seed (Anethum graveolens) on performance, serum biochemicals and gut health of broiler chickens. J Anim Physiol Anim Nutr (Berl) 103:525–533. https://doi.org/10.1111/jpn.13052

Vlaicu PA, Panaite TD, Untea AE, Idriceanu L, Cornescu GM (2021) Herbal plants as feed additives in broiler chicken diets. Archiva Zootechnica 24:76–95. https://doi.org/10.2478/azibna-2021-0015

Wang J, Deng L, Chen M, Che Y, Li L, Zhu L, Chen G, Feng T (2024) Phytogenic feed additives as natural antibiotic alternatives in animal health and production: a review of the literature of the last decade. Anim Nutr S2405654524000301:244. https://doi.org/10.1016/j.aninu.2024.01.012

Wang C, Zhao F, Li Z, Jin X, Chen X, Geng Z, Hu H, Zhang C (2021) Effects of resveratrol on growth performance, intestinal development, and antioxidant status of broilers under heat stress. Animals (Basel) 11:1427. https://doi.org/10.3390/ani11051427

Weeks A, Daly DC, Simpson BB (2005) The phylogenetic history and biogeography of the frankincense and myrrh family (Burseraceae) based on nuclear and chloroplast sequence data. Mol Phylogenet Evol 35:85–101. https://doi.org/10.1016/j.ympev.2004.12.021

Xiang YD, He Z, Pouton C, Hoerr FJ, Xiao Z (2017) Target animal safety and residual study for Berberine and other phytogenic compounds in broiler chickens. Arch Clin Microbiol 08. https://doi.org/10.4172/1989-8436.100069

Xie Z, Shen G, Wang Y, Wu C (2019) Curcumin supplementation regulates lipid metabolism in broiler chickens. Poult Sci 98:422–429. https://doi.org/10.3382/ps/pey315

Xu X, Yi H, Wu J, Kuang T, Zhang J, Li Q, Du H, Xu T, Jiang G, Fan G (2021) Therapeutic effect of berberine on metabolic diseases: both pharmacological data and clinical evidence. Biomed Pharmacother 133:110984. https://doi.org/10.1016/j.biopha.2020.110984

Yadav S, Teng P-Y, Souza dos Santos T, Gould RL, Craig SW, Lorraine Fuller A, Pazdro R, Kim WK (2020) The effects of different doses of curcumin compound on growth performance, antioxidant status, and gut health of broiler chickens challenged with Eimeria species. Poult Sci 99:5936–5945. https://doi.org/10.1016/j.psj.2020.08.046

Yalçin S, Eser H, Onbaşilar İ, Yalçin S (2020) Effects of dried thyme (Thymus vulgaris L.) leaves on performance, some egg quality traits and immunity in laying hens. Ankara Üniversitesi Veteriner Fakültesi Dergisi. https://doi.org/10.33988/auvfd.677150

Yang L, Liu G, Liang X, Wang M, Zhu X, Luo Y, Shang Y, Yang J, Zhou P, Gu X (2019) Effects of berberine on the growth performance, antioxidative capacity and immune response to lipopolysaccharide challenge in broilers. Anim Sci J 90:1229–1238. https://doi.org/10.1111/asj.13255

Yoshikawa M, Murakami T, Kishi A, Kageura T, Matsuda H (2001) Medicinal flowers. III. Marigold. (1): hypoglycemic, gastric emptying inhibitory, and gastroprotective principles and new oleanane-type triterpene oligoglycosides, calendasaponins a, B, C, and D, from Egyptian Calendula officinalis. Chem Pharm Bull (Tokyo) 49:863–870. https://doi.org/10.1248/cpb.49.863

Zamljen T, Jakopič J, Hudina M, Veberič R, Slatnar A (2021) Influence of intra and inter species variation in chilies (capsicum spp.) on metabolite composition of three fruit segments. Sci Rep 11:4932. https://doi.org/10.1038/s41598-021-84458-5

Zeng Q, Deng H, Li Y, Fan T, Liu Y, Tang S, Wei W, Liu X, Guo X, Jiang J, Wang Y, Song D (2021) Berberine directly targets the NEK7 protein to block the NEK7-NLRP3 interaction and exert anti-inflammatory activity. J Med Chem 64:768–781. https://doi.org/10.1021/acs.jmedchem.0c01743

Zeng Z, Zhang S, Wang H, Piao X (2015) Essential oil and aromatic plants as feed additives in non-ruminant nutrition: a review. J Anim Sci Biotechnol 6:7. https://doi.org/10.1186/s40104-015-0004-5

Zhang Y, Guo L, Huang J, Sun Y, He F, Zloh M, Wang L (2019) Inhibitory effect of Berberine on broiler P-glycoprotein expression and function: in situ and in vitro studies. IJMS 20:1966. https://doi.org/10.3390/ijms20081966

Zhang M, Viennois E, Prasad M, Zhang Y, Wang L, Zhang Z, Han MK, Xiao B, Xu C, Srinivasan S, Merlin D (2016) Edible ginger-derived nanoparticles: a novel therapeutic approach for the prevention and treatment of inflammatory bowel disease and colitis-associated cancer. Biomaterials 101:321–340. https://doi.org/10.1016/j.biomaterials.2016.06.018

Zhang C, Wang L, Zhao XH, Chen XY, Yang L, Geng ZY (2017a) Dietary resveratrol supplementation prevents transport-stress-impaired meat quality of broilers through maintaining muscle energy metabolism and antioxidant status. Poult Sci 96:2219–2225. https://doi.org/10.3382/ps/pex004

Zhang C, Zhao X, Wang L, Yang L, Chen X, Geng Z (2017b) Resveratrol beneficially affects meat quality of heat-stressed broilers which is associated with changes in muscle antioxidant status. Anim Sci J 88:1569–1574. https://doi.org/10.1111/asj.12812

Zhang C, Zhao XH, Yang L, Chen XY, Jiang RS, Jin SH, Geng ZY (2017c) Resveratrol alleviates heat stress-induced impairment of intestinal morphology, microflora, and barrier integrity in broilers. Poult Sci 96:4325–4332. https://doi.org/10.3382/ps/pex266

Zhao W, Huang X, Han X, Hu D, Hu X, Li Y, Huang P, Yao W (2018) Resveratrol suppresses gut-derived NLRP3 Inflammasome partly through stabilizing mast cells in a rat model. Mediat Inflamm 2018:1–10. https://doi.org/10.1155/2018/6158671

Zhao X, Shao T, Wang YQ, Lu XL, Luo JB, Zhou WD (2013) The phytoestrogen daidzein may affect reproductive performance of Zhedong white geese by regulating gene mRNA levels in the HPG axis. Br Poult Sci 54:252–258. https://doi.org/10.1080/00071668.2013.767439

Zhao X, Yang ZB, Yang WR, Wang Y, Jiang SZ, Zhang GG (2011) Effects of ginger root (Zingiber officinale) on laying performance and antioxidant status of laying hens and on dietary oxidation stability. Poult Sci 90:1720–1727. https://doi.org/10.3382/ps.2010-01280

Zhao RQ, Zhou YC, Ni YD, Lu LZ, Tao ZR, Chen WH, Chen J (2005) Effect of daidzein on egg-laying performance in Shaoxing duck breeders during different stages of the egg production cycle. Br Poult Sci 46:175–181. https://doi.org/10.1080/00071660500064808

Zitterl-Eglseer K, Sosa S, Jurenitsch J, Schubert-Zsilavecz M, Della Loggia R, Tubaro A, Bertoldi M, Franz C (1997) Anti-oedematous activities of the main triterpendiol esters of marigold (Calendula officinalis L.). J Ethnopharmacol 57:139–144. https://doi.org/10.1016/s0378-8741(97)00061-5

Effect of *Lippia origanoides* Essential Oil Extract on *Salmonella* serovar *Enteritidis* as a Natural Alternative to Antibiotics in the Poultry Industry

Abdiel Atencio-Vega, Dante J. Bueno, Juan D. Latorre,
Jesus A. Maguey-Gonzalez, Awad A. Shehata,
Wolfgang Eisenreich, Billy M. Hargis,
and Guillermo Tellez-Isaias

Contents

A. Atencio-Vega · J. D. Latorre · J. A. Maguey-Gonzalez · B. M. Hargis
Department of Poultry Science, University of Arkansas, Fayetteville, AR, USA

D. J. Bueno (✉)
Instituto Nacional de Tecnología Agropecuaria EEA Concepción del Uruguay,
Concepción del Uruguay, Entre Ríos, Argentina

Facultad de Ciencia y Tecnología, sede Basavilbaso, Universidad Autónoma de Entre Ríos,
Basavilbaso, Entre Ríos, Argentina
e-mail: bueno.dante@inta.gob.ar

A. A. Shehata · W. Eisenreich
Structural Membrane Biochemistry, Bavarian NMR Center, Technical University of Munich
(TUM), Garching, Bayern, Germany

G. Tellez-Isaias
Department of Poultry Science, University of Arkansas, Fayetteville, AR, USA

Gut Health LLC, Fayetteville, AR, USA

© The Author(s), under exclusive license to Springer Nature Switzerland AG 2024
A. A. Shehata et al. (eds.), *Alternatives to Antibiotics against Pathogens in
Poultry*, https://doi.org/10.1007/978-3-031-70480-2_9

Abstract

Salmonella serotypes Enteritidis (SE) and Typhimurium are thought to be the most significant serovars in terms of human infections, having the most effects on public health. In addition, chicken products continue to be the main source of illness for humans. The global chicken supply is getting increasingly interconnected, and the infection of SE is spreading globally as a result of breeding stocks that are contaminated being traded internationally. In addition to foodborne transmission, cross-contamination in kitchens or food processing areas and intimate contact with animals or habitats that are contaminated can also transmit SE. The poultry industry has been under constant pressure to reduce or stop using antibiotics in production for more than a decade due to concerns about food safety (including antibiotic resistance and consumer expectations) and the actual effects of pathogen contamination on food safety. Essential oils are well-known for being organic, odorous, volatile substances with potential medicinal benefits that have attracted a lot of interest from scientists, particularly in the chicken sector. This chapter explains *Salmonella* ser. Enteritidis as a unique foodborne pathogen, provides updates on the use of essential oils as natural antibiotic substitutes in the poultry industry, and discusses the use of *Lippia origanoides* to lessen the effects of SE in broiler chickens. This plant exhibits a wide range of chemotypes, with the majority having distinct essential oil compositions. Furthermore, it produces essential oils with a wide range of chemical compositions, including the so-called chemotypes of thymol and carvacrol. It is an alternate method of controlling SE in poultry production.

Keywords

Salmonella ser. Enteritidis · Essential oils · Quinta essentia · Foodborne pathogen · *Lippia origanoides*

1   Introduction

It is estimated that *Salmonella* spp. infects 93.8 million humans worldwide annually, and some isolates are multidrug resistant (Billah and Rahman 2024). Regarding human infections, *Salmonella* serotypes Enteritidis (SE) and Typhimurium are considered the most important serovars with the greatest impact on public health in several countries (Galán-Relaño et al. 2023). Moreover, poultry products are still the most important source of human infection (Mattock et al. 2024). The first line of treatment for this bacterial illness is antimicrobial therapy; however, due to antibiotic abuse in both human medicine and animal husbandry, antimicrobial resistance has become an issue (Castro-Vargas et al. 2020). Antibiotics have played a crucial role in the poultry industry for several decades, primarily for disease prevention, growth promotion, and improving intestinal health microbiota (Xu et al. 2021), and it has sparked questions about antibiotic resistance, the impact on the environment, and possible health risks not only to humans but also for animals. Consumers' being aware of the possible hazards and issues linked to antibiotic use in chicken farms has prompted researchers and poultry producers to attempt to employ alternatives to antibiotics (AGPs).

Essential oils (EOs) are known as natural, volatile, and fragrant compounds with putative therapeutic characteristics that have drawn a lot of attention from researchers, especially in the poultry industry. These oils have been used in aromatherapy, conventional medicine, and culinary arts and they can be derived from a variety of plant components, including leaves, flowers, bark, roots, seeds, and fruits (Yammine et al. 2024).

But their place in the chicken business is a modern one, with a range of possible benefits that line up with the industry's goals of increased output, improved animal welfare, and decreased need for traditional treatments, such as antibiotics. Herbs and aromatic plants were the first putative pharmaceuticals used in antiquity to heal ailments and other maladies. Their historical significance remains since they are still used today, at least in adjuvant therapy in both animal and human medicine. According to de Sousa et al. (2023), in the sixteenth century the alchemist and physician Theophrastus von Hohenheim, better known as Paracelsus, introduced the theory "Quinta essentia," which basically stated that extracting the potent healing essences from plant materials through distillation can be used as a medicinal alternative. There are over 3000 known EOs, 300 of which are highly important commercially (Abd El-Hack et al. 2022). These extracts from plants are used in diverse industries such as food, pharmaceutical, agronomic, sanitary, and cosmetic. They bring potential value to many aspects of chicken management and could provide a flexible and natural solution to health and hygiene issues in poultry production. Antioxidant, anti-inflammatory, and immune-modulating qualities of EOs are well recognized to offer protection against infections brought on by a variety of microbes such as bacteria, viruses, and fungi just to mention a few of them (de Sousa et al. 2023). The objective of this literature review was to discuss the potential of EOs as a non-chemical strategy to control SE in broiler chickens.

2   Essential Oils as Natural Alternatives to Antibiotics in the Poultry Industry

Due to worries about food safety (including antibiotic resistance and consumer expectations) and the real consequences of pathogen contamination on food safety, the poultry industry has been under constant pressure to cut back on or completely stop using antibiotics in production for more than the past 10 years (Elsayed et al. 2024; Manjankattil et al. 2024). As part of the one-health concept, the use of antibiotics in the chicken industry has been connected as one of the factors contributing to the worldwide antibiotic resistance dilemma (de Mesquita Souza Saraiva et al. 2022). Beyond unintentionally adding to the public health burden of antibiotic resistance, rearing poultry, and other livestock with antibiotics also leads to environmental pollution, with concerns about the role that the poultry industry could play in the emergence of multidrug-resistant pathogens (Mitra et al. 2024; Parajuli et al. 2024).

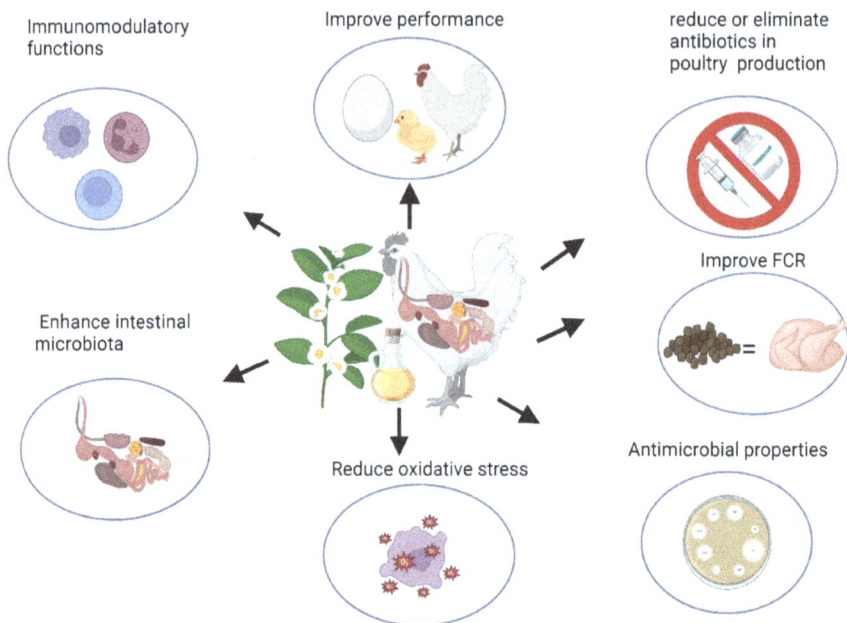

Fig. 9.1 Effects of essential oils on poultry production as a natural alternative to antibiotics. Created with BioRender.com

Essential oils (Fig. 9.1) are secondary metabolites from several parts of plants that have been used in Ayurvedic medicine and food spices for thousands of years (Qneibi et al. 2024). Due to their medicinal and industrial uses, EOs from the following plant families have been thoroughly studied: Alliaceae (onion), Apiaceae (celery), Asteraceae (aster), Lamiaceae (oregano, thyme, lavender, peppermint, sage oils), Verbenaceae (*Lippia*), Lauraceae (cinnamon oil), Liliaceae (garlic oil), Myrtaceae (tea tree oil), Poaceae (grass), and Rutaceae (rue). The chemical compositions of oregano, cinnamon, garlic, thyme, black pepper, lavender, mint, sage, and tea tree EOs make them widely utilized around the world (Mucha and Witkowska 2021).

Essential oils have been used to increase body weight, growth, egg production, and improve feed conversion ratio in poultry species (Adaszyńska-Skwirzyńska and Szczerbińska 2017). EOs have antibacterial (Mohamed et al. 2023; Pezantes-Orellana et al. 2024), immunomodulatory and antioxidant properties, which have been reported to protect against a range of illnesses, (Bhirich et al., 2024). Moreover, these oils are considered generally as safe (GRAS; Prakash et al. 2024). When applied at appropriate doses, regulators consider EOs safe and free from any adverse effects on poultry performance and health. Furthermore, compared with antibiotics, EOs can form a less persistent residue in poultry products, which could otherwise present food safety problems (Rodilla et al., 2024). In addition, EOs naturally occurring organic residue in poultry products is deemed safe for consumers and the environment because they are considered a naturally occurring and rapidly biodegradable chemical (Altay et al. 2024).

Dietary supplementation of EOs in broiler are a safe alternative to antibiotic growth promoters (Thirumeignanam et al. 2024). Although, EOs improve the performance of broilers subjected to heat stress conditions (Señas-Cuesta et al. 2023), there are still some concerns to analyze before introducing them to the poultry industry, like the inherent variability in compositions, the endless potential in formulations, the lack of information regarding the best dosages, and the most efficient application (Huang et al. 2024). Additional research and development are required to address the obstacles associated with these natural pharmacological techniques and fully harness their promise in poultry production.

3   *Salmonella* ser. Enteritidis: An Exceptional Foodborne Pathogen

Salmonella ser. Enteritidis is a Gram-negative bacterium of the Enterobacteriaceae family with significant relevance as a foodborne pathogen (Putturu et al. 2015). This bacterium is characterized by its straight, rod shape, motile due to its flagellum, Gram-negative bacterium with an approximate length of $0.7–1.5 \times 2.0–5.0\ \mu m$. The optimal temperature environment for growing *Salmonella* is 37 ° C and a pH of around 7.0 (Yang et al. 2014). An infection with SE can have several negative health effects. The bacterium frequently causes gastroenteritis, which manifests as fever, cramping in the abdomen, nausea, vomiting, and diarrhea in humans (Zha et al. 2019). Symptoms usually develop 6–72 h after infected food or drink is consumed. Dehydration can happen under extreme circumstances, especially in susceptible groups including small children, the elderly, and people with compromised immune systems (Food Safety and Inspection Service 2020). Additionally, SE can cause bacteremia, a condition in which the germs enter the bloodstream and may cause complications including sepsis (Mileva et al. 2016).

Although normally asymptomatic, in poultry species SE can show clinical signs such as depression, poor performance weakness, diarrhea, and dehydration (Shoaib et al. 2017). A host's age, immunity, the presence of coinfections, environmental stress, managerial traits, and infective dose are only a few of the variables that affect how severe is SE infection (Shaji et al. 2023). In poultry species, SE can be transmitted in both routes, vertical and horizontal. Vertical transmission refers to the spread of *Salmonella* spp. from parent birds that are infected with the pathogen to the chicks. The infection directly affects the yolk, vitelline membrane, and albumen. SE causes the infection to start in reproductive organs such as the ovary and oviduct (Shaji et al. 2023).

The pathogenic bacteria can break through the eggshell and infect the developing chick within. Chicks may also contract SE in the hatchery or brooding environment if they come into touch with contaminated surfaces, tools, or other diseased birds during or soon after hatching. Conversely, horizontal transmission occurs when birds are exposed to contaminated surfaces such as feed, water, and litter. SE spreads rapidly via fecal-oral transmission as soon as one of the birds carries the bacterium (Thomas et al. 2009; Gast et al. 2020).

Poultry products tainted with SE gave rise to a pandemic in the 1980s. This bacterium continues to produce outbreaks and the chicken supply is becoming more globalized, and the infection spreads worldwide due to the international commerce of contaminated breeding stocks (Li et al. 2021). The complete genome sequence of strains of SE related to poultry has been reported (Wang et al. 2019; Xu et al. 2021). This intracellular lifestyle helps make the organism virulent and enables it to migrate within the host (Dandekar et al. 2012). Human infections with this bacterium almost exclusively occur via ingesting contaminated food products, particularly animal products—poultry, eggs, meat, and unpasteurized dairy foods are common sources of SE contamination (Ehuwa et al. 2021). Because SE colonizes the ceca, shedding SE into the feces allows eggs or poultry meat contamination during processing (Van Immerseel et al. 2004).

Besides foodborne transmission, SE can spread through close contact with infected animals or environments and cross-contamination in kitchens or food processing environments (Ehuwa et al. 2021). Virulent strains of SE are associated with gastrointestinal disorders and in some children, elderly, or immunocompromised persons can be lethal (Liu et al. 2024; Mkangara 2023). A systemic and chronic infection may occur when SE translocates the intestinal barrier (Liu et al. 2022).

The most important steps to improve safety prevention include:

Strict biosecurity measures to minimize the possibility of SE infection of animals and their environment (Totton et al. 2012); legislation and standards for the production, processing and storage of food products, and particularly prepackaged foods that are prone to SE contamination to prevent or reduce contamination (Carrique-Mas and Davies 2008).

Consumers health education campaigns teach and encourage good food hygiene practices such as cooking food to kill pathogens (Tiozzo et al. 2011); washing hands regularly when handling these foods or any food exposed to the environment (Chai et al. 2012); preventing cross-contamination of other foods when preparing and eating solid or liquid food (van Velsen et al. 2014). In relation to eggs and layers, egg pasteurization or vaccination of the poultry flocks (Sakha and Fujikawa 2012). Furthermore, enhance surveillance systems for outbreak detection and investigation so that early steps can be taken to limit the spread of infection (Wagenaar et al. 2013).

As one of the main foodborne pathogens, SE control requires a multidisciplinary effort from farm to fork (Nazari Moghadam et al. 2023; Travis et al. 2014; Russell et al. 2024). By targeting the contamination and transmission of SE, the burden of SE-related illness will be greatly reduced, and public health and poultry welfare will be well protected (Chattaway et al. 2019) (Fig. 9.2).

Transmission routes of *Salmonella Enteritidis*

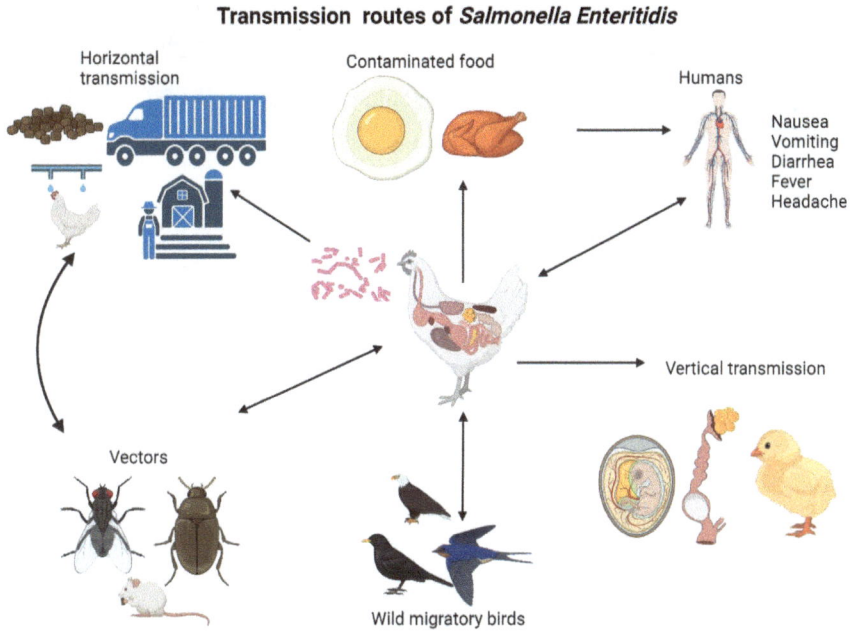

Fig. 9.2 Transmission routes of *Salmonella* Enteritidis. Created with BioRender.com

4 The Use of *Lippia origanoides* to Mitigate the Effect of *Salmonella ser.* Enteritidis in Broiler Chickens

There are about 800 species and 32 genera in the Verbenaceae family. V*erbena, Lippia, Citharexylum, Stachytarpheta, Glandularia*, and *Duranta* are the most representative genera (Cardoso et al. 2021). *Lipia origanoides* (common names: Mexican oregano or Mexican-sauge), is high in antibacterial EOs (Guillín et al. 2023). Since the use of antibiotics as growth promoters is banned in the EU, consumer pressures, and strategic interest to dominate the export market of poultry products, EOs, those of *Lippia origanoides*, have become an object of investigation as potential antimicrobials in the control of *Salmonella* infections in broiler chickens (Rubio Ortega et al. 2023; Orimaye et al. 2024, Fig. 9.3).

Moreover, EOs provide an eco-friendly and non-toxic substitute for antibiotics, all while decreasing antibiotic residues, improving gut health, decreasing the colonization of pathogens, and improving the immune response in broiler chickens (Bautista-Hernández et al. 2021). Several investigators have shown that the EOs from *Lippia origanoides* reduce *Salmonella* spp. contamination in poultry products (Hernandes et al. 2017).

Studies show that *Lippia origanoides* thymol and carvacrol, the most abundant EO present in the plant, have strong antimicrobial activity against SE (Madrid-Garcés et al. 2022; Betancourt et al. 2012). Furthermore, *Lippia origanoides* have

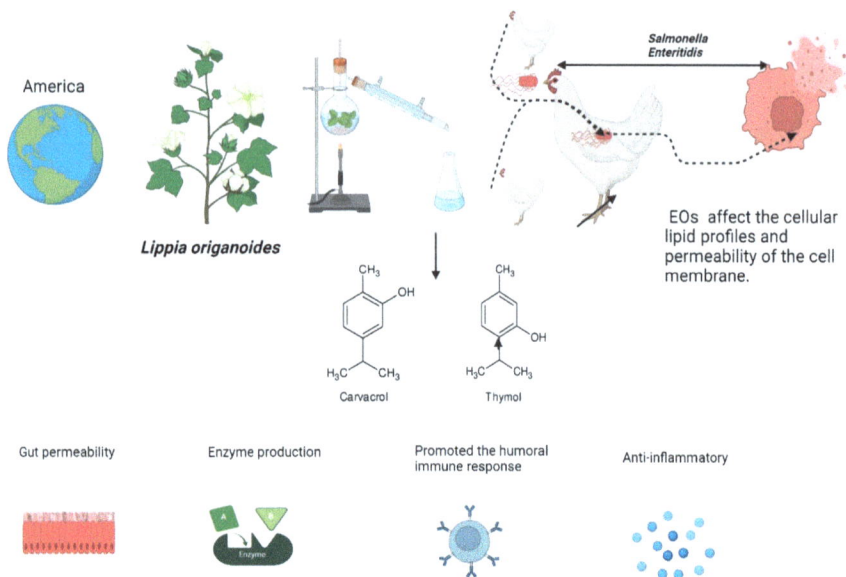

Fig. 9.3 The use of Lippia organoids to combat *Salmonella ser.* Enteritidis. Created with BioRender.com

several diversities of chemotypes, with a predominance of different essential oil compositions. Essential oils obtained from this plant can have large chemical composition variations, including the so-called thymol and carvacrol chemotypes (Stashenko et al. 2010). The main EO component of the thymol chemotype is thymol, which may correspond to more than 40% of the EOs (Betancur-Galvis et al. 2011).

Thymol content is highest in *Lippia origanoides* Kunth, an indigenous species found in the Patia region of southwest Colombia, where it has adapted to dry soil (Betancourt López 2024). Thymol has a strong, spicy, phenolic odor, vaguely resembling thyme intermingled with mint and cloves. Moreover, anti-inflammatory properties and the ability to protect the bird against respiratory infections are some of the benefits that this compound (Benitez-Llano et al. 2024). In addition to thymol, this chemotype (and others) may have smaller proportions of other compounds, such as p-cymene, γ-terpinene, and carvacrol (Castillo et al. 2024). In contrast, the carvacrol chemotype primarily consists of carvacrol EO, often found in quantities exceeding 50% of EOs (Sarrazin et al. 2015). Carvacrol produces a strong, herbal, peppery odor, and the oil is also known for its antimicrobial and antioxidant effects; oils high in carvacrol find frequent use in culinary applications and natural remedies for certain ailments (Castilho et al. 2019). Thymol, p-cymene, and γ-terpinene are other EOs found in this chemotype, but in lower concentrations (Sahu et al. 2024).

References

Abd El-Hack ME, El-Saadony MT, Salem HM, El-Tahan AM, Soliman MM, Youssef GBA, Taha AE, Soliman SM, Ahmed AE, El-kott AF, Al Syaad KM, Swelum AA (2022) Alternatives to antibiotics for organic poultry production: types, modes of action and impacts on bird's health and production. Poult Sci 101

Adaszyńska-Skwirzyńska M, Szczerbińska D (2017) Use of essential oils in broiler chicken production–A review. Ann Anim Sci 17:317–335

Altay Ö, Köprüalan Ö, İlter I, Koç M, Ertekin FK, Jafari SM (2024) Spray drying encapsulation of essential oils; process efficiency, formulation strategies, and applications. Crit Rev Food Sci Nutr 64(4):1139–1157

Bautista-Hernández I, Aguilar CN, Martínez-Ávila GC, Torres-León C, Ilina A, Flores-Gallegos AC, Chávez-González ML (2021) Mexican oregano (Lippia graveolens Kunth) as source of bioactive compounds: A review. Molecules 26(17):5156

Benitez-Llano CA, Florez-Acosta OA, Velasquez-Polo DD, Mesa-Arango AC, Zapata-Zapata C (2024) Preparation, physicochemical characterization, and stability study of Lippia origanoides essential oil-based Nanoemulsion as a topical delivery system. Pharm Nanotechnol 12(3):251–261

Betancourt López L (2024) Oregano essential oils as a nutraceutical additive in poultry diets. IntechOpen. https://doi.org/10.5772/intechopen

Betancourt L, Phandanauvong V, Patiño R, Ariza-Nieto C, Afanador-Téllez G (2012) Composition and bactericidal activity against beneficial and pathogenic bacteria of oregano essential oils from four chemotypes of Origanum and Lippia genus. Revista de la Facultad de Medicina Veterinaria y de Zootecnia 59(1):21–31

Betancur-Galvis L, Zapata B, Baena A, Bueno J, Ruíz-Nova CA, Stashenko E, Mesa-Arango AC (2011) Antifungal, cytotoxic and chemical analyses of essential oils of Lippia origanoides HBK grown in Colombia. Revista de la Universidad Industrial de Santander. Salud 43(2):141–148

Bhirich N, Meknassi GS, Yafout M, Elhorr H, Said AAH, Mojemmi B (2024) A review of plants and essential oils effective as natural remedies against bedbugs. Indian J Pharm Drug Studies 3(1)

Billah MM, Rahman MS (2024) Salmonella in the environment: A review on ecology, antimicrobial resistance, seafood contaminations, and human health implications. Journal of Hazardous Materials. Advances:100407

Cardoso PH, O'Leary N, Olmstead RG, Moroni P, Thode VA (2021) An update of the Verbenaceae genera and species numbers. Plant Ecol Evol 154:80–86

Carrique-Mas JJ, Davies RH (2008) Salmonella Enteritidis in commercial layer flocks in Europe: legislative background, on-farm sampling and main challenges. Braz J Poult Sci 10:1–9

Castilho CV, Leitão SG, Silva VD, Miranda CDO, Santos MCDS, Bizzo HR, da Silva NC (2019) In vitro propagation of a carvacrol-producing type of Lippia origanoides Kunth: A promising oregano-like herb. Ind Crop Prod 130:491–498

Castillo LN, Calva J, Ramírez J, Vidari G, Armijos C (2024) Chemical analysis of the essential oils from three populations of Lippia dulcis Trevir. Grown at different locations in southern ecuador. Plants 13(2):253

Castro-Vargas RE, Herrera-Sánchez MP, Rodríguez-Hernández R, Rondón-Barragán IS (2020) Antibiotic resistance in Salmonella spp. isolated from poultry: A global overview. Vet World 13:2070–2084

Chai SJ, White PL, Lathrop SL, Solghan SM, Medus C, McGlinchey BM, Mahon BE (2012) Salmonella enterica serotype Enteritidis: increasing incidence of domestically acquired infections. Clin Infect Dis 54(suppl_5):S488–S497

Chattaway MA, Dallman TJ, Larkin L, Nair S, McCormick J, Mikhail A et al (2019) The transformation of reference microbiology methods and surveillance for salmonella with the use of whole genome sequencing in England and Wales. Front Public Health 7:317

Dandekar T, Astrid F, Jasmin P, Hensel M (2012) Salmonella enterica: a surprisingly well-adapted intracellular lifestyle. Front Microbiol 3:164

de Mesquita Souza Saraiva M, Lim K, do Monte DFM, Givisiez PEN, Alves LBR, de Freitas Neto OC, Gebreyes WA (2022) Antimicrobial resistance in the globalized food chain: A one health perspective applied to the poultry industry. Braz J Microbiol 53:1–22

de Sousa DP, Damasceno ROS, Amorati R, Elshabrawy HA, de Castro RD, Bezerra DP, Lima TC (2023) Essential oils: Chemistry and pharmacological activities. Biomolecules 2023(13):1144

Ehuwa O, Jaiswal AK, Jaiswal S (2021) Salmonella, food safety and food handling practices. Food Secur 10(5):907

Elsayed MM, El-Basrey YF, El-Baz AH, Dowidar HA, Shami A, Al-Saeed FA, Khedr MH (2024) Ecological prevalence, genetic diversity, and multidrug resistance of salmonella enteritidis recovered from broiler and layer chicken farms. Poult Sci 103(2):103320

Galán-Relaño, Á; Valero Díaz, A; Huerta Lorenzo, B; Gómez-Gascón, L; Mena Rodríguez, MÁ; Carrasco Jiménez, E; Pérez Rodríguez, F; Astorga Márquez, RJ (2023) Salmonella and salmonellosis: an update on public health implications and control strategies. Animals 13, 3666. https://doi.org/10.3390/ ani13233666

Gast RK, Jones DR, Guraya R, Anderson KE, Karcher DM (2020) Research note: horizontal transmission and internal organ colonization by salmonella Enteritidis and salmonella Kentucky in experimentally infected laying hens in indoor cage-free housing. Poult Sci 99(11):6071–6074. https://doi.org/10.1016/j.psj.2020.08.006.

Guillín Y, Cáceres M, Stashenko EE, Hidalgo W, Ortiz C (2023) Untargeted metabolomics for unraveling the metabolic changes in planktonic and sessile cells of salmonella Enteritidis ATCC 13076 after treatment with Lippia origanoides essential oil. Antibiotics 12(5):899

Hernandes C, Pina ES, Taleb-Contini SH, Bertoni BW, Cestari IM, Espanha LG, Pereira AMS (2017) Lippia origanoides essential oil: an efficient and safe alternative to preserve food, cosmetic and pharmaceutical products. J Appl Microbiol 122(4):900–910

Huang J, Guo F, Abbas W, Hu Z, Liu L, Qiao J, Wang Z (2024) Effects of microencapsulated essential oils and organic acids preparation on growth performance, slaughter characteristics, nutrient digestibility and intestinal microenvironment of broiler chickens. Poult Sci 103655:103655

Li S, He Y, Mann DA, Deng X (2021) Global spread of *salmonella* Enteritidis via centralized sourcing and international trade of poultry breeding stocks. Nat Commun 12(1):5109

Liu B, Zhang X, Ding X, Bin P, Zhu G (2022) The vertical transmission of *salmonella* Enteritidis in a one-health context. One Health 16:100469

Liu J, Hou Y, Zhao L, Chen G, Chen J, Zhao Q et al (2024) Antimicrobial resistance and the genomic epidemiology of multidrug-resistant salmonella enterica serovar Enteritidis ST11 in China. Front Biosci (Landmark Ed) 29(3):112

Madrid-Garcés TA, González-Herrera LG, López-Herrera A, Parra-Suescún JE (2022) Improvement of productive and metabolic indicators of broiler by the application of Lippia origanoides essential oil in an in vivo intestinal inflammation model. Revista Facultad Nacional de Agronomía Medellín 75(2):9971–9981

Manjankattil S, Dewi G, Peichel C, Creek M, Bina P, Cox R, Johny AK (2024) Effect of Pimenta essential oil against salmonella Agona and salmonella Saintpaul in ground Turkey meat and nonprocessed Turkey breast meat. Poult Sci 103(2):103279

Mattock J, Chattaway MA, Hartman H, Dallman TJ, Smith AM, Keddy K, Langridge GC (2024) A One Health Perspective on Salmonella enterica Serovar Infantis, an emerging human multidrug-resistant pathogen. Emerging Infectious Diseases 30(4):701

Mileva S, Gospodinova M, Todorov I (2016) Salmonella enteritidis primary bacteremia in previously healthy patient from Taiwan: case report. Ther Adv Infect Dis 3:128–132

Mitra SD, Shome R, Bandopadhyay S, Geddam S, Kumar AP, Murugesan D, Shome BR (2024) Genetic insights of antibiotic resistance, pathogenicity (virulence) and phylogenetic relationship of Escherichia coli strains isolated from livestock, poultry and their handlers-a one health snapshot. Mol Biol Rep 51(1):404

Mkangara M (2023) Prevention and control of human salmonella enterica infections: an implication in food safety. Int J Food Sci *2023*

Mohamed AA, Alotaibi BM (2023) Essential oils of some medicinal plants and their biological activities: a mini review. J Umm Al-Qura Univ Appl Sci 9(1):40–49

Mucha W, Witkowska D (2021) The applicability of essential oils in different stages of production of animal-based foods. Molecules 26(13):3798. https://doi.org/10.3390/molecules26133798

Nazari Moghadam M, Rahimi E, Shakerian A, Momtaz H (2023) Prevalence of salmonella typhimurium and salmonella Enteritidis isolated from poultry meat: virulence and antimicrobial-resistant genes. BMC Microbiol 23(1):168

Orimaye OE, Ekunseitan DA, Omaliko PC, Fasina YO (2024) Mitigation potential of herbal extracts and constituent bioactive compounds on salmonella in meat-type poultry. Animals 14(7):1087

Parajuli A, Mitchell J, King N, Arjyal A, Latham S, King R, Baral S (2024) Drivers of antimicrobial resistance within the communities of Nepal from one health perspective: a scoping review. Front Public Health 12:1384779

Pezantes-Orellana C, German Bermúdez F, Matías De la Cruz C, Montalvo JL, Orellana-Manzano A (2024) Essential oils: a systematic review on revolutionizing health, nutrition, and omics for optimal Well-being. Front Med 11:1337785

Prakash B, Singh PP, Gupta V, Raghuvanshi TS (2024) Essential oils as green promising alternatives to chemical preservatives for Agri-food products: new insight into molecular mechanism, toxicity assessment, and safety profile. Food Chem Toxicol 183:114241

Putturu R, Eevuri T, Ch B, Nelapati K (2015) Salmonella enteritidis-foodborne pathogen-a review. Int J Pharm Biol Sci 5(1):86–95

Qneibi M, Bdir S, Maayeh C, Bdair M, Sandouka D, Basit D, Hallak M (2024) A comprehensive review of essential oils and their pharmacological activities in neurological disorders: exploring neuroprotective potential. Neurochem Res 49(2):258–289

Rodilla JM, Rosado T, Gallardo E (2024) Essential oils: chemistry and food applications. Food Secur 13(7):1074

Rubio Ortega A, Guinoiseau E, Poli JP, Quilichini Y, de Rocca Serra D, del Carmen Travieso Novelles M et al (2023) The primary mode of action of Lippia graveolens essential oil on salmonella enterica subsp. Enterica Serovar Typhimurium. Microorganisms 11(12):2943

Russell G, Nenov A, Hancock JT (2024) How hydrogen (H2) can support food security: from farm to fork. Appl Sci 14(7):2877

Sahu A, Parai D, Choudhary HR, Singh DD (2024) Essential oils as alternative antimicrobials: current status. Recent Advances in Anti-Infective Drug Discovery Formerly Recent Patents on Anti-Infective Drug Discovery 19(1):56–72

Sakha MZ, Fujikawa H (2012) Growth characteristics of salmonella Enteritidis in pasteurized and unpasteurized liquid egg products. Biocontrol Sci 17(4):183–190

Sarrazin SLF, Da Silva LA, De Assunção APF, Oliveira RB, Calao VY, Da Silva R, Mourão RHV (2015) Antimicrobial and seasonal evaluation of the carvacrol-chemotype oil from Lippia origanoides kunth. Molecules 20(2):1860–1871

Señas-Cuesta R, Stein A, Latorre JD, Maynard CJ, Hernandez-Velasco X, Petrone-Garcia V, Greene ES, Coles M, Gray L, Laverty L, Martin K, Loeza I, Uribe AJ, Martínez BC, Angel-Isaza JA, Graham D, Owens CM, Hargis BM, Tellez-Isaias G (2023) The effects of essential oil from Lippia origanoides and herbal betaine on performance, intestinal integrity, bone mineralization and meat quality in broiler chickens subjected to cyclic heat stress. Front Physiol 14:1184636

Shaji S, Selvaraj RK, Shanmugasundaram R (2023) Salmonella infection in poultry: A review on the pathogen and control strategies. Microorganisms 11

Shoaib M, Zafar MA, Riaz A (2017) Salmonellosis in poultry, new prospects of an old disease: a review

Stashenko EE, Martínez JR, Ruíz CA, Arias G, Durán C, Salgar W, Cala M (2010) Lippia origanoides chemotype differentiation based on essential oil GC-MS and principal component analysis. J Sep Sci 33(1):93–103

Thirumeignanam D, Chellapandian M, Arulnathan N, Parthiban S, Kumar V, Vijayakumar MP, Chauhan S (2024) Evaluation of natural antimicrobial substances blend as a replacement for antibiotic growth promoters in broiler chickens: enhancing growth and managing intestinal bacterial diseases. Curr Microbiol 81(2):55

Thomas ME, Klinkenberg D, Ejeta G, Van Knapen F, Bergwerff AA, Stegeman JA, Bouma A (2009) Quantification of horizontal transmission of salmonella enterica serovar enteritidis bacteria in pair-housed groups of laying hens. Appl Environ Microbiol 75:6361–6366

Tiozzo B, Mari S, Magaudda P, Arzenton V, Capozza D, Neresini F, Ravarotto L (2011) Development and evaluation of a risk-communication campaign on salmonellosis. Food Control 22(1):109–117

Totton SC, Farrar AM, Wilkins W, Bucher O, Waddell LA, Wilhelm BJ et al (2012) A systematic review and meta-analysis of the effectiveness of biosecurity and vaccination in reducing salmonella spp. in broiler chickens. Food Res Int 45(2):617–627

Travis DA, Sriramarao P, Cardona C, Steer CJ, Kennedy S, Sreevatsan S, Murtaugh MP (2014) One medicine one science: a framework for exploring challenges at the intersection of animals, humans, and the environment. Ann N Y Acad Sci 1334(1):26–44

Van Immerseel F, De Buck J, Pasmans F, Bohez L, Boyen F, Haesebrouck F, Ducatelle R (2004) Intermittent long-term shedding and induction of carrier birds after infection of chickens early posthatch with a low or high dose of salmonella Enteritidis. Poult Sci 83(11):1911–1916

van Velsen L, Beaujean DJ, van Gemert-Pijnen JE, van Steenbergen JE, Timen A (2014) Public knowledge and preventive behavior during a large-scale salmonella outbreak: results from an online survey in The Netherlands. BMC Public Health 14:1–9

Wagenaar JA, Hendriksen RS, Carrique-Mas J (2013) Practical considerations of surveillance of salmonella serovars other than Enteritidis and typhimurium. Revue Scientifique et Technique-Office International des Epizooties 32(2):509–519

Wang H, Cai L, Hu H, Xu X, Zhou G (2019) Complete genome sequence of salmonella enterica Serovar Enteritidis NCM 61, with high potential for biofilm formation, isolated from meat-related sources. Microbiol Resour Announc. 8(2):e01434–e01418. https://doi.org/10.1128/MRA.01434-18

Xu Y, Abdelhamid AG, Yousef AE (2021) Draft genome sequence of salmonella enterica subsp. *enterica* Serovar Enteritidis ODA 99-30581-13, a heat-resistant strain Isolated from Shell eggs. Microbiol Resour Announc 10(9):e01461–e01420. https://doi.org/10.1128/MRA.01461-20

Yammine J, Chihib NE, Gharsallaoui A, Ismail A, Karam L (2024) Advances in essential oils encapsulation: development, characterization and release mechanisms. Polymer Bulletin 81(5):3837–3882

Yang Y, Khoo WJ, Zheng Q, Chung HJ, Yuk HG (2014) Growth temperature alters salmonella Enteritidis heat/acid resistance, membrane lipid composition and stress/virulence related gene expression. Int J Food Microbiol 172:102–109. https://doi.org/10.1016/j.ijfoodmicro.2013.12.006. Epub 2013 Dec 11

Zha L, Garrett S, Sun J (2019) Salmonella infection in chronic inflammation and gastrointestinal cancer. Diseases 7:28

Taming the Tiny Titans: Exploring Bacteriophages as an Alternative to Antibiotics in the Poultry Industry

10

Guillermo Tellez, Dante J. Bueno, Inkar Castellanos-Huerta, Jesus A. Maguey-Gonzalez, Lauren Laverty, Abdil Atancio, Victor M. Petrone, Juan D. Latorre, Awad A. Shehata, Wolfgang Eisenreich, Hafez M. Hafez, and Billy M. Hargis

Contents

G. Tellez (✉) · I. Castellanos-Huerta · J. A. Maguey-Gonzalez · L. Laverty · A. Atancio ·
J. D. Latorre · B. M. Hargis
Department of Poultry Science, University of Arkansas, Fayetteville, AR, USA
e-mail: gtellez@uark.edu; hafez.mohamed@fu-berlin.de

D. J. Bueno
Instituto Nacional de Tecnología Agropecuaria EEA Concepción del Uruguay,
Concepción del Uruguay, Entre Ríos, Argentina

V. M. Petrone
Departamento de Ciencias Pecuarias, Universidad Nacional Autónoma de México (UNAM),
Cuautitlan Izcalli, Estado de México, Mexico

A. A. Shehata · W. Eisenreich
Structural Membrane Biochemistry, Bavarian NMR Center, Technical University of Munich
(TUM), Garching, Bayern, Germany

H. M. Hafez
Institute of Poultry Diseases, Faculty of Veterinary Medicine, Free University of Berlin,
Berlin, Germany

197

Abstract

In recent years, the global poultry industry has faced unprecedented challenges in maintaining the health and productivity of poultry flocks, compounded by the escalating concerns surrounding the widespread use of antibiotics. As the industry strives to address these challenges, an innovative and promising alternative has emerged: the use of bacteriophages. This chapter delves into the pivotal role of bacteriophages as a groundbreaking alternative to antibiotics in the poultry industry, exploring their potential to revolutionize disease management, enhance biosecurity, and safeguard both animal and human health. With growing apprehensions regarding antibiotic resistance and its far-reaching implications for public health, there is an urgent need for sustainable and effective alternatives. Bacteriophages, or phages, are viruses that specifically target and infect bacteria, offering a precise and targeted approach to combating bacterial infections in poultry. This chapter sets the stage for an in-depth exploration of the science behind bacteriophages, their mechanisms of action, the potential benefits they bring to the poultry farming landscape, and unfolds the historical context of antibiotic use in the poultry industry, shedding light on the factors that have fuelled the quest for alternative strategies. The focus will then shift to the fundamental characteristics of bacteriophages, illustrating how these viral predators can be harnessed to control bacterial pathogens in poultry production. Furthermore, the chapter will examine key studies and advancements in the field, providing a comprehensive overview of the current state of research and application of bacteriophages in poultry farming.

Keywords

Bacteriophages · Safety · Advantages · Mechanism of action · Challenges

1 Concern about Antibiotic Resistance in Poultry and Its Public Health Implications

The idea that superbugs could become the leading cause of death in humans by the year 2050 is a concerning scenario often discussed in the context of antibiotic resistance (Painuli et al. 2023). Superbugs are bacteria that have developed resistance to multiple antibiotics, making it challenging to treat infections caused by them. Several factors contribute to the emergence of antibiotic-resistant bacteria, including overuse and misuse of antibiotics in humans and animals, inadequate sanitation and hygiene, and the global spread of resistant strains (Salam et al. 2023). If antibiotic resistance continues to rise at the current rate, it could lead to a situation where common infections become difficult or impossible to treat and increase mortality rates. The World Health Organization (WHO), as well as various health organizations and experts worldwide, have been emphasizing the importance of addressing antibiotic resistance to prevent such a scenario (WHO 2022). Efforts include promoting responsible antibiotic use, developing new antibiotics, and implementing measures to prevent the spread of resistant strains. It is essential for governments, healthcare professionals, researchers, veterinarians, and the general public to work together to combat antibiotic resistance, as the consequences could be severe. By promoting proper antibiotic stewardship, investing in research and development of new antibiotics, and implementing measures to prevent infections, it is possible to mitigate the risk of superbugs becoming a leading cause of death in the future (Medina et al. 2020).

The concern about antibiotic resistance in poultry is a significant issue with potential public health implications. Antibiotics have been widely used in the poultry industry for growth promotion, disease prevention, and treatment. However, the misuse and overuse of antibiotics in livestock, including poultry, contributes to the development of antibiotic-resistant bacteria (Emes et al. 2024).

Development of resistant bacteria:

- Continuous exposure of bacteria to low doses of antibiotics in poultry farming can lead to the development of antibiotic-resistant strains (Khan and Asaduzzaman 2023).
- Resistant bacteria can spread to humans through the food supply chain, direct contact with animals, or environmental contamination.

Transmission to humans:

- People can be exposed to antibiotic-resistant bacteria through the consumption of contaminated poultry products, such as meat and eggs.
- Handling raw poultry or contact with farm environments may also pose a risk of transmission (Hannan et al. 2023).

Reduced effectiveness of antibiotics:

- When antibiotic-resistant bacteria are transmitted to humans, it can limit the effectiveness of commonly used antibiotics, making infections harder to treat.
- This could lead to increased healthcare costs, longer hospital stays, and higher mortality rates for some infections (Collignon 2013).

One-Health approach:

- Addressing antibiotic resistance in poultry requires a "One Health" approach, recognizing the interconnectedness of human, animal, and environmental health.
- Collaboration between the healthcare, veterinary, and agricultural sectors is essential to develop and implement strategies to mitigate antibiotic resistance (Aslam et al. 2021).

Regulatory measures:

- Many countries have implemented regulations to restrict the use of certain antibiotics in animal feed for growth promotion.
- Monitoring and surveillance programs are essential to track antibiotic use in poultry and the prevalence of antibiotic-resistant bacteria (Baral and Mozafari 2020).

Alternative farming practices:

- Adopting alternative farming practices, such as improving hygiene, vaccination, and disease prevention measures, can help reduce the reliance on antibiotics in poultry farming (Adegoke et al. 2016).

Consumer awareness:

- Raising awareness among consumers about the importance of responsible antibiotic use in agriculture can influence purchasing behaviors and encourage demand for antibiotic-free poultry products (Suri et al. 2021).
- Antibiotic resistance in poultry is a complex issue with implications for public health. Efforts to address this concern involve a combination of regulatory measures, sustainable farming practices, and public awareness campaigns to ensure the responsible use of antibiotics in the poultry industry.

2 Implications of Using Bacteriophages in the Poultry Industry

Bacteriophages, or phages, are viruses that specifically target and infect bacteria. Protein or lipoprotein capsids, typically affixed to a tail, make up bacteriophages. In single (ss) or double (ds) strains, the phage nucleic acid (DNA or RNA), which can be organized as linear or circular molecules, is enclosed by the capsid. A small number of bacteriophages also have envelopes (Salmond and Fineran 2015; Ackermann 2009). The classification of bacteriophages is regularly updated and approved by the International Committee on the Taxonomy of Viruses (ICTV), according to host range, size, shape and morphology, nucleic acid, and genomic similarity (Lefkowitz et al. 2018). The phage classification was changed, and the morphology-based families Podovirus, Myovirus, and Siphovirus were eliminated along with the order Caudovirales, which was replaced by a new class named "Caudoviricetes," according to the most recent taxonomy update published by the ICTV in August 2022. In the class Caudoviricetes, there are now 14 families spread across four orders: *Kirjokansivirales*, *Thumleimavirales*, *Methanobavirales*, and *Crispvirales* (Table 10.1). There are currently 631 genera that have not yet been classified at the family or order level, along with 37 subfamilies and 33 more families that have been formed but not assigned to an order (Turner et al. 2023).

The use of bacteriophages in the poultry industry carries implications for the well-being of the birds, food safety, and the broader goal of creating a sustainable

Table 10.1 Current taxonomy of the class Caudoviricetes following ICTV*

Order	Family	Subfamily	Genus	Species
Crassvirales	Crevaviridae	2	3	4
	Intestiviridae	3	11	18
	Steigviridae	1	12	15
	Suoliviridae	5	16	36
Kirjokansivirales	Graaviviridae		2	2
	Haloferuviridae		3	3
	Pyrstoviridae		1	1
	Shortaselviridae		1	1
Methanobavirales	Anaerodiviridae		1	1
	Leisingerviridae		1	1
Thumleimavirales	Druskaviridae		2	2
	Hafunaviridae		4	10
	Halomagnusviridae		1	1
	Soleiviridae		1	1
Unidentified	33	11	59	96
	Unidentified	37	631	2060

ICTV taxonomy (https://talk.ictvonline.org/taxonomy/). Abd-El Wahab et al. (2023)

and environment-friendly agricultural system (Żbikowska et al. 2020). The benefits of bacteriophage application include:

(i) *Targeted approach*: Bacteriophages offer a highly targeted approach to combat bacterial infections in poultry. Unlike broad-spectrum antibiotics, phages selectively infect and destroy specific bacterial strains, minimizing the disruption of the natural microbiota in the birds' digestive systems (Ngiam et al. 2021).

(ii) *Reduced antibiotic dependency*: Bacteriophage therapy provides an alternative to antibiotics, reducing the dependency on these drugs and contributing to the mitigation of antibiotic resistance (Burrowes et al. 2011).

(iii) *Food safety*: Contaminated poultry products can pose a risk to human health. Bacteriophages can be applied directly to poultry or through feed to control bacterial pathogens, thereby enhancing the safety of poultry products and reducing the risk of foodborne illnesses (Schmelcher and Loessner 2016).

The implications of using bacteriophages in the poultry industry extend beyond the immediate benefits of targeted bacterial control. With the potential to reduce antibiotic dependency, enhance food safety, and contribute to sustainable farming practices, bacteriophage therapy holds promise for creating a healthier and more resilient poultry industry. However, addressing challenges related to phage specificity, regulatory approval, and public perception is crucial for the successful integration of this innovative approach into mainstream poultry production; these and other challenges are discussed later (Łobocka et al. 2021). As ongoing research continues to refine our understanding of bacteriophage applications, the poultry industry stands to benefit from a more sustainable and environmentally conscious future.

3 Effectiveness in Combating Common Poultry Pathogens

They exhibit a high degree of specificity, often targeting particular strains of bacteria. This specificity is advantageous as it allows for the precise targeting of pathogenic bacteria without affecting beneficial microbiota. The potential of bacteriophages in controlling bacterial infections in poultry has gained attention as a sustainable and eco-friendly alternative to antibiotics.

Several studies have shown the effectiveness against common poultry pathogens, including:

– *Salmonella* spp.: *Salmonella* spp. is a major concern in poultry farming due to its ability to cause foodborne illnesses in humans. Studies have shown that certain bacteriophages can effectively target and reduce *Salmonella* spp. populations in poultry, offering a potential means of controlling the spread of this pathogen (Shahdadi et al. 2024).
– *Escherichia coli*: Another prevalent poultry pathogen, *E. coli*, is responsible for various infections, leading to high economic losses in the industry. Bacteriophages have demonstrated success in controlling *E. coli* in poultry, showing promise as a targeted and efficient intervention (Huff et al. 2003).
– *Campylobacter* spp.: Campylobacteriosis is a common foodborne illness associated with poultry meat consumption. Bacteriophages targeting *Campylobacter* have shown potential in reducing the prevalence of this pathogen in poultry, offering a novel approach to improve food safety (Peh et al. 2023).

The use of bacteriophages in combating common poultry pathogens shows great promise as a sustainable and targeted alternative to traditional antibiotics. As research continues to explore the potential of phage therapy, addressing challenges related to specificity, regulatory approval, and understanding phage–host interactions will be critical. The integration of bacteriophages into poultry farming practices could contribute to a more resilient and sustainable poultry industry while addressing concerns associated with antibiotic resistance.

4 Safety Profile for Both Birds and Consumers

Bacteriophages have gained attention in the poultry industry due to their ability to specifically target and destroy harmful bacteria without affecting beneficial bacteria. In the context of poultry farming, these viruses have been employed to combat bacterial infections such as *Salmonella* spp. and *Escherichia coli*, which pose significant risks to both avian health and food safety (Chaudhary et al. 2024).

Bacteriophages offer a targeted approach to combat bacterial infections in birds. Unlike broad-spectrum antibiotics, which can disrupt the balance of the gut microbiota and lead to the development of antibiotic-resistant strains, bacteriophages selectively target and eliminate specific pathogens. This targeted action can help maintain the overall health and well-being of the avian population (Kamiński and Paczesny 2024).

Furthermore, bacteriophages have shown effectiveness in controlling bacterial infections in various poultry species. Research has demonstrated their potential to reduce the prevalence of *Salmonella* and other harmful bacteria, contributing to improved flock health and productivity (Li et al. 2022).

The use of bacteriophages in poultry farming not only benefits avian health but also enhances consumer safety. Contamination of poultry products with pathogenic bacteria is a significant concern in the food industry. Bacteriophage-based interventions offer a natural and specific means of reducing bacterial contamination in poultry products (Żbikowska et al. 2020).

The targeted action of bacteriophages ensures that non-pathogenic bacteria crucial for maintaining the balance of the gut microbiome remain unaffected, addressing concerns about the potential development of antibiotic-resistant strains and preserving the overall safety of poultry products (Ranveer et al. 2024).

While the use of bacteriophages presents a promising alternative to antibiotics in poultry farming, challenges exist. One such challenge is the need for comprehensive research to establish optimal dosage, application methods, and potential interactions with other interventions. Additionally, regulatory frameworks and public perception may influence the widespread adoption of bacteriophage-based solutions in poultry farming (Chaudhary et al. 2024).

Bacteriophages are effective in maintaining a safe profile for both birds and consumers in the poultry industry. Their targeted approach to bacterial infections offers a viable alternative to traditional antibiotics, addressing concerns related to antibiotic resistance and preserving the delicate balance of the avian microbiome. As research and development in this field continue, bacteriophages have the potential to revolutionize poultry farming practices, ensuring both avian health and consumer safety in a sustainable and environmentally friendly manner (Jaglan et al. 2024).

5   Advantages over Traditional Antibiotics

Attributes of bacteriophages that make them attractive antibacterial agents compared with antibiotics are illustrated in Table 10.2. One of the primary advantages of bacteriophages lies in their remarkable specificity. Unlike broad-spectrum antibiotics, which can inadvertently harm beneficial bacteria, bacteriophages are highly selective in their targeting. Each bacteriophage is designed to infect a particular bacterial strain, ensuring precision in eliminating the harmful pathogen while sparing the beneficial microbiota. This specificity minimizes collateral damage to the host organism and reduces the risk of disrupting the delicate balance of the microbiome (Gao et al. 2024).

Bacteriophages exhibit a remarkable ability to adapt and evolve, a characteristic that distinguishes them from traditional antibiotics. As bacteria evolve and develop resistance mechanisms against antibiotics, bacteriophages can coevolve to counteract these defenses. This dynamic interaction between bacteriophages and bacteria

Table 10.2 Comparison between bacteriophages and antibiotics as antibiotics (Sarhan and Azzazy 2015)

Bacteriophages	Antibiotics
Very specific, affecting only the targeted bacterial species with no disruption of normal flora, therefore minimizing the possibility of secondary infections	Target both pathogenic microorganisms and normal microflora, which may lead to serious secondary infections
Autodosing through replication at the site of infection (repeated administration may not be needed)	Metabolized and eliminated and do not accumulate at the site of infection
No serious side effects have been described Minor side effects may be due to the liberation of endotoxins from bacteria lysed in vivo by the phages	Multiple side effects, including allergies, intestinal disorders, secondary infections, adverse effects on the kidney and the liver
Phage-resistant bacteria are not resistant to other phages having a similar target range	Resistance to antibiotics extends over targeted bacteria
Finding new phages against developed bacterial resistance can be achieved in days	Developing a new antibiotic against antibiotic-resistant bacteria is a very lengthy and expensive process
Ability to clear biofilms	Limited ability of biofilm clearance

provides a sustainable approach to combating infections, with the potential to stay ahead of bacterial resistance. The adaptability of bacteriophages offers a personalized and evolving therapeutic strategy capable of responding to the ever-changing landscape of bacterial pathogens (Brenner et al. 2024).

The escalating threat of antibiotic resistance has become a global health crisis, rendering many conventional antibiotics ineffective. Bacteriophages present a promising solution to this challenge. Since bacteriophages target bacteria through mechanisms distinct from antibiotics, they can be effective against antibiotic-resistant strains. Moreover, the ability of bacteriophages to evolve rapidly allows them to circumvent resistance mechanisms employed by bacteria, offering a dynamic and innovative approach to addressing the growing problem of antibiotic resistance (Jiang et al. 2024).

Traditional antibiotics often come with a host of side effects, ranging from gastrointestinal discomfort to more severe reactions, such as allergic responses. Bacteriophages, due to their specificity, are less likely to cause unintended side effects. By selectively targeting the offending bacteria, bacteriophages spare the beneficial microbiota, minimizing disruptions to the host organism and reducing the likelihood of adverse reactions. This characteristic enhances the safety profile of bacteriophage therapy compared to traditional antibiotics (Torres-Barceló 2018).

The advantages of bacteriophages over traditional antibiotics position them as a promising alternative in the fight against bacterial infections. Their specificity, adaptability, and ability to overcome antibiotic resistance make bacteriophages a compelling option for the future of infectious disease treatment. While challenges remain in terms of standardization, regulatory approval, and widespread application, ongoing research and clinical trials suggest that bacteriophages have the potential to revolutionize the field of antibacterial therapy, offering new hope in the face of an evolving and increasingly resistant microbial landscape (Burrowes et al. 2011).

6   Mechanisms of Action of Bacteriophages and their Potential to Reshape the Landscape of Poultry Health Management

Since bacteriophages are highly specific to their bacterial hosts, each type of phage typically targets a particular species or even a strain of bacteria. This specificity allows for precise targeting of pathogenic bacteria without affecting beneficial bacteria in the poultry microbiota.

Bacteriophages possess two lifecycles: lysogenic or temperate, lytic or virulent (Fig. 10.1). The bacteriophage attaches itself to the surface of the bacteria and injects its genetic material into the cytoplasm to initiate the infection. Adsorption can involve components of the bacterial cell wall or projecting structures such as flagella, pili, and capsules. Lytic bacteriophages quickly seize control of the host cell's metabolic machinery in order to produce new virions. The host cell is quickly lysed to end the growth cycle. As long as the bacterial host is present in sufficient numbers to support multiplication, cell lysis will release hundreds of additional

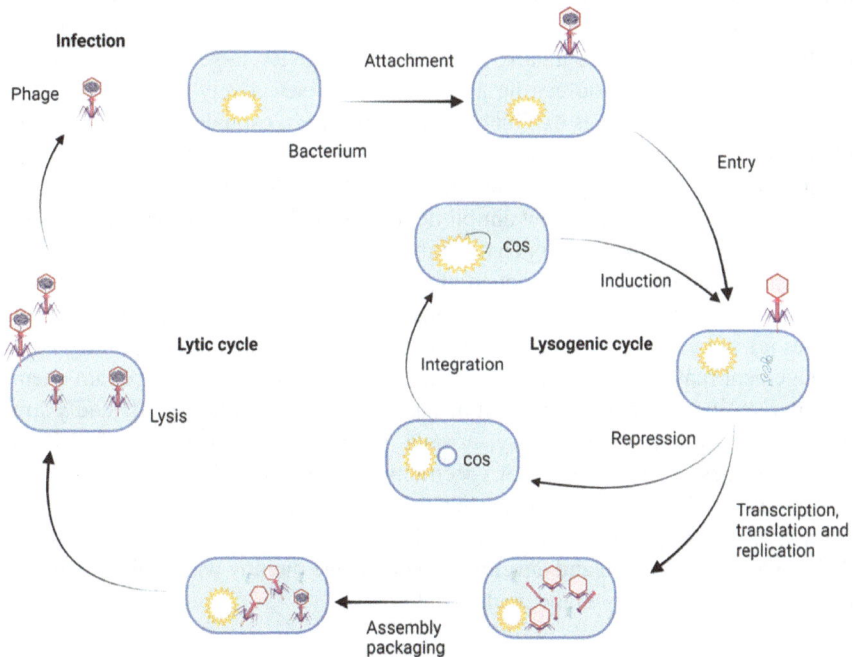

Fig. 10.1 Life cycle of bacteriophage. © Abd-El Wahab et al. (2023)

virus particles within minutes or hours. This process can be repeated. This lytic activity can significantly reduce the population of pathogenic bacteria in the poultry gut or environment. On the other hand, lysogenic bacteriophages are able to insert their genetic material into the host cell's genome and stay in a dormant state (as prophage) until the bacteria is exposed to specific stimuli that cause the lytic cycle to begin. The lysogenic interaction that exists between a temperate phage and its host bacteria gives the phage genome a safe haven, prevents the replication of non-virulent homologous phages, and may even change the host cell's phenotype (Sarhan and Azzazy 2015; Kazi and Annapure 2016). Even though the phage may eventually lyse the cell, this will not stop bacterial infection right away and could even propagate antimicrobial resistance and other virulence genes. Therefore, it is imperative to make sure that phage therapies only employ lytic phages (Barron 2022).

Furthermore, phages can penetrate and disrupt biofilms formed by bacteria. Biofilms are complex structures that bacteria use to adhere to surfaces, making them more resistant to antibiotics and other antimicrobial agents. By, phages can make pathogenic bacteria more susceptible to other control measures (Liu et al. 2016).

Bacteriophages and bacteria are engaged in an ongoing evolutionary arms race. As bacteria evolve to resist phage infection, phages can, in turn, evolve to overcome these bacterial defenses.

Bacteria encode hundreds of diverse defense systems that protect them from viral infection and inhibit phage propagation (Antine et al. 2024), while phages encode an arsenal of anti-defense proteins that can disable a variety of bacterial defense mechanisms (Yirmiya et al. 2024). This dynamic interaction can potentially provide a sustainable and evolving solution for controlling bacterial infections in poultry. Some phages have a narrow spectrum of activity, targeting specific strains of bacteria. This precision can be advantageous in selectively eliminating pathogenic strains. Other phages exhibit a broader spectrum, capable of infecting multiple bacterial species. This versatility can be useful in addressing complex bacterial communities in the poultry environment (Koskella and Meaden 2013).

With growing concerns about antibiotic resistance, bacteriophages offer an alternative or complementary approach to antibiotics in managing bacterial infections in poultry. Phage therapy has the potential to reduce reliance on antibiotics and mitigate the development of antibiotic-resistant bacteria. In addition, since bacteriophages are natural entities that exist in the environment. Their use in poultry health management may have minimal environmental impact compared to certain chemical interventions. Hence, phages can be part of a sustainable and eco-friendly strategy for disease control in poultry farming. The unique mechanisms of action of bacteriophages, coupled with their specificity and adaptability, make them promising candidates for reshaping poultry health management practices, especially in the context of reducing antibiotic usage and addressing bacterial infections (Bisesi et al. 2024). Ongoing research and developments in phage therapy will likely continue to enhance its applicability in the poultry industry.

7 Challenges Associated with Phage Therapy, Phage Selection, and Delivery Methods

Overall, phage therapy, the use of bacteriophages to treat bacterial infections, has gained attention as a potential alternative or complementary approach to conventional antibiotics. However, like any medical intervention, phage therapy comes with its own set of challenges. Here are some key challenges associated with phage therapy, particularly in terms of phage selection and delivery methods:

Phage specificity and diversity:

- Narrow host range: Phages are often highly specific to particular bacterial strains, which may limit their effectiveness against a broad range of pathogens.
- Genetic diversity: Bacteria can evolve and develop resistance to specific phages, potentially leading to treatment failures. Maintaining a diverse phage bank is crucial to address this challenge (Bollback and Huelsenbeck 2001).

Phage pharmacokinetics:

- *Stability and shelf life*: Phages can be sensitive to environmental conditions, and their stability during storage is a concern. Developing methods to ensure phage stability and a longer shelf life is essential for practical application.
- *Phage clearance*: Phages may be rapidly cleared from the body by the immune system before they can effectively target and eliminate bacterial infections. Strategies to improve phage circulation and persistence in the body need to be developed (Lin et al. 2020).

Bioavailability and tissue penetration:

- *Phage accessibility to infection sites*: Ensuring that phages reach the target site of infection in sufficient quantities is challenging, especially in the case of deep-seated infections or biofilm-associated infections.
- *Phage penetration through mucus and tissues*: Mucus barriers and tissues can impede the effective delivery of phages to the infection site. Overcoming these barriers is critical for successful phage therapy (Bichet et al. 2021).

Regulatory challenges:

- *Lack of standardized protocols*: Developing standardized protocols for phage therapy, including phage isolation, characterization, and production, is necessary for regulatory approval.
- *Regulatory approval*: Regulatory agencies may have limited experience with phage therapy, and the approval process can be complex. Establishing clear regulatory pathways is crucial for the integration of phage therapy into mainstream medicine (Furfaro et al. 2018).

Ethical and safety considerations:

- *Phage safety*: Ensuring the safety of phage therapy, including the absence of harmful side effects and potential horizontal gene transfer, is a significant concern.
- *Ethical considerations*: The ethical implications of using live viruses as therapeutic agents need to be carefully considered, including issues related to informed consent and patient understanding (Verbeken et al. 2014).

Phage resistance development:

- *Rapid adaptation*: Bacteria can evolve mechanisms to resist phage infection, potentially reducing the effectiveness of phage therapy over time. Monitoring and addressing the development of phage resistance are ongoing challenges (Egido et al. 2022).

Optimal dosing and treatment regimens:

- *Dose optimization*: Determining the optimal phage dosage and treatment duration for different infections is challenging and may require a personalized approach.
- *Combination therapy*: Exploring the potential benefits of combining phage therapy with other antimicrobial agents to enhance efficacy and reduce the likelihood of resistance (Lin et al. 2022).

Addressing these challenges requires ongoing research, collaboration between scientists and clinicians, and the development of innovative strategies to optimize phage therapy for various clinical scenarios.

8   Potential Economic Benefits of Using Phages in the Poultry Industry

The poultry industry plays a crucial role in global food security, providing a significant source of protein for a growing population. However, the industry faces challenges, with bacterial infections causing substantial economic losses. Traditional antibiotic use has been a cornerstone in mitigating bacterial infections, but the rise of antibiotic resistance has prompted the exploration of alternative solutions. Bacteriophages, viruses that specifically target bacteria, offer a promising avenue for combating bacterial infections in poultry. This essay explores the potential economic benefits of incorporating phages into the poultry industry. As has been discussed in this chapter, perhaps one of the primary economic advantages of phage therapy in poultry is the potential to reduce antibiotic dependency. Antibiotic resistance is a global health concern, and the agricultural sector is a significant contributor to the problem. By incorporating bacteriophages as an alternative to antibiotics, the poultry industry can contribute to the global effort to curb antibiotic resistance. This reduction in antibiotic usage not only helps preserve the efficacy of existing antibiotics but also addresses consumer concerns about antibiotic residues in poultry products.

Moreover, bacteriophages can provide targeted and specific control over bacterial infections in poultry, leading to improved overall animal health. Healthy birds are more productive, with enhanced growth rates, improved feed conversion ratios, and higher egg production. The reduction in disease-related mortality and morbidity can result in substantial economic gains for poultry producers as they experience increased yields and a more efficient production cycle (Żbikowska et al. 2020).

Phage therapy offers a potentially cost-efficient and sustainable solution for controlling bacterial infections in poultry. Unlike antibiotics, which may require a lengthy and costly approval process, phages can be relatively easily isolated and characterized for specific bacterial strains. Additionally, the self-replicating nature of phages allows for the production of high volumes at a relatively low cost. This affordability can make phage therapy an economically viable option for poultry producers, particularly in regions with limited access to expensive antibiotics (Torres-Acosta et al. 2020).

As consumer awareness of antibiotic resistance and food safety grows, there is an increasing demand for poultry products raised using sustainable and responsible practices. Incorporating phage therapy in poultry production aligns with these consumer preferences, offering a competitive advantage in the marketplace. Poultry producers adopting phage-based interventions can differentiate their products, potentially gaining market access and commanding premium prices for antibiotic-free and phage-treated poultry products (Warner et al. 2014).

The economic benefits of integrating bacteriophages into the poultry industry are promising. By reducing antibiotic dependency, improving animal health and productivity, ensuring cost-efficiency, and enhancing food safety, phage therapy emerges as a sustainable and economically viable solution. As the poultry industry navigates the challenges posed by bacterial infections, harnessing the potential of phages represents a proactive and forward-thinking approach that aligns with both economic interests and broader societal concerns. The adoption of phage therapy in poultry production not only contributes to the industry's economic resilience but also promotes a more sustainable and responsible model for meeting the growing global demand for poultry products (García et al. 2019).

9   Regulatory Landscape Surrounding Phage Therapy

The regulatory pathway for phage therapy in poultry involves gaining approval from relevant authorities. Different countries may have varying procedures for evaluating and approving novel therapies.

In the United States, for example, the Food and Drug Administration (FDA) oversees the approval of phage-based products. The European Medicines Agency (EMA) plays a similar role in the European Union. Navigating these approval processes requires robust scientific evidence demonstrating the safety and efficacy of phage products in poultry (Pelfrene et al. 2016; Yang et al. 2023).

Establishing clear definitions and standards for phage therapy is crucial for regulatory clarity. Defining the characteristics of phage products, including their composition, concentration, and intended use, helps regulators evaluate their safety and effectiveness. Collaborative efforts between industry stakeholders, regulatory agencies, and scientific communities are essential to develop standardized guidelines that ensure consistency in assessing phage therapy products (Plaut and Stibitz 2019).

Regulatory submissions for phage therapy in poultry must include comprehensive data on safety, efficacy, and quality. This may involve rigorous testing in controlled environments and field trials. Gathering data on the specificity of phages, potential resistance development, and long-term effects on poultry health is essential. Transparency in reporting data strengthens the regulatory case for phage therapy and fosters confidence among regulatory agencies and consumers (Naanwaab et al. 2014).

Phage therapy represents a relatively novel approach in comparison to traditional antibiotics. The absence of well-established precedents for evaluating phage products poses a challenge for regulatory bodies. Establishing a framework that accommodates the unique characteristics of phage therapy while ensuring safety and efficacy is a delicate balancing act. Bacteriophages evolve rapidly, leading to concerns about the stability and consistency of phage-based products over time. Regulatory frameworks must adapt to accommodate the dynamic nature of phages, ensuring that products remain effective and safe throughout their shelf life (Saleh 2020).

Achieving global harmonization in the regulation of phage therapy for poultry is a complex endeavor. Divergent regulatory standards across countries may impede the international trade of phage-based products. Efforts to align regulatory approaches globally will facilitate the development and distribution of phage therapy solutions on a broader scale (Leung et al. 2017).

In this context, there are several opportunities. *Industry collaboration* with academia and regulatory agencies can accelerate the development of robust scientific data on phage therapy. *Establishing research consortia* can facilitate the generation of standardized protocols for testing and evaluating phage products, addressing regulatory concerns. *Educating regulators, poultry producers, and consumers* about the benefits and risks of phage therapy is crucial. Increased awareness fosters a better understanding of the regulatory process and promotes acceptance of phage-based solutions in the poultry industry. *Recognizing the dynamic nature of phage therapy*, regulatory frameworks should be designed to adapt to scientific advancements. *Regular updates and revisions* to guidelines will ensure that the regulatory landscape remains relevant and responsive to emerging challenges (Greer 2005).

The regulatory landscape surrounding phage therapy in the poultry industry is multifaceted, with challenges and opportunities intertwined. While navigating the approval process and addressing concerns about the evolving nature of phages present hurdles, collaborative efforts and adaptive regulatory frameworks offer promising solutions. As the poultry industry seeks sustainable alternatives to antibiotics, a *well-defined and globally harmonized regulatory* pathway for phage therapy is essential for realizing its potential benefits in safeguarding poultry health and ensuring food safety (Chaudhary et al. 2024).

10 Importance of Further Research and Development

One of the primary reasons for intensifying research in the poultry industry is the pressing need for sustainability. The environmental impact of poultry farming, including issues such as waste management and resource consumption, demands innovative solutions. Research can identify and develop sustainable practices, such as improved feed formulations, waste recycling, and energy-efficient production systems. These advancements not only reduce the industry's environmental footprint but also contribute to long-term viability and resilience in the face of climate change (Kittler et al. 2017).

The poultry industry is susceptible to various diseases that can have devastating effects on both animal welfare and economic stability. Investing in research and development allows for the discovery of new and improved vaccines, diagnostic tools, as well as biosecurity measures. By staying ahead of emerging diseases and enhancing preventive strategies, the industry can safeguard flocks, minimize economic losses, and maintain consumer confidence in the safety of poultry products (Fiorilla et al. 2024).

Advancements in genetics, nutrition, and management practices through research contribute significantly to improving the efficiency and productivity of poultry production. *Genetic research* helps develop breeds that exhibit enhanced growth rates, disease resistance, and feed efficiency. *Nutritional research* refines feed formulations to optimize growth, while minimizing environmental impact. *Management innovations*, guided by research findings, lead to more effective production processes, reducing costs and increasing overall productivity (Najafi and Zolfagharinia 2024).

As consumer preferences evolve, the poultry industry must adapt to meet changing demands. Research enables the development of new products, such as value-added and functional foods, addressing health and sustainability concerns. Investigating consumer trends also allows the industry to enhance animal welfare practices, responding to growing interest in ethically produced and humanely raised poultry (Leinonen and Kyriazakis 2016).

Research and development in the poultry industry play a crucial role in addressing global food security challenges. Poultry represents an efficient and resource-effective means of converting feed into protein, making it a vital component of a sustainable food system. By continually improving production practices and addressing constraints, the industry contributes to a more resilient and secure global food supply (Mottet and Tempio 2017).

The importance of ongoing research and development in the poultry industry cannot be overstated. Sustainable practices, disease management, efficiency improvements, consumer-driven innovations, and contributions to global food security all hinge on continuous scientific exploration and application. The future of the poultry industry lies in the hands of researchers committed to driving positive change, ensuring that poultry farming remains not only economically viable but also environmentally responsible and capable of meeting the nutritional needs of a growing global population (Manning and Baines 2004).

11   The Use of Bacteriophages in Poultry Feed

Bacteriophages in poultry feed have shown promise in enhancing overall performance and feed efficiency. By reducing the prevalence of pathogenic bacteria, phages contribute to a healthier gut environment, improving nutrient absorption and optimizing feed conversion rates. This, in turn, can lead to better growth rates, increased weight gain, and improved feed utilization, ultimately benefiting both poultry producers and consumers (Jiang et al. 2024).

While the use of bacteriophages in poultry feed presents a novel and promising approach, there are challenges and considerations that must be addressed. *The selection and engineering of phages* for specific pathogens require careful consideration, as does the potential for the development of phage-resistant bacteria. Additionally, regulatory approval and widespread adoption of this technology in the poultry industry may take time, necessitating further research and collaboration between scientists, industry stakeholders, and regulatory bodies. The use of bacteriophages in poultry feed represents a groundbreaking approach to enhancing poultry health and productivity. This alternative to traditional antibiotics aligns with the growing demand for sustainable and eco-friendly farming practices while addressing concerns related to antibiotic resistance. As research and development in this field continue, bacteriophages hold great promise as a valuable tool for promoting the well-being of poultry flocks and meeting the challenges of modern poultry production.

References

Abd-El Wahab A, Basiouni S, El-Seedi HR, Ahmed MFE, Bielke LR, Hargis B, Tellez-Isaias G, Eisenreich W, Lehnherr H, Kittler S, Shehata AA, Visscher C (2023) An overview of the use of bacteriophages in the poultry industry: successes, challenges, and possibilities for overcoming breakdowns. Front Microbiol 14:1136638

Ackermann HW (2009) Phage classification and characterization. Methods Mol Biol 501:127–140

Adegoke AA, Faleye AC, Singh G, Stenström TA (2016) Antibiotic-resistant superbugs: assessment of the interrelationship of occurrence in clinical settings and environmental niches. Molecules 22(1):29

Antine SP, Johnson AG, Mooney SE, Leavitt A, Mayer ML, Yirmiya E, Amitai G, Sorek R, Kranzusch PJ (2024) Structural basis of Gabija anti-phage defence and viral immune evasion. Nature 625(7994):360–365

Aslam B, Khurshid M, Arshad MI, Muzammil S, Rasool M, Yasmeen N, Shah T, Chaudhry TH, Rasool MH, Shahid A, Xueshan X (2021) Antibiotic resistance: one health one world outlook. Front Cell Infect Microbiol 11:1153

Barron M (2022) Phage therapy: past, present and future. American Society for Microbiology. https://asm.org/articles/2022/august/phage-therapy-past,-present-and-future

Baral B, Mozafari MR (2020) Strategic moves of "superbugs" against available chemical scaffolds: signaling, regulation, and challenges. ACS Pharmacol Transl Sci 3(3):373–400

Bichet MC, Chin WH, Richards W, Lin YW, Avellaneda-Franco L, Hernandez CA, Oddo A, Chernyavskiy O, Hilsenstein V, Neild A, Li J (2021) Bacteriophage uptake by mammalian cell layers represents a potential sink that may impact phage therapy. Iscience 24(4):102287

Bisesi AT, Möbius W, Nadell CD, Hansen EG, Bowden SD, Harcombe WR (2024) Bacteriophage specificity is impacted by interactions between bacteria. Msystems:e01177–e01123

Bollback JP, Huelsenbeck JP (2001) Phylogeny, genome evolution, and host specificity of single-stranded RNA bacteriophage (family Leviviridae). J Mol Evol 52:117–128

Brenner T, Schultze DM, Mahoney D, Wang S (2024) Reduction of Nontyphoidal salmonella enterica in broth and on raw chicken breast by a broad-spectrum bacteriophage cocktail. J Food Prot 87(1):100207

Burrowes B, Harper DR, Anderson J, McConville M, Enright MC (2011) Bacteriophage therapy: potential uses in the control of antibiotic-resistant pathogens. Expert Rev Anti-Infect Ther 9(9):775–785

Chaudhary V, Kajla P, Lather D, Chaudhary N, Dangi P, Singh P, Pandiselvam R (2024) Bacteriophages: a potential game changer in food processing industry. Crit Rev Biotechnol:1–25

Collignon P (2013) Superbugs in food: a severe public health concern. Lancet Infect Dis 13(8):641–643

Egido JE, Costa AR, Aparicio-Maldonado C, Haas PJ, Brouns SJ (2022) Mechanisms and clinical importance of bacteriophage resistance. FEMS Microbiol Rev 46(1):fuab048

Emes E, Guitian J, Knight GM Naylor N (2024) The relationship between antibiotic use in humans and poultry and antibiotic resistance prevalence in humans: an ecological regression study of campylobacter in the UK. medRxiv, pp. 2024-01

Fiorilla E, Gariglio M, Martinez-Miro S, Rosique C, Madrid J, Montalban A, Biasato I, Bongiorno V, Cappone EE, Soglia D, Schiavone A (2024) Improving sustainability in autochthonous slow-growing chicken farming: exploring new frontiers through the use of alternative dietary proteins. J Clean Prod 434:140041

Furfaro LL, Payne MS, Chang BJ (2018) Bacteriophage therapy: clinical trials and regulatory hurdles. Front Cell Infect Microbiol 8:376

Gao FZ, He LY, Liu YS, Zhao JL, Zhang T, Ying GG (2024) Integrating global microbiome data into antibiotic resistance assessment in large rivers. Water Res 250:121030

García R, Latz S, Romero J, Higuera G, García K, Bastías R (2019) Bacteriophage production models: An overview. Front Microbiol 10:1187

Greer GG (2005) Bacteriophage control of foodborne bacteria. J Food Prot 68(5):1102–1111

Hannan A, Ihsan M, Haque MA, Du X (2023) Zoonoses and AMR: silent spreader of superbug pandemic. Zoonosis 4:186–201

Huff WE, Huff GR, Rath NC, Balog JM, Donoghue AM (2003) Bacteriophage treatment of a severe Escherichia coli respiratory infection in broiler chickens. Avian Dis 47(4):1399–1405

Jaglan AB, Vashisth M, Sharma P, Verma R, Virmani N, Bera BC, Vaid RK, Singh RK, Anand T (2024) Phage mediated biocontrol: a promising green solution for sustainable agriculture. Indian J Microbiol:1–10

Jiang A, Liu Z, Lv X, Zhou C, Ran T, Tan Z (2024) Prospects and challenges of bacteriophage substitution for antibiotics in livestock and poultry production. Biology 13(1):28

Kamiński B, Paczesny J (2024) Bacteriophage challenges in industrial processes: a historical unveiling and future outlook. Pathogens 13(2):152

Kazi M, Annapure US (2016) Bacteriophage biocontrol of foodborne pathogens. J Food Sci Technol 53(3):1355–1362

Khan A, Asaduzzaman M (2023) Use of antibiotics in poultry and poultry farmers-a cross-sectional survey in Pakistan. Front Public Health 11:1154668

Kittler S, Witt mann J, Mengden RALP, Klein G, Rohde C, Lehnherr H (2017) The use of bacteriophages as one-health approach to reduce multidrug-resistant bacteria. Sustain Chem Pharm 5:80–83

Koskella B, Meaden S (2013) Understanding bacteriophage specificity in natural microbial communities. Viruses 5(3):806–823

Leinonen I, Kyriazakis I (2016) How can we improve the environmental sustainability of poultry production? Proc Nutr Soc 75(3):265–273

Lefkowitz EJ, Dempsey DM, Hendrickson RC, Orton RJ, Siddell SG, Smith DB (2018) Virus taxonomy: the database of the international committee on taxonomy of viruses (ICTV). Nucleic Acids Res 46:D708–D717

Leung V, Szewczyk A, Chau J, Hosseinidoust Z, Groves L, Hawsawi H, Anany H, Griffiths MW, Ali MM, Filipe CD (2017) Long-term preservation of bacteriophage antimicrobials using sugar glasses. ACS Biomater Sci Eng 4(11):3802–3808

Li J, Zhao F, Zhan W, Li Z, Zou L, Zhao Q (2022) Challenges for the application of bacteriophages as effective antibacterial agents in the food industry. J Sci Food Agric 102(2):461–471

Lin J, Du F, Long M, Li P (2022) Limitations of phage therapy and corresponding optimization strategies: a review. Molecules 27(6):1857

Lin YW, Chang RY, Rao GG, Jermain B, Han ML, Zhao JX, Chen K, Wang JP, Barr JJ, Schooley RT, Kutter E (2020) Pharmacokinetics/pharmacodynamics of antipseudomonal bacteriophage therapy in rats: a proof-of-concept study. Clin Microbiol Infect 26(9):1229–1235

Liu Y, Mi Z, Niu W, An X, Yuan X, Liu H, Wang Y, Feng Y, Huang Y, Zhang X, Zhang Z (2016) Potential of a lytic bacteriophage to disrupt Acinetobacter baumannii biofilms in vitro. Future Microbiol 11(11):1383–1393

Łobocka M, Dąbrowska K, Górski A (2021) Engineered bacteriophage therapeutics: rationale, challenges and future. BioDrugs 35(3):255–280

Manning L, Baines RN (2004) Globalisation: a study of the poultry-meat supply chain. Br Food J 106(10/11):819–836

Medina MJ, Legido-Quigley H Hsu LY (2020) Antimicrobial resistance in one health. Global health security: recognizing vulnerabilities, creating opportunities, pp. 209–229

Mottet A, Tempio G (2017) Global poultry production: current state and future outlook and challenges. Worlds Poult Sci J 73(2):245–256

Naanwaab C, Yeboah OA, Ofori Kyei F, Sulakvelidze A, Goktepe I (2014) Evaluation of consumers' perception and willingness to pay for bacteriophage treated fresh produce. Bacteriophage 4(4):e979662

Najafi M, Zolfagharinia H (2024) A multi-objective integrated approach to address sustainability in a meat supply chain. Omega 124:103011

Ngiam L, Schembri MA, Weynberg K, Guo J (2021) Bacteriophage isolated from non-target bacteria demonstrates broad host range infectivity against multidrug-resistant bacteria. Environ Microbiol 23(9):5569–5586

Painuli S, Semwal P, Sharma R, Akash S (2023) Superbugs or multidrug-resistant microbes: a new threat to the society. Health Sci Rep 6(8):e1480

Peh E, Szott V, Reichelt B, Friese A, Rösler U, Plötz M, Kittler S (2023) Bacteriophage cocktail application for campylobacter mitigation-from in vitro to in vivo. BMC Microbiol 23(1):209

Pelfrene E, Willebrand E, Cavaleiro Sanches A, Sebris Z, Cavaleri M (2016) Bacteriophage therapy: a regulatory perspective. J Antimicrob Chemother 71(8):2071–2074

Plaut RD, Stibitz S (2019) Regulatory considerations for bacteriophage therapy products. Phage therapy: a practical approach, pp. 337–349

Ranveer SA, Dasriya V, Ahmad MF, Dhillon HS, Samtiya M, Shama E, Anand T, Dhewa T, Chaudhary V, Chaudhary P, Behare P (2024) Positive and negative aspects of bacteriophages and their immense role in the food chain. npj Sci Food 8(1):1

Salam MA, Al-Amin MY, Salam MT, Pawar JS, Akhter N, Rabaan AA, Alqumber MA (2023) Antimicrobial resistance: a growing serious threat for global public health. In: Healthcare. (Vol. 11, No. 13, p. 1946). MDPI

Saleh FR (2020) Bacteriophage as bio-preserving to enhance shelf life of fruit and vegetables. Plant Arch 20(1):359–362

Salmond GP, Fineran PC (2015) A century of the phage: past, present and future. Nat Rev Microbiol 13(12):777–786

Sarhan WA, Azzazy HME (2015) Phage approved in food, why not as a therapeutic? Expert Rev Anti-Infect Ther 13:91–101

Schmelcher M, Loessner MJ (2016) Bacteriophage endolysins: applications for food safety. Curr Opin Biotechnol 37:76–87

Shahdadi M, Safarirad M, Berizi E, Hosseinzadeh S, Phimolsiripol Y, Khaneghah AM (2024) Investigating the effect of phage on reducing salmonella spp. in poultry meat: a systematic review and meta-analysis. Food Control 160:110380

Suri M, Aggarwal S, Saini A, Bhardwaj M, Singh P, Shukla H (2021) Attitudes and awareness about antimicrobials usage and resistance in Delhi, INDIA. J Adv Sci Res 12(01 Suppl 1):317–325

Torres-Acosta M, González-Mora A, Ruiz-Ruiz F, Rito-Palomares M, Benavides J (2020) Economic evaluation of M13 bacteriophage production at large-scale for therapeutic applications using aqueous two-phase systems. J Chem Technol Biotechnol 95(11):2822–2833

Torres-Barceló C (2018) The disparate effects of bacteriophages on antibiotic-resistant bacteria. Emerg Microbes Infect 7(1):1–12

Turner D, Shkoporov AN, Lood C, Millard AD, Dutilh BE, Alfenas-Zerbini P, van Zyl LJ, Aziz RK, Oksanen HM, Poranen MM, Kropinski AM, Barylski J, Brister JR, Chanisvili N, Edwards RA, Enault F, Gillis A, Knezevic P, Krupovic M, Kurtböke I, Kushkina A, Lavigne R, Lehman S, Lobocka M, Moraru C, Moreno Switt A, Morozova V, Nakavuma J, Reyes Muñoz A, Rūmnieks J, Sarkar BL, Sullivan MB, Uchiyama J, Wittmann J, Yigang T, Adriaenssens EM (2023) Abolishment of morphology-based taxa and change to binomial species names: 2022 taxonomy update of the ICTV bacterial viruses subcommittee. Arch Virol 23;168(2):74

Verbeken G, Huys I, Pirnay JP, Jennes S, Chanishvili N, Scheres J, Górski A, De Vos D, Ceulemans C (2014) Taking bacteriophage therapy seriously: a moral argument. BioMed Res Int 2014

Warner CM, Barker N, Lee SW, Perkins EJ (2014) M13 bacteriophage production for large-scale applications. Bioprocess Biosyst Eng 37:2067–2072

WHO (2022) Antimicrobial resistance. Available online: https://www.who.int/news-room/fact-sheets/detail/antimicrobial-resistance. Accessed 2 Nov 2022

Yang Q, Le S, Zhu T, Wu N (2023) Regulations of phage therapy across the world. Front Microbiol 14:1250848

Yirmiya E, Leavitt A, Lu A, Ragucci AE, Avraham C, Osterman I, Garb J, Antine SP, Mooney SE, Hobbs SJ, Kranzusch PJ, Amitai G, Sorek R (2024) Phages overcome bacterial immunity via diverse anti-defence proteins. Nature 625(7994):352–359

Żbikowska K, Michalczuk M, Dolka B (2020) The use of bacteriophages in the poultry industry. Animals 10(5):872

Peptides as Alternatives to Antibiotics in Poultry Health Management

11

Inkar Castellanos-Huerta, Abdiel Atencio Vega,
Jesus A. Maguey-Gonzalez, Lauren Laverty,
Awad A. Shehata, Wolfgang Eisenreich, Billy M. Hargis,
and Guillermo Tellez-Isaias

Contents

I. Castellanos-Huerta · A. A. Vega · J. A. Maguey-Gonzalez · L. Laverty · B. M. Hargis ·
G. Tellez-Isaias (✉)
Division of Agriculture, Department of Poultry Science, University of Arkansas,
Fayetteville, AR, USA
e-mail: gtellez.gf@uark.edu

A. A. Shehata · W. Eisenreich
Structural Membrane Biochemistry, Bavarian NMR Center, Technical University of Munich
(TUM), Garching, Bayern, Germany

Abstract

The poultry industry faces a critical challenge: antibiotic resistance. Once considered miraculous, antibiotics now breed resilient pathogens, endangering poultry health and global food security. Widespread antibiotic use has created resistant strains, threatening poultry and public health. Peptides, with diverse structures and functions, emerge as promising alternatives for sustainable poultry health. These small yet potent molecules offer a unique approach to disease control, targeting a broad spectrum of pathogens. As short amino acid chains, Peptides possess distinctive properties crucial for combating infectious agents. Antimicrobial Peptides (AMPs) play a central role in defense mechanisms, disrupting microbial membranes. The transition to peptides signifies a paradigm shift in disease control, addressing threats from *Salmonella*, *Escherichia coli*, and viruses. Peptides exhibit multifaceted functionalities, tackling resistance concerns and preventing future strains. Peptides offer hope in crisis, with antimicrobial, immunomodulatory, and wound-healing capabilities. Their comprehensive approach uniquely targets invaders, disrupting membranes, inhibiting processes, and orchestrating immune responses. This exploration unveils the potential of peptides in mitigating threats from antibiotic-resistant pathogens. Peptides as antibiotic alternatives promise effective and sustainable poultry health management, marking a crucial step toward resilient farming practices.

Keywords
Peptides · Antibiotic resistance · Poultry health management · Alternative strategies · Antimicrobial Peptides (AMPs)

1   Introduction

The poultry industry is at a critical crossroads, contending with the daunting specter of antibiotic resistance (Agyare et al. 2018). This challenge jeopardizes poultry health and carries far-reaching implications for global food security and human welfare (Gržinić et al. 2023). Antibiotics, long regarded as essential in managing prevalent diseases among poultry flocks, have inadvertently fueled the rise of resilient pathogens due to their widespread and indiscriminate use (Agyare et al. 2018; de Mesquita Souza Saraiva et al. 2022; Gržinić et al. 2023).

The consequence of this overreliance and misuse—antibiotic resistance—has emerged as a pressing concern within poultry farming (Agyare et al. 2018). Pathogens, spanning bacteria, viruses, and fungi, have evolved mechanisms to evade the impact of antibiotics and drugs, resulting in the proliferation of strains impervious to conventional treatments (de Mesquita Souza Saraiva et al. 2022). This disconcerting trend not only endangers the well-being of poultry populations but also presents significant public health risks through the potential transmission of antibiotic-resistant pathogens via the food chain (Bacanlı and Başaran 2019; de Mesquita Souza Saraiva et al. 2022). In response to this escalating crisis, exploring alternative strategies has become imperative. Among these potential alternatives, peptides have surfaced as promising candidates for sustainable solutions in poultry health management. Distinguished by their diverse structures and multifaceted functionalities, peptides present a novel avenue for combatting the challenges posed by infectious agents in poultry farming (Bagley 2014).

This chapter explores peptides to shed light on their prospective role as alternatives to antibiotics in safeguarding poultry health. By delving into the intricacies of peptides, understanding their mechanisms of action, and elucidating their diverse functions, this exploration aims to offer insights into how peptides might effectively address the burgeoning concerns surrounding antibiotic resistance in poultry. As the discussion delves deeper into peptides, the focus will shift toward unraveling their antimicrobial activities, immunomodulatory properties, wound-healing capabilities, and the challenges linked to their integration into poultry health management. This comprehensive exploration aims to illuminate the potential of peptides as invaluable assets in combating the threats posed by antibiotic-resistant pathogens in the poultry industry. Exploring alternative solutions has become imperative in pursuing sustainable and effective strategies against the escalating threat of antibiotic resistance and its impact on poultry health. With their diverse structures and multifaceted functionalities, peptides have emerged as promising candidates in combating pathogenic challenges in poultry farming. This chapter dives into peptides, highlighting their potential as antibiotic alternatives in mitigating the threats posed by infectious agents in poultry.

2 Antimicrobial Peptides (AMPs) and their Mechanisms

Antimicrobial peptides (AMPs) stand as nature's guardians, possessing a remarkable arsenal of defense mechanisms that render them potent agents against various pathogens threatening health (Mourtada 2018). These peptides, distinguished by their diversity in structure and function, are integral components of the innate immune system, orchestrating a multifaceted response against microbial invaders (Hemshekhar et al. 2016). AMPs, also known as host defense peptides, are a diverse group of naturally occurring molecules that play a crucial role in the innate immune response of organisms, including humans, animals, and plants (Broekaert et al. 1995; Hemshekhar et al. 2016; Andrés et al. 2022; Alhhazmi et al. 2024). Several features characterize these peptides:

(i) Molecular weight (1–10 kDa).
(ii) Diverse structures (commonly adopted amphipathic structures) are essential for their interaction with and disruption of microbial membranes.
(iii) Cationic nature, which is believed to contribute to their interaction with the negatively charged microbial membranes. However, not all AMPs are positively charged; some are neutral or even negatively charged; this positive charge allows them to interact with the negatively charged microbial membranes, disrupting the membrane integrity and leading to the death of the microorganism (Hemshekhar et al. 2016).
(iv) Hydrophobic amino acid residues contribute to their ability to insert into and disrupt lipid bilayers, such as those found in microbial cell membranes.
(v) Secondary structures, including alpha-helices, beta-sheets, and extended structures. The specific secondary structure often depends on the surrounding environment and the target microorganism.
(vi) Broad-spectrum activity: AMPs often exhibit a broad spectrum of antimicrobial activity, meaning they can target a wide range of microorganisms, including bacteria, fungi, viruses, and even parasites.
(vii) Fast action: One of the notable characteristics of AMPs is their rapid action. They can act quickly to disrupt microbial membranes, leading to the death of the microorganism.
(viii) Modularity, AMPs are often modular, meaning that different peptide regions can contribute to different functions, such as membrane disruption, immunomodulation, or even the direct killing of microorganisms (Dwivedi et al. 2024).

2.1 Understanding the Structure and Diversity of Antimicrobial Peptides (AMPs)

AMPs, often small and cationic, exhibit remarkable structure diversity. They encompass a spectrum of sequences and configurations, allowing them to interact with microbial membranes, intracellular components, and immune receptors (Hemshekhar et al. 2016; Dwivedi et al. 2024). Their varied structures contribute to their versatility in targeting a broad range of pathogens, including bacteria, fungi, and viruses that afflict poultry flocks (Coustard et al. 2024).

2.2 Mechanisms of Action of AMPs against Various Pathogens Affecting Poultry

AMPs execute their antimicrobial activities through multiple mechanisms, exerting lethal effects on pathogens while evading resistance mechanisms (Mourtada 2018).

Mechanisms of Action of Antimicrobial peptides (AMPs) against Various Pathogens Affecting Poultry

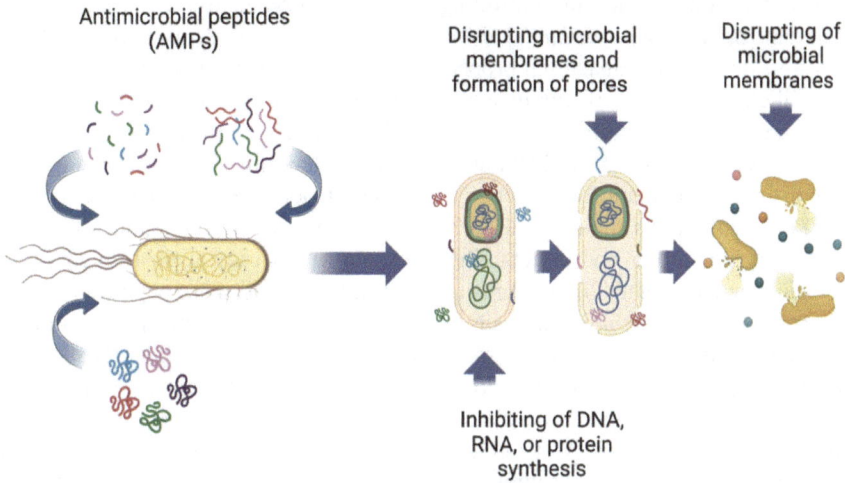

Fig. 11.1 Principal mechanisms of action of antimicrobial peptides (AMPs), Created by BioRender

One such mechanism involves disrupting microbial membranes by forming pores or causing membrane destabilization, leading to cell lysis. Additionally, AMPs can interfere with intracellular processes vital for pathogen survival, such as inhibiting DNA, RNA, or protein synthesis (Kuhn and Di 2024), ultimately impeding microbial growth and propagation. In poultry health, AMPs showcase their prowess against a myriad of pathogens. They combat notorious bacterial strains like *Salmonella* spp., (Mulukutla et al. 2024). *Escherichia coli* and *Clostridium perfringens* commonly afflict poultry populations (Gu et al. 2024; Mulukutla et al. 2024). Moreover, these peptides exhibit efficacy against fungal pathogens such as *Aspergillus* spp. (Deokar et al. 2024) and various viruses, including retrovirus (Mulukutla et al. 2024) (Fig. 11.1).

2.3 Examples of Specific AMPs and their Efficacy in Controlling Poultry Diseases

Specific AMPs have demonstrated remarkable efficacy in controlling poultry diseases (Liu et al. 2024). For instance, cathelicidins, a class of AMPs found in avian species (Cui et al. 2024), exhibit broad-spectrum antimicrobial activities against Gram-positive and Gram-negative bacteria, including *Salmonella* spp. and *E. coli* (Mulukutla et al. 2024; Xia et al. 2024). Additionally, β-defensins, another group of AMPs, showcase potent antimicrobial effects against various pathogens commonly associated with poultry infections.

Peptides exhibit diverse antimicrobial activities against various pathogens, including bacteria, fungi, parasites, and viruses. Here are examples of peptides known for their efficacy in controlling these microorganisms: Bacterial Control. Derived from *Lactococcus lactis*, nisin is a well-known peptide used as a food preservative due to its antibacterial properties, particularly against Gram-positive bacteria like *Staphylococcus aureus* and *Listeria monocytogenes* (Guo et al. 2024b; Sharafi et al. 2024) and bacitracin as an antibiotic peptide effective against Gram-positive bacteria (Li et al. 2024); Fungal Control. Plectasin, derived from a fungus, demonstrates potent antifungal activity, making it a potential candidate for combating fungal infections (Dulta et al. 2024). Cecropins are natural peptides found in insects such as moths and insects. They have shown efficacy against fungal infections in plants (Lu et al. 2024); Parasite Control. Defensins, these peptides found in various organisms, including humans and plants, exhibit broad-spectrum antimicrobial activity, including effectiveness against certain parasites such as *Plasmodium* species responsible for malaria (Yang et al. 2024); Viral Control. Cathelicidin exhibited antiviral properties against various viruses, including influenza viruses and HIV (Cui et al. 2024; Watts et al. 2024). LL-37, a human cathelicidin-derived peptide, has shown antiviral activity against herpes simplex virus (HSV) and human papillomavirus (HPV) (Cui et al. 2024; Kalia and Sarkar 2024).

These examples showcase a wide range of peptides with remarkable antimicrobial properties effective against bacteria, fungi, parasites, and viruses. They provide a glimpse into the extensive library of peptides under scrutiny for their potential in combating infectious diseases across diverse microorganisms. Furthermore, synthetic peptides based on naturally occurring AMPs display encouraging prospects in addressing poultry diseases (Nazeer et al. 2021). Specifically, some synthetic peptides derived from natural templates exhibit enhanced stability and heightened efficacy, introducing innovative therapeutic options to tackle antibiotic-resistant pathogens.

The multifaceted mechanisms of action demonstrated by AMPs highlight their efficacy against a diverse spectrum of pathogens, underscoring their potential as viable alternatives to traditional antibiotics in poultry disease management. Delving into the complexities of AMPs and their diverse functionalities unveils promising pathways for utilizing these peptides as essential assets in preserving poultry health.

3   Immunomodulatory Properties of Peptides: Strengthening Poultry Health through Immune Regulation

Beyond their direct antimicrobial actions, peptides serve as pivotal regulators in modulating the intricate immune system of poultry, orchestrating a symphony of responses that bolster disease resistance and fortify overall poultry health (Cui et al. 2024). These peptides exert profound influences by fine-tuning immune responses and enhancing the resilience of poultry against various diseases. Peptides' Roles in Modulating the Immune System of Poultry. Peptides, endowed with diverse structures and functionalities, play a multifaceted role in modulating the immune system of poultry (Dong et al. 2024). Their interactions with immune cells and signaling molecules instigate events that amplify the body's defense mechanisms. These molecules act as signaling agents, communicating with immune cells to activate or suppress specific immune responses crucial for combating infections (Liu et al. 2024b).

3.1 Stimulation of Immune Responses and Enhancement of Disease Resistance

Peptides stimulate the immune system by triggering the production of cytokines, the molecular messengers orchestrating immune cell functions (Millan-Linares et al. 2024). Cytokines, including interleukins and interferons, play pivotal roles in regulating immune responses, fostering the activation of immune cells, and promoting their coordinated actions against pathogens (Kuhn and Di 2024). Additionally, peptides enhance phagocytosis (Mei et al. 2024), the process by which immune cells engulf and eliminate invading microbes, further fortifying the body's defense mechanisms against infectious agents. Furthermore, peptides regulate inflammatory responses, balancing the necessary inflammatory reactions to combat infections and preventing excessive inflammation that can lead to tissue damage (Cui et al. 2024). This regulatory function ensures that the immune system responds efficiently to threats without causing collateral damage to healthy tissues, thereby promoting faster recovery from infections.

3.2 Impact of Immunomodulation by Peptides on Overall Poultry Health

The immunomodulatory properties exhibited by peptides contribute significantly to overall poultry health and resilience against diseases (Liu et al. 2024b). By fine-tuning immune responses, peptides aid in combating infections more effectively while mitigating the risk of excessive immune reactions that might lead to immunopathology, for instance, decreasing intestinal inflammation (Abreu et al. 2023). This delicate balance achieved by peptides ensures that the immune system remains primed to respond to microbial challenges without causing undue harm to the host. Moreover, the enhanced disease resistance conferred by peptides reduces disease-related economic losses in poultry farming (Liu et al. 2024b). Poultry flocks fortified with robust immune responses facilitated by peptide-mediated immunomodulation exhibit improved productivity, reduced morbidity, and enhanced overall well-being. Beyond their pivotal role as antimicrobial agents, peptides emerge as vital regulators of the poultry immune system. Their ability to stimulate immune responses, enhance disease resistance, and regulate inflammatory reactions not only aids in combating infections but also fosters robust overall poultry health and resilience against diseases.

4   Peptides and Wound Healing in Poultry: Accelerating Tissue Repair and Recovery

Peptides, renowned for their multifaceted functionalities, emerge as promising allies in expediting wound-healing processes within poultry (Zhang et al. 2024). These molecules exhibit remarkable potential in managing injuries, fostering tissue repair, and promoting swift recovery in poultry farming settings.

4.1   Examining Peptides' Potential in Facilitating Wound-Healing Processes

Peptides showcase significant promise in expediting the intricate wound-healing process in poultry (Mangoni et al. 2016). Their diverse structures and functionalities contribute to their pivotal roles in orchestrating cellular responses vital for tissue repair. Peptides interact with various cell types involved in wound healing, stimulating their activities and promoting the coordinated events crucial for restoring tissue integrity (Ramos et al. 2011; Mangoni et al. 2016; Haidari et al. 2023).

4.2   Specific Peptides Known for their Wound-Healing Capabilities and Mechanisms of Action

Certain peptides have garnered recognition for their profound wound-healing capabilities in poultry (Mingfu et al. 2023). For instance, certain growth factor-mimicking peptides stimulate cell proliferation, angiogenesis, and collagen synthesis, essential processes for tissue regeneration (Hu et al. 2023). Additionally, peptides that mimic extracellular matrix components facilitate cell adhesion and migration, aiding in the reconstruction of damaged tissues (Rasouli et al. 2023). These peptides operate through intricate mechanisms, such as binding to specific cell receptors or signaling pathways, triggering events that accelerate wound closure and tissue regeneration. Some peptides also possess anti-inflammatory properties, reducing excessive inflammation at the wound site and promoting a conducive environment for healing (Haidari et al. 2023).

Furthermore, peptides' ability to modulate inflammatory responses aids in creating an optimal microenvironment conducive to healing. By mitigating excessive inflammation, peptides facilitate a balanced healing process, minimizing tissue damage and promoting more efficient recovery in poultry affected by injuries or wounds, especially at the intestinal level (Saravanan et al. 2023). Integrating peptides into wound management strategies in poultry farming heralds a new era of enhanced tissue repair and recovery. The application of specific peptides known for their wound-healing capabilities, mechanisms of action, and potential to manage injuries highlight a promising avenue for fostering optimal health and welfare intestinal standards within poultry flocks.

5 Challenges in Implementing Peptides as Alternatives in Poultry Health Management

Integrating peptides as alternatives to conventional treatments in poultry health management presents a promising yet intricate endeavor. Despite their potential, several hurdles and limitations impede the seamless utilization of peptides within poultry farming practices.

Addressing the Hurdles and Limitations in Utilizing Peptides. One of the primary challenges involves overcoming hurdles related to the intricate nature of peptides. Peptides are susceptible to enzymatic degradation, necessitating innovative formulations or delivery systems that ensure their stability and sustained efficacy within the poultry system (Ali et al. 2023). Achieving a delicate balance between maintaining peptide integrity and ensuring effective delivery to target sites remains a significant obstacle.

5.1 Challenges Related to Peptide Stability, Production Scalability, and Cost-Effectiveness

Peptide stability, both in storage and within the biological milieu of poultry, is a critical concern. It is a priority for their therapeutic effectiveness to ensure that peptides retain their structural integrity and bioactivity under varying environmental conditions (Selvaraj and Chen 2023). Moreover, the scalability of peptide production poses a challenge, as large-scale synthesis methods that maintain quality standards while remaining cost-effective are yet to be fully optimized (Bellavita et al. 2023). The cost of peptide production remains a significant factor inhibiting widespread adoption. Synthesizing peptides using sophisticated techniques often incurs high production costs, hindering their cost-effectiveness as viable alternatives to traditional treatments (dos Santos and Franco 2023).

5.2 Regulatory Aspects and Safety Considerations Associated with Peptide Use in Poultry

Regulatory frameworks governing the use of peptides in poultry health management pose another hurdle (Arun et al. 2023). Ensuring compliance with stringent regulations, including establishing safety profiles, defining appropriate dosages, and evaluating potential risks, demands rigorous assessment and validation. Furthermore, addressing concerns about residues and withdrawal periods in poultry products requires meticulous scrutiny to ensure consumer safety. Safety considerations encompass not only the direct effects of peptides on poultry health but also potential environmental impacts and implications for human consumption (Chen et al. 2023; Kamal et al. 2023; Singh et al. 2023). Comprehensive safety assessments are imperative to ascertain the absence of adverse effects on poultry, humans, and the environment (Chen et al. 2023).

Mitigating these challenges necessitates concerted efforts from researchers, industry stakeholders, and regulatory bodies. Innovations in peptide stabilization techniques, advancements in scalable production methods, and comprehensive safety evaluations are pivotal in surmounting these obstacles and unlocking the full potential of peptides as viable alternatives in poultry health management.

6 Advancements in Peptide Synthesis and Delivery: Revolutionizing Peptide Applications in Poultry Health

The peptide synthesis and delivery realm has witnessed remarkable strides, paving the way for enhanced stability, efficacy, and innovative administration methods, thereby revolutionizing their applications in poultry health management.

6.1 Innovations in Peptide Synthesis Techniques for Enhancing Stability and Efficacy

Recently, novel peptide synthesis techniques have emerged to bolster peptide stability and efficacy in poultry settings. Innovations such as solid-phase peptide synthesis (SPPS) and liquid-phase peptide synthesis (LPPS) have enabled the production of peptides with improved structural integrity, thereby enhancing their stability against enzymatic degradation and environmental factors (Guo et al. 2024a; Schüttel et al. 2024). Furthermore, advancements in peptide modification strategies, such as cyclization or incorporation of non-natural amino acids, have increased resistance to degradation, prolonging peptide activity within the avian system (Enninful et al. 2024). These innovations ensure peptides retain their bioactivity and therapeutic potential, even under challenging conditions in poultry environments. A possible new approach for this problem with clear advantages is using microalgae to express peptides as alternative antibiotics (Dávalos-Guzmán et al. 2023). It is an innovative and promising approach in the biotechnology field. Microalgae, such as *Chlamydomonas reinhardtii* and others, offer several advantages in producing therapeutic peptides regarding its efficient and cost-effective platform, offering a relatively simple and scalable system for producing heterologous proteins and proper post-translational modifications of proteins (Sun et al. 2024). Providing a crucial correct folding, stability, and functionality of therapeutic peptides.

6.2 Novel Delivery Systems Facilitating the Administration of Peptides in Poultry

Developing novel delivery systems is a cornerstone in optimizing peptide administration within poultry farming. Nanotechnology-based delivery systems, such as nanoparticles and liposomes, offer promising avenues for efficient and targeted delivery of peptides to specific sites of action within poultry organisms (Luo et al. 2024). Nano-formulations encapsulating peptides protect them from degradation and facilitate controlled release, ensuring sustained and optimal therapeutic concentrations at the desired sites. Additionally, encapsulation within carrier molecules or matrices enhances peptides' bioavailability and cellular uptake, maximizing their efficacy in managing poultry health (Primo et al. 2024).

6.3 Prospects and Developments in Improving Peptide Applications

The future of peptide applications in poultry health management holds exciting prospects for further innovations. Ongoing research focuses on refining existing peptide synthesis methods, seeking cost-effective and scalable approaches that maintain high quality and bioactivity. Integrating cutting-edge technologies, such as gene-editing techniques like CRISPR/Cas9, into peptide research offers prospects for designing novel peptides with enhanced functionalities tailored to address specific health challenges (Sen and Mukhopadhyay 2024).

The continuous evolution of peptide synthesis techniques, innovative delivery systems, and future-oriented research endeavors promise a transformative impact on poultry health management (García et al. 2024; Liu et al. 2024b). Harnessing these advancements can elevate the efficacy, accessibility, and applicability of peptides as indispensable tools in ensuring optimal health and well-being within poultry flocks.

7 Case Studies and Successful Applications: Peptides Transforming Poultry Health Management

Examining real-world case studies is a testament to the remarkable success of strategically implementing peptides in poultry farming. These instances underscore the efficacy of peptides and highlight their potential to replace antibiotics, offering invaluable lessons and insights for broader adoption in the industry.

7.1 Showcasing Successful Use of Peptides in Poultry Farming

Real-world case studies reveal compelling success stories where peptides have important roles in transforming poultry health management (Liu et al. 2024b). Instances abound where peptides, leveraging their diverse functionalities, have exhibited remarkable efficacy in controlling and preventing diseases among poultry flocks (García et al. 2024). For instance, in a case study involving the management of bacterial infections in poultry, specific antimicrobial peptides demonstrated superior efficacy against *Clostridium perfringens*, non-clostridial strains, including *Listeria monocytogenes*, methicillin-resistant *Staphylococcus aureus*, *Streptococcus suis*, *Streptococcus pyogenes*, *Enterococcus cecorum*, and *Enterococcus faecalis* (García et al. 2024). These peptides effectively suppressed bacterial growth and mitigated the emergence of antibiotic-resistant strains, ensuring sustained disease control without the pitfalls of resistance development.

7.2   Peptides Effectively Replacing Antibiotics in Disease Management

A notable highlight in these case studies is the successful replacement of antibiotics with peptides in disease management within poultry farming. Instances where peptides have outperformed antibiotics in combating pathogens showcase their potential as viable alternatives (Grace et al. 2024). In a striking case, a poultry farm grappling with recurrent bacterial infections successfully transitioned from antibiotic-based treatments to peptide-based solutions (Lamichhane et al. 2024). This shift resolved the persistent infections and reduced antibiotic usage, mitigating the risk of antibiotic resistance while ensuring effective disease control.

7.3   Lessons Learned and Implications for Broader Adoption in the Industry

The lessons from these case studies carry profound implications for the broader adoption of peptides in the poultry industry. Firstly, they underscore the efficacy and versatility of peptides as potent alternatives to antibiotics, offering a sustainable approach to disease management (Ioannou et al. 2023). Moreover, these case studies highlight the importance of proactive measures in mitigating the risks associated with antibiotic resistance (Luo et al. 2023). The successful adoption of peptides addresses immediate disease challenges and contributes to the long-term goal of preserving antibiotic efficacy for both animal and human health. Additionally, the cases emphasize the need for further research, innovation, and stakeholder collaboration to optimize peptide applications in poultry health management (Xie et al. 2023). Addressing challenges related to peptide stability, production scalability, and regulatory frameworks emerges as critical areas for advancement.

The successful case studies in which peptides have effectively transformed disease management in poultry farming are compelling evidence of their efficacy and potential. These instances offer valuable lessons and insights, advocating for broader adoption of peptides as indispensable tools in ensuring sustainable and resilient poultry health.

8   Future Perspectives and Recommendations: Paving the Way for Peptides in Poultry Health

The potential of peptides as alternatives to antibiotics in poultry health management heralds a promising future, underscoring the need for concerted efforts to leverage their efficacy. Recommendations for further research, development, and integration into standard poultry health protocols carry profound implications for sustainable farming practices.

8.1 Summarizing the Potential of Peptides as Alternatives to Antibiotics in Poultry Health

Peptides emerge as formidable contenders in the quest for sustainable solutions to combat the escalating challenges of antibiotic resistance in poultry farming. Their diverse functionalities, ranging from antimicrobial action to immunomodulation and wound-healing capabilities, showcase a breadth of applications that position peptides as invaluable assets in safeguarding poultry health. Furthermore, the ability of peptides to address disease challenges while minimizing the risks associated with antibiotic resistance underscores their potential to revolutionize poultry health management. Their multifaceted mechanisms of action and versatility render them pivotal in promoting optimal health and welfare standards within poultry flocks.

8.2 Recommendations for Further Research, Development, and Integration

Moving forward, recommendations for extensive research, development, and integration of peptides into standard poultry health protocols are imperative. This fosters collaborative efforts among researchers, industry stakeholders, and regulatory bodies to advance peptide-based interventions. Research endeavors should focus on elucidating the nuances of peptide functions, optimizing synthesis techniques, and developing innovative delivery systems. Exploring the potential of engineered peptides tailored to specific poultry health challenges is a promising avenue for enhancing therapeutic efficacy and minimizing potential limitations. Integrating peptides into standard poultry health protocols necessitates the establishment of guidelines, protocols, and regulations governing their use. Rigorous safety assessments, dosage optimization, and residue monitoring protocols are essential for ensuring their efficacy, safety, and compliance within the poultry farming industry.

8.3 Insights into Long-Term Implications and Role in Sustainable Farming Practices

Adopting peptides in poultry health management holds profound implications for sustainable farming practices. Beyond their immediate benefits in disease control, peptides contribute to fostering sustainable practices by reducing reliance on antibiotics, mitigating risks associated with resistance development, and promoting overall animal welfare. The long-term implications of peptide adoption extend to preserving antibiotic efficacy and safeguarding animal and human health. By embracing peptides as alternatives, poultry farming can transition toward more resilient, environmentally friendly, and socially responsible practices.

The outlook for peptides in poultry health management is promising, contingent upon concerted efforts in research, development, integration, and regulatory frameworks. Leveraging the potential of peptides as alternative antibiotics addresses immediate disease challenges and contributes to sustainable poultry farming practices, ensuring the health, welfare, and safety of poultry and consumers alike.

References

Abreu R, Semedo-Lemsaddek T, Cunha E, Tavares L, Oliveira M (2023) Antimicrobial drug resistance in poultry production: current status and innovative strategies for bacterial control. Microorganisms 11:953

Agyare C, Boamah VE, Zumbi CN, Osei FB (2018) Antibiotic use in poultry production and its effects on bacterial resistance. Antimicrob Resist Glob Threat:33–51

Alhhazmi AA, Alluhibi SS, Alhujaily R, Alenazi ME, Aljohani TL, Al-Jazzar A-AT, Aljabri AD, Albaqami R, Almutairi D, Alhelali LK, others (2024) Novel antimicrobial peptides identified in legume plant, Medicago truncatula. Microbiol Spectr 12:e01827–e01823

Ali M, van Gent ME, de Waal AM, van Doodewaerd BR, Bos E, Koning RI, Cordfunke RA, Drijfhout JW, Nibbering PH (2023) Physical and functional characterization of PLGA nanoparticles containing the antimicrobial peptide SAAP-148. Int J Mol Sci 24:2867

Andrés CMC, Pérez de la Lastra JM, Juan CA, Plou FJ, Pérez-Lebeña E (2022) The role of reactive species on innate immunity. Vaccine 10:1735

Arun B, Rejeesh E, Rani NM (2023) Future perspective of peptide antibiotic market. In: Antimicrobial peptides. Elsevier, pp 311–320

Bacanlı M, Başaran N (2019) Importance of antibiotic residues in animal food. Food Chem Toxicol 125:462–466

Bagley C (2014) Potential role of synthetic antimicrobial peptides in animal health to combat growing concerns of antibiotic resistance-a review. Wyno Acad J Agric Sci 2:19–28

Bellavita R, Braccia S, Galdiero S, Falanga A (2023) Glycosylation and lipidation strategies: approaches for improving antimicrobial peptide efficacy. Pharmaceuticals 16:439

Broekaert WF, Terras F, Cammue B, Osborn RW (1995) Plant defensins: novel antimicrobial peptides as components of the host defense system. Plant Physiol 108:1353

Chen B, Zhang Z, Zhang Q, Xu N, Lu T, Wang T, Hong W, Fu Z, Penuelas J, Gillings M (2023) Antimicrobial peptides in the global microbiome: biosynthetic genes and resistance determinants. Environ Sci Technol 57:7698

Coustard SM, Rossignol C, Collin A, Blanc F, Lallier N, Schouler C, Duval ELB, Travel A, Lalmanach A-C (2024) Research note: intestinal avian defensin 2 and robustness of chicks. Poult Sci 103:103175

Cui X, Huang Y, Peng Z, Li Z, Cen S (2024) Mammalian antimicrobial peptides: defensins and cathelicidins. In: Molecular medical microbiology. Elsevier, pp 551–573

Dávalos-Guzmán SD, Martinez-Gutierrez F, Martínez-González L, Quezada-Rivera JJ, Lorenzo-Leal AC, Bach H, Morales-Domínguez JF, Soria-Guerra RE (2023) Antimicrobial activity of the Flo peptide produced in Scenedesmus acutus and Nannochloropsis oculata. World J Microbiol Biotechnol 39:1–10

de Mesquita Souza Saraiva M, Lim K, do Monte DFM, Givisiez PEN, Alves LBR, de Freitas Neto OC, Kariuki S, Júnior AB, de Oliveira CJB, Gebreyes WA (2022) Antimicrobial resistance in the globalized food chain: a one health perspective applied to the poultry industry. Braz J Microbiol 1–22

Deokar GS, Nirmal NP, Kshirsagar SJ (2024) Plant seeds: a potential bioresource for isolation of nutraceutical and bioactive compounds. In: Bioactive extraction and application in food and nutraceutical industries. Springer, pp 333–372

Dong X, Bie J, Liu X (2024) Research note: isolation and immunomodulatory activity of bursal peptide, a novel peptide from avian immune system developments. Poult Sci 103:103294

dos Santos C, Franco OL (2023) Advances in the use of plants as potential biofactories in the production of antimicrobial peptides. Pept Sci 115:e24290

Dulta K, Kumar K, Thakur A, Singh S, Ağçeli GK, Singh D (2024) 171Laetiporus sulphureus (Bull.). In: Murrill, in: edible and medicinal mushrooms of the Himalayas. CRC Press, pp 171–187

Dwivedi M, Parmar MD, Mukherjee D, Yadava A, Yadav H, Saini NP (2024) Biochemistry, mechanistic intricacies, and therapeutic potential of antimicrobial peptides: an alternative to traditional antibiotics. Curr Med Chem 31:6110

Enninful GN, Kuppusamy R, Tiburu EK, Kumar N, Willcox MD (2024) Non-canonical amino acid bioincorporation into antimicrobial peptides and its challenges. J Pept Sci 30:e3560

García-Vela S, Guay L-D, Rahman MRT, Biron E, Torres C, Fliss I (2024) Antimicrobial activity of synthetic enterocins A, B, P, SEK4, and L50, alone and in combinations, against Clostridium perfringens. Int J Mol Sci 25:1597

Grace D, Knight-Jones TJ, Melaku A, Alders R, Jemberu WT (2024) The public health importance and management of infectious poultry diseases in smallholder systems in Africa. Food Secur 13:411

Gržinić G, Piotrowicz-Cieślak A, Klimkowicz-Pawlas A, Górny RL, Lawniczek-Walczyk A, Piechowicz L, Olkowska E, Potrykus M, Tankiewicz M, Krupka M, others (2023) Intensive poultry farming: a review of the impact on the environment and human health. Sci Total Environ 858:160014

Gu Q, Yan J, Lou Y, Zhang Z, Li Y, Zhu Z, Liu M, Wu D, Liang Y, Pu J, others (2024) Bacteriocins: curial guardians of gastrointestinal tract. Compr Rev Food Sci Food Saf 23:1–27

Guo D, Xiong Y, Fu B, Sha Z, Li B, Wu H (2024a) Liquid-liquid phase separation in bacteria. Microbiol Res 127627:127627

Guo L, Wambui J, Wang C, Broos J, Stephan R, Kuipers OP (2024b) Rombocin, a short stable natural Nisin variant, displays selective antimicrobial activity against listeria monocytogenes and employs a dual mode of action to kill target bacterial strains, vol 13. ACS Synth Biol, p 370

Haidari H, Melguizo-Rodríguez L, Cowin AJ, Kopecki Z (2023) Therapeutic potential of antimicrobial peptides for treatment of wound infection. Am J Physiol-Cell Physiol 324:C29–C38

Hemshekhar M, Anaparti V, Mookherjee N (2016) Functions of cationic host defense peptides in immunity. Pharmaceuticals 9:40

Hu Q, Chen C, Lin Z, Zhang L, Guan S, Zhuang X, Dong G, Shen J (2023) The antimicrobial peptide Esculentin-1a (1–21) NH2 stimulates wound healing by promoting angiogenesis through the PI3K/AKT pathway. Biol Pharm Bull 46:382–393

Ioannou P, Baliou S, Kofteridis DP (2023) Antimicrobial peptides in infectious diseases and beyond—a narrative review. Life 13:1651

Kalia V, Sarkar S (2024) Vitamin D and antiviral immunity. In: Feldman and Pike's Vitamin D. Elsevier, pp 1011–1034

Kamal I, Ashfaq UA, Hayat S, Aslam B, Sarfraz MH, Yaseen H, Rajoka MSR, Shah AA, Khurshid M (2023) Prospects of antimicrobial peptides as an alternative to chemical preservatives for food safety. Biotechnol Lett 45:137–162

Kuhn JM, Di YP (2024) Antimicrobial peptides in lung health and pulmonary diseases. In: Lung biology and pathophysiology. CRC Press, pp 3–22

Lamichhane B, Mawad AM, Saleh M, Kelley WG, Harrington PJ, Lovestad CW, Amezcua J, Sarhan MM, El Zowalaty ME, Ramadan H, others (2024) Salmonellosis: an overview of epidemiology, pathogenesis, and innovative approaches to mitigate the antimicrobial resistant infections. Antibiotics 13:76

Li C, Zhou Z, Wang W, Zhao Y, Yin X, Meng Y, Zhao P, Wang M, Liu X, Wang X, others (2024) Development of antibacterial peptides with membrane disruption and folate pathway inhibitory activities against methicillin-resistant Staphylococcus aureus. J Med Chem 67:1044

Liu X, Hong H, Wang J, Huang J, Li J, Tao Y, Liu M, Pang H, Li J, Bo R (2024a) Mucosal immune responses and protective efficacy elicited by oral administration AMP-ZnONPs-adjuvanted inactivated H9N2 virus in chickens. Poult Sci 103:103496

Liu X, Wang X, Shi X, Wang S, Shao K (2024b) The immune enhancing effect of antimicrobial peptide LLv on broilers chickens. Poult Sci 103:103235

Lu T, Ji Y, Chang M, Zhang X, Wang Y, Zou Z (2024) The accumulation of modular serine protease mediated by a novel circRNA sponging miRNA increases Aedes aegypti immunity to fungus. BMC Biol 22:1–18

Luo L, Huang W, Zhang J, Yu Y, Sun T (2024) Metal-based nanoparticles as antimicrobial agents: a review. ACS Appl Nano Mater 7:2529

Luo X, Chen H, Song Y, Qin Z, Xu L, He N, Tan Y, Dessie W (2023) Advancements, challenges and future perspectives on peptide-based drugs: focus on antimicrobial peptides. Eur J Pharm Sci 181:106363

Mangoni ML, McDermott AM, Zasloff M (2016) Antimicrobial peptides and wound healing: biological and therapeutic considerations. Exp Dermatol 25:167–173

Mei K-C, Thota N, Wei P-S, Yi B, Bonacquisti EE, Nguyen J (2024) Calreticulin P-domain-derived "eat-me" peptides for enhancing liposomal uptake in dendritic cells. Int J Pharm 653:123844

Millan-Linares MC, Rivero-Pino F, Gonzalez-de la Rosa T, Villanueva A, Montserrat-de la Paz S (2024) Identification, characterization, and molecular docking of immunomodulatory oligopeptides from bioavailable hempseed protein hydrolysates. Food Res Int 176:113712

Mingfu N, Qiang G, Yang L, Ying H, Chengshui L, Cuili Q (2023) The antimicrobial peptide MetchnikowinII enhances Ptfa antigen immune responses against avian Pasteurella multocida in chickens. J Vet Med Sci 85:964–971

Mourtada R (2018) Engineering membrane-selective antibiotic peptides to combat multidrug resistance. (PhD Thesis). Massachusetts Institute of Technology

Mulukutla A, Shreshtha R, Deb VK, Chatterjee P, Jain U, Chauhan N (2024) Recent advances in antimicrobial peptide-based therapy. Bioorg Chem 107151:107151

Nazeer N, Rodriguez-Lecompte JC, Ahmed M (2021) Bacterial-specific aggregation and killing of immunomodulatory host defense peptides. Pharmaceuticals 14:839

Primo LMDG, Roque-Borda CA, Canales CSC, Caruso IP, de Lourenço IO, Colturato VMM, Sábio RM, de Melo FA, Vicente EF, Chorilli M (2024) Antimicrobial peptides grafted onto the surface of N-acetylcysteine-chitosan nanoparticles can revitalize drugs against clinical isolates of mycobacterium tuberculosis. Carbohydr Polym 323:121449

Ramos R, Silva JP, Rodrigues AC, Costa R, Guardão L, Schmitt F, Soares R, Vilanova M, Domingues L, Gama M (2011) Wound healing activity of the human antimicrobial peptide LL37. Peptides 32:1469–1476

Rasouli M, Soleimani M, Hosseinzadeh S, Ranjbari J (2023) Bacterial cellulose as potential dressing and scaffold material: toward improving the antibacterial and cell adhesion properties. J Polym Environ 31:1–20

Saravanan P, Balachander N, Kesav Ram Singh K (2023) Anti-inflammatory and wound healing properties of lactic acid bacteria and its peptides. Folia Microbiol. (Praha):1–17

Schüttel M, Will E, Sangouard G, Zarda A, Habeshian S, Nielsen AL, Heinis C (2024) Solid-phase peptide synthesis in 384-well plates. J Pept Sci 30:e3555

Selvaraj SP, Chen J-Y (2023) Conjugation of antimicrobial peptides to enhance therapeutic efficacy. Eur J Med Chem 259:115680

Sen D, Mukhopadhyay P (2024) Antimicrobial resistance (AMR) management using CRISPR-Cas based genome editing. Gene Genome Ed 100031:100031

Sharafi T, Ghaemi EA, Rafiee M, Ardebili A (2024) Combination antimicrobial therapy: in vitro synergistic effect of anti-staphylococcal drug oxacillin with antimicrobial peptide nisin against Staphylococcus epidermidis clinical isolates and Staphylococcus aureus biofilms. Ann Clin Microbiol Antimicrob 23:1–12

Singh A, Duche RT, Wandhare AG, Sian JK, Singh BP, Sihag MK, Singh KS, Sangwan V, Talan S, Panwar H (2023) Milk-derived antimicrobial peptides: overview, applications, and future perspectives. Probiotics Antimicrob. Proteins 15:44–62

Sun S-N, Fan LL, Diao A, Fan Z-C (2024) Chlamydomonas reinhardtii-derived triple BmKbpp distorts membrane integrity for inhibiting bacterial growth. Process Biochem 137:10–19

Watts S, Hänni E, Smith GN, Mahmoudi N, Freire RV, Lim S, Salentinig S (2024) Human antimicrobial peptide inactivation mechanism of enveloped viruses. J Colloid Interface Sci 657:971–981

Xia R, Xiao H, Xu M, Hou L, Han Y, Zhou Z (2024) Insight into the inhibitory activity and mechanism of bovine cathelicidin BMAP 27 against salmonella typhimurium. Microb Pathog 187:106540

Xie W-Y, Yuan Y, Wang Y-T, Liu D-Y, Shen Q, Zhao F-J (2023) Hazard reduction and persistence of risk of antibiotic resistance during thermophilic composting of animal waste. J Environ Manag 330:117249

Yang J, Xu Q, Shen W, Jiang Z, Gu X, Li F, Li B, Wei J (2024) The toll/IMD pathways mediate host protection against dipteran parasitoids. J Insect Physiol 104614:104614

Zhang G, Wang Y, Qiu H, Lu L (2024) Facile one-pot synthesis of flower-like ellagic acid microparticles incorporating anti-microbial peptides for enhanced wound healing. J Mater Chem B 12:500–507

Advances in Anti-Mycotoxins

12

Jesus A. Maguey-Gonzalez, Juan D. Latorre,
Lauren Laverty, Inkar Castellanos-Huerta,
Awad A. Shehata, Wolfgang Eisenreich,
and Guillermo Tellez-Isaias

Contents

J. A. Maguey-Gonzalez · J. D. Latorre · L. Laverty · I. Castellanos-Huerta · G. Tellez-Isaias
(✉)
Department of Poultry Science, University of Arkansas, Fayetteville, AR, USA
e-mail: gtellez@uark.edu

A. A. Shehata · W. Eisenreich
Structural Membrane Biochemistry, Bavarian NMR Center, Technical University of Munich
(TUM), Garching, Bayern, Germany

© The Author(s), under exclusive license to Springer Nature Switzerland AG 2024
A. A. Shehata et al. (eds.), *Alternatives to Antibiotics against Pathogens in Poultry*, https://doi.org/10.1007/978-3-031-70480-2_12

Abstract

Mycotoxins are toxic chemicals produced by certain fungi and pose a significant threat to human and animal health through contaminated food and feed. These toxins, primarily produced by *Aspergillus*, *Claviceps*, *Penicillium*, *Fusarium*, and *Alternaria* fungi, can cause acute or chronic illness. Despite the identification of more than 500 mycotoxins, six major groups (aflatoxins, fumonisins, ochratoxins, trichothecenes, zearalenone, and patulin) pose the most significant food safety concerns due to their frequent occurrence. Mycotoxins weaken the immune system, making animals more susceptible to infections and diseases, ultimately impacting their health and performance. These toxins can cause a range of harmful effects, including damage to the liver (hepatotoxic), suppress the immune system (immunotoxic), cause genetic mutations (mutagenic), increase cancer risk (carcinogenic), and disrupt fetal development (teratogenic) in several animal species. Controlling mycotoxin contamination remains a significant global challenge. Even with the implementation of the best practices in agriculture, storage, and processing, their unpredictable presence and exceptional resistance to different treatments pose a permanent threat to food safety. Therefore, this chapter explores the latest advances in anti-mycotoxin strategies for poultry, focusing on innovative approaches to counteract mycotoxin contamination.

Keywords

Mycotoxins · Poultry · Health · Anti-mycotoxins · Additives

1 Introduction

Mycotoxins are low molecular weight secondary metabolites produced by filamentous fungi of the genera: *Aspergillus*, *Fusarium*, *Penicillium*, *Claviceps*, and *Alternaria*. Secondary metabolites have no relevance in terms of the biochemical processes essential for the growth or development of fungi; however, they are capable of causing significant diseases in humans and animals (Janik et al. 2020). Mycotoxins can be produced simultaneously by several fungal genera, for example, aflatoxins are produced by species of the genera *Aspergillus*; species of the genera *Fusarium* produce fumonisins; ochratoxin A produced by species of the *Aspergillus* and *Penicillium* genera, trichothecenes produced by different species of the Fusarium genera; and zearalenone produced by species of the genus *Fusarium* (Zain 2011). The mycotoxins are considered the most frequent in food products, therefore they can cause significant economic losses and health problems (Ochieng et al. 2021). The consumption of diets contaminated with mycotoxins causes the development of a disease called mycotoxicosis, which can be acute or chronic. The course of the mycotoxicosis disease will depend on the dose and time of exposure to mycotoxins. In addition, to the species, age, sex, immunological and nutritional status of the individual. The effects of mycotoxins among animals are very similar (Awuchi et al.

2022). Some effects caused by mycotoxins in birds are: hepatotoxic, hepatocarcino-genic, teratogenic, and mutagenic (Ali 2020). It also causes immunosuppression, anorexia, decreased growth, poor absorption of food, increased morbidity and mortality, changes in certain biochemical values of blood serum, capillary fragility, and hemorrhages (Okasha et al. 2024). Due to the severe effects caused by mycotoxins, numerous researchers have developed various methods to reduce their toxic effects on animals (based on physical, chemical, and biological strategies) (Agriopoulou et al. 2020). Among these, physical methods are the most used in the animal production industry, specifically adsorbents, because they are considered one of the most effective methods in the elimination of mycotoxins (Kamle et al. 2019). Adsorbents are divided into inorganic and organic (also called bio-adsorbents). Inorganic adsorbents are the most studied and used to reduce the toxic effects caused by mycotoxins. However, bio-adsorbents have gained relevance importance, since they are safe, profitable, and effective. Furthermore, they are capable of adsorbing mycotoxins without compromising the micronutrients of the animal diet and releasing compounds toxic to animals, as inorganic adsorbents do (Karlovsky et al. 2016). The mycotoxins adsorption effectiveness of various bio-adsorbents has been studied, which resemble or exceed inorganic adsorbents in their effectiveness. Consequently, this chapter explores the most recent developments in poultry anti-mycotoxin methods for feed decontamination.

2 Mycotoxins

2.1 Fungi and Mycotoxins

Fungi are eukaryotic, heterotrophic, and ubiquitous microorganisms, which present two types of metabolic processes: primary and secondary. Primary metabolites are essential compounds for growth and include proteins, carbohydrates, nucleic acids, and lipids. These primary products must be synthesized if they cannot be obtained from the growth medium. Secondary metabolites can be defined as chemical compounds resulting from specific biosynthetic pathways, whose production is not necessary for the normal growth and development of the fungus (Daou et al. 2021). However, they are present in numerous species, and therefore their persistence in evolution implies a competitive benefit in nature (Nigam and Singh 2014). Mycotoxins constitute a structurally diverse group of toxic and low molecular weight compounds, which are generally less than 1000 Da. Approximately 500 potentially toxic mycotoxins produced by more than 100 fungi species have been identified. *Aspergillus*, *Fusarium*, *Penicillium*, *Claviceps*, and *Alternaria* are some species of fungi capable of producing various mycotoxins simultaneously, such as aflatoxins produced by *Aspergillus* species, fumonisins produced by *Fusarium* species, ochratoxin A produced by *Aspergillus* and *Penicillium* species, trichothecenes produced by different *Fusarium* species, and zearalenone produced by *Fusarium* species (Haque et al. 2020; Hoerr 2020) (Fig. 12.1).

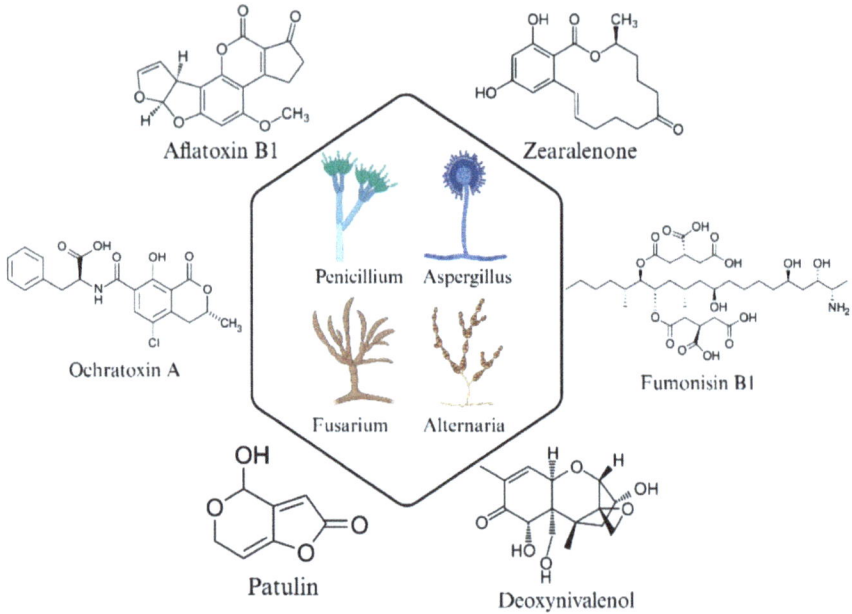

Fig. 12.1 Main fungi and mycotoxins "Created with BioRender.com"

The presence of mycotoxins in grains and rations has been investigated worldwide and it has been reported that it is possible to have grains with a high mycotoxin content, which are frequently contaminated with more than one mycotoxin (Omotayo et al. 2019). Furthermore, it has been observed that inhalation, skin contact, and consumption of a diet contaminated with mycotoxins can cause the development of an acute or chronic health disease in animals and humans (Smith et al. 2016). This disease is called mycotoxicosis and its severity depends on the toxicity of the mycotoxin, dose, and exposure time; as well as the age, sex, immunological and nutritional status of the individual, causing significant economic losses for the animal production industry (Bennett 1987).

In general, mycotoxins are odorless, tasteless, and colorless; they are also molecules that are poorly soluble in water and soluble in moderately polar organic solvents, so they are easily extracted with methanol, chloroform, acetone, acetonitrile, and dimethyl sulfoxide (Steyn 1995). Chemically they are stable in foods and resistant to the temperatures of cooking, ultra-pasteurization, frying, boiling, fermentation, and nixtamalization, making them difficult to eliminate once they are produced (Carvajal-Moreno 2022; Kumar et al. 2017). Although they can be destroyed in an autoclave with ammonium or sodium hypochlorite, mycotoxins are difficult to eliminate because they are not destroyed by most of the processes used in animal food production (Kumar et al. 2017).

2.2 Mycotoxins Production

Several factors influence fungal growth and the subsequent synthesis of mycotoxins in grains, the main factors (Fig. 12.2) include the environmental conditions for the colonization of the fungus, such as water activity, temperature, pH, oxygen, substrate compositions (Aldars-García et al. 2018), and factors that are based on interactions between the species of the fungus and its substrate (Plant susceptibility, spore load, presence of insects, stress conditions, and physical integrity of grains) (Kumar et al. 2021). The production of mycotoxins in the field occurs when there is drought, stress, and high temperatures. Healthy plants are less likely to be colonized; however, under conditions of stress plants are highly prone to contamination by fungi and subsequent synthesis of mycotoxins. Furthermore, mycotoxins are synthesized exponentially mainly in inadequate grain storage conditions (heat and high humidity) (Marin et al. 2013).

The optimal temperature for mold growth varies from 20 °C to 37 °C, and for the production of toxins, it is 25.5 °C, a significant factor for mycotoxin production is water activity, the optimal range of this parameter varies from 0.83 to above 0.9 (Aldars-García et al. 2018). Additional factors increasing mold growth and mycotoxin production are high relative humidity (70–90%) and moisture content (20–25%). Fungi can colonize and produce mycotoxins during two stages (Field and storage) (Bhat et al. 2010). Field fungi that colonize grains and produce preharvest mycotoxins, typically require high levels of humidity in grain (20–22%), such as the genera *Alternaria* and *Fusarium*. Storage fungi that colonize grains and produce post-harvest mycotoxins according to their storage conditions, such as the genera *Aspergillus* and *Penicillium*. In general, the conditions for the development of pre-harvest mycotoxins are high temperatures, high levels of humidity, and damage by insects, which cause stress in the plants. On the other hand, inadequate post-harvest conditions are inadequate transportation, storage, drying, handling, and packaging. During the pre-harvest and post-harvest stages, mycotoxins are capable

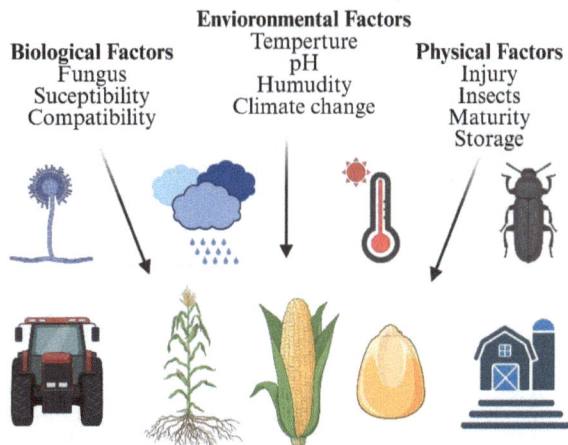

Fig. 12.2 Fundamental variables that affect the formation of mycotoxins "Created with BioRender.com"

Enviornmental Factors

Biological Factors
Fungus
Suceptibility
Compatibility

Temperture
pH
Humudity
Climate change

Physical Factors
Injury
Insects
Maturity
Storage

of greatly reducing the value and marketability of grains due to their negative effects (Cleveland et al. 2003; Dowd 2003; Magan and Aldred 2007).

2.3   Occurrence in Feed Ingredients

Feed ingredients that are the most susceptible to being contaminated by mycotoxins are barley, corn, rice, sorghum, and wheat; and oilseeds such as cotton, peanuts, rapeseed, sunflowers, peanuts, and nuts (Mahato et al. 2019). According to the Food and Agriculture Organization of the United Nations (FAO) report, 25% of global agricultural products are contaminated by mycotoxins, resulting in economic losses (Eskola et al. 2020). It is estimated that between 25 and 40% of cereals are contaminated with more than two different mycotoxins (Wan et al. 2020). However, the occurrence of mycotoxins is changing as average temperatures increase due to climate change in the world (Perrone et al. 2020). Mycotoxin contamination is considered an unpredictable and inevitable problem due to its resistance to high temperatures and chemical or physical treatments. Therefore, management, prevention measures, and adequate control strategies are required to combat the problems generated by mycotoxins (Zain 2011).

2.4   Effects on Poultry

The effects of mycotoxins are very similar in all animals. However, it was previously mentioned that susceptibility varies between species, age, breed, sex, immunological, and nutritional status. The most common effects on health (Fig. 12.3) are nephrotoxicity, hepatotoxicity, immunosuppression, carcinogenicity, and teratogenicity (Filazi et al. 2017). The susceptibility of birds to aflatoxins has a great variation, recognizing ducks as the most susceptible, followed by turkeys, quails, and chickens as the least susceptible (Rawal et al. 2010). One of the most important factors in determining susceptibility to mycotoxins is age, with young animals being more vulnerable than adults. Mycotoxins are generally considered lipophilic,

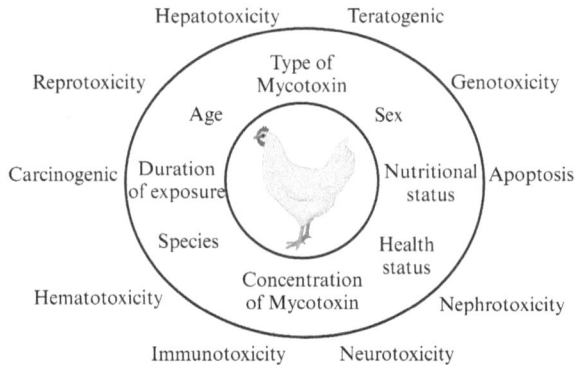

Fig. 12.3 The most common effects of mycotoxins on poultry health "Created with BioRender.com"

like the liver, which is why they are stored and concentrated in hepatocytes. The organ that is severely affected is the liver, and the primary lesions include hemorrhage, necrosis, fatty changes in hepatocytes, and hepatomegaly (Coulombe 1993). These effects can occur in the centrilobular region or the portal region, depending on the species. Microscopic liver changes include hydropic degeneration, bile duct hyperplasia, and periportal fibrosis (Jand et al. 2005). Mycotoxicosis in poultry is characterized by metabolic disorders, rejection of feed, alteration of the feed conversion rate, decrease in weight gain, alteration in nutrient adsorption, decrease in body weight, and immunosuppression, which leads to death (Vörösházi et al. 2024). Therefore, feed contaminated with mycotoxins is still considered an economically important problem in the poultry sector. Furthermore, prolonged consumption of mycotoxins is reflected in the low production quality of foods such as eggs and meat, which leads to economic losses in the production sector (El-Sayed et al. 2022). Due to these global problems, several decontamination and detoxification practices have been developed to prevent and reduce the risks caused by mycotoxins in feed (Jamil et al. 2024).

2.5 Mycotoxins Control

There are multiple origins for the presentation of fungal contamination, so the methods to control exposure to fungi and mycotoxins are mainly preventive and of a comprehensive nature throughout the entire food production chain (plant, growth, harvest, storage, and distribution) (Vila-Donat et al. 2018) (Fig. 12.4). The Codex Alimentarius has developed several codes of practice for the prevention and

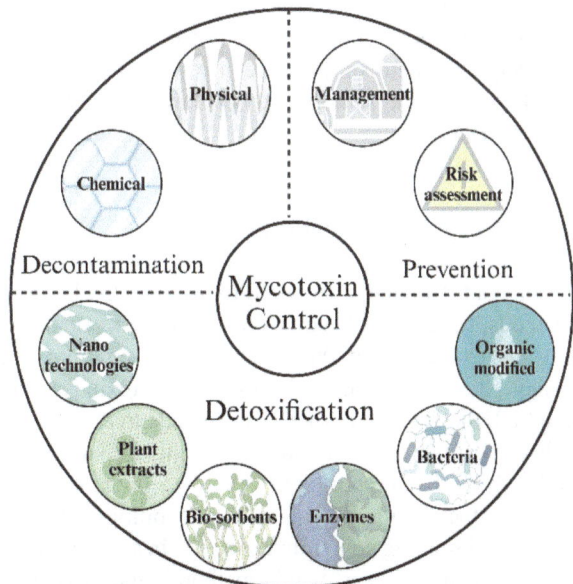

Fig. 12.4 Main strategies for mycotoxin control "Created with BioRender.com"

reduction of mycotoxins, which are divided into two: recommended practices based on good manufacturing practices, and the use of Hazard Analysis and Critical Control Points (Chen et al. 2021). However, these methods are not sufficient, so other strategies are required to decontaminate or detoxify foods contaminated by mycotoxins (Dänicke 2002). Mycotoxin control and prevention strategies include pre-harvest and post-harvest strategies, in addition to physical, chemical, and biological methods (Agriopoulou et al. 2020).

Pre-harvest strategies include the use of pesticides, crop rotation, irrigation, planting time, physiological stage of the plant, selection of high-quality seeds, and use of genetically modified seeds that are tolerant to *Aspergillus* infection and environmental stressors (Janssen et al. 2019).

Post-harvest strategies include good management in storage, drying, distribution, control of humidity, temperature, quantity of stored grains and the use of physical, chemical, and biological methods. If adequate drying of crops is achieved before storage, can prevent the production of mycotoxins (Magan and Aldred 2007).

3   Decontamination Methods

When prevention is not achieved at the field level or during harvest, employ myco-toxin decontamination procedures using physical, chemical, or biological methods, although some methods use combinations of the different principles of action (Čolović et al. 2019). Decontamination is defined as the program of post-harvest treatments aimed at eliminating or minimizing the toxic effects of mycotoxins in animals. The characteristics of an ideal decontamination method are: easy to use, economical, does not form toxic compounds, does not alter the nutritional or organoleptic properties of foods, be capable effective in eliminating, destroying, and inactivating various mycotoxins (Čolović et al. 2019). Numerous biological, chemical, and physical methods have been effective in destroying or inactivating several mycotoxins at the same time, but they do not cover the other equally important requirements.

3.1   Chemical Methods

Feeds contaminated with mycotoxins can also be decontaminated using chemical methods, which have become one of the most popular strategies since they are studied and evaluated for the optimum of changing the molecular structures of the mycotoxins. However, in most cases, its toxicity is not necessarily reduced (Luo et al. 2018). Chemical procedures include treatment with acid/base solutions, use of substances such as ammonia, ozonation, inorganic salts, and chlorination (Liu et al. 2022). It has been reported that these chemical agents are ineffective, and they also decrease the nutritional value or palatability of the feed ingredients. Acidification consists of soaking contaminated foods in acid solutions for a certain period of time. One of the advantages of this method is that it is simple to perform and does not require special equipment or specific skills. This method has been effective when using citric, lactic, tartaric, and hydrochloric acid. However, when succinic, acetic, ascorbic, or formic acids are used, decontamination success has been minimal (Rushing and Selim 2019; Shanakhat et al. 2018).

Ozonation is another method commonly used for mycotoxin decontamination. Ozone disrupts fungal cells through oxidizing sulfhydryl and amino acid groups of enzymes or attacks the polyunsaturated fatty acids of the cell wall. Although the mechanism of detoxification is not very clear for some mycotoxins, it is believed that ozone reacts with the functional groups in the mycotoxin molecules, changes their molecular structures, and forms products with lower molecular weight, less double bonds, and less toxicity (Afsah-Hejri et al. 2020; Conte et al. 2020).

Electrolyzed oxidizing water treatment is a technique that has recently been developed to decontaminate feeds with mycotoxins due to its biocidal activity and its physicochemical properties: pH, oxidation-reduction potential, and available chlorine concentration (Villarreal-Barajas et al. 2022; Xiong et al. 2014). Electrolyzed oxidizing water contains many OH groups that play an important role against conidia and mycelia structures due to the destruction of these cellular structures and reducing mycotoxin molecules (Castro-Ríos et al. 2021). The disadvantage of this method is that electrolyzed water quickly loses its activity when it comes into contact with organic matter (Gómez-Espinosa et al. 2017). These chemical methods have had moderate to high success. However, if combined, decontamination levels can become higher, either by reducing the time in which they are carried out or by increasing the percentage of mycotoxin decontamination (Sipos et al. 2021).

3.2 Physical Methods

Most physical methods have variable effectiveness and depend on the level of contamination and distribution of mycotoxins in the grains (Peng et al. 2018). Physical procedures are mainly based on the elimination of contaminated products or the inactivation of mycotoxins, which include mechanical classification, separation, segregation by density, washing, polishing, flotation, grinding, extraction with solvents, supplementation of adsorbent agents, thermal inactivation, autoclaving, extrusion, irradiation, ultrasound treatment, and roasting and the use of microwaves (Aiko and Mehta 2015). Furthermore, some of the physical methods are expensive, can cause large losses of the product, and eliminate or destroy essential nutrients from the feed (Kaushik 2015). Solvent extraction (95% ethanol, 90% aqueous acetone, 80% isopropanol, hexane-methanol, methanol-water, acetonitrile-water, hexane-ethanol-water, acetone-hexane-water) can remove traces of mycotoxin; however, its large-scale application is limited by high costs and the disposal of toxic waste (Goldblatt 1971; Stahr and Obioha 1982).

Milling has no direct effect on the elimination of mycotoxins in grains, this strategy is aimed at changing the distribution of toxins between grains and different resulting fractions (Peng et al. 2018). The fractions with mycotoxin depend on the penetration of the fungus into the endosperm of the grain, so in products where the distribution of aflatoxins is on the surface of the grain, the lowest level of mycotoxin is found in flour and the highest level is found in germ and bran (Kolosova and Stroka 2011). The application of physical methods simultaneously, such as sorting, washing, grinding, and hulling the grains, was effective significant decontamination of mycotoxin (Luo et al. 2018). It is important to note that the effectiveness of these methods depends on many factors such as humidity, the level of contamination, the distribution of the mycotoxin, and the presence of additives in the feed (Kolosova and Stroka 2011).

Heat treatment can obtain positive results, the advantages of this method are that it is economical, fast (2 h or less), and easy to perform. However, mycotoxins are sufficiently resistant to heat in a temperature range of conventional processing (80–121 °C), especially aflatoxins, so this method is not effective enough for the decontamination of AFB1. Mycotoxin sensitivity to heat treatment is affected by moisture, pH, and ionic strength of the feed ingredients. Furthermore, longer processing periods can affect the quality of the feed, even if the reduction of mycotoxins is achieved. It has been reported that pressure cooking largely reduces mycotoxin contamination in feed compared to other thermal methods. Autoclaving can reduce the presence of mycotoxins in feed ingredients (Bullerman and Bianchini 2007; Lee et al. 2015).

Gamma (ÿ) irradiation is a common method to decontaminate animal feed (Di Stefano et al. 2014). It consists of irradiating food products with a gamma ray source (60 Co), this technique showed a reduction effectiveness of 65% mycotoxins (Kolosova and Stroka 2011). The presence of water in the gamma radiation for the degradation of mycotoxin plays an important role, since the radiolysis that water undergoes results in the formation of highly reactive free radicals, generating compounds of low biological activity that have the potential to cause further harm (Calado et al. 2014) (Kolosova and Stroka 2011). They are also sensitive to UV radiation, which can lead to the formation of less toxic products, through photooxidation (Diao et al. 2015) (Kolosova and Stroka 2011). This technique is considered favorable compared to gamma radiation as it does not cause changes in nutrients (Wang et al. 2023). Pulsed light technology has also been studied for the reduction of mycotoxins in feeds by applying a series of high-intensity, short bursts of light pulse that reduce the number of bacteria and their metabolites (Mandal et al. 2020). Conversely, pulsed light technology induces sensory changes, and also has a higher investment cost (Mir et al. 2021).

4   Detoxification Methods

4.1   Adsorbents

Additives have been the method of choice for the protection of animals (mainly pigs, poultry, and cattle) against the consumption of contaminated ingredients, to reduce the absorption of mycotoxins in the gastrointestinal tract and their subsequent distribution in blood and target organs (Di Stefano et al. 2014). Currently, major efforts have been directed to eliminate or reduce the effect of mycotoxins through the use of different products that can suppress or reduce absorption, promote excretion, or modify the mode of action of mycotoxins. The use of adsorbents is one of the most practical methods for the decontamination of mycotoxins within the animal industry (Kolosova and Stroka 2011). Adsorbents are high molecular weight compounds that can bind mycotoxins, and their mode of action is based on preventing the absorption of the mycotoxin present in the food by not dissociating in the gastrointestinal tract of the animal after ingestion, limiting its bioavailability (Huwig et al. 2001). Adsorbents bind mycotoxins by different mechanisms that may be involved in the same adsorption process, including hydrogen bonds, hydrophobic bonding, electrostatic attraction, ion exchange, complexation, chelation, and precipitation, which lead to the immobilization of the mycotoxins molecules (Puvača et al. 2023). Depending on their mode of action, adsorbents act by binding mycotoxins to their surface (adsorption) or by degrading or transforming mycotoxins into less toxic metabolites (biotransformation).

4.2   Adsorbents Interactions

Depending on their mode of action, adsorbents act by binding mycotoxins physically, chemically, or by the interaction of both (Physicochemical interactions) or by degrading or transforming mycotoxins into less toxic metabolites (biotransformation). Physical adsorption is normally a reversible phenomenon, as are Van der Waals interactions, and electrostatic attraction interactions, which include polarization interactions. These interactions are important in the case of adsorbents that have an ionic structure (zeolites) (Di Gregorio et al. 2014). Hydrophobic binding is a complex process that involves more than one type of bond, but the most important characteristic is the lipophilic–hydrophilic balance which allows quantitative and qualitative prediction of mycotoxin-bonding. Electrostatic binding is a process in which the dominant characteristic is the electrostatic attraction between an ionized molecule and an adsorbent (Boudergue et al. 2009). Whether electrostatic binding or hydrophobic binding can be affected by pH: if the molecule and adsorbent are anionic above a certain pH, electrostatic repulsion will prevent hydrophobic binding, which would be possible at a lower pH (Di Gregorio et al. 2014).

In chemical adsorption, there is an exchange of electrons between the adsorbent and the mycotoxin, forming a single layer on a solid surface in an irreversible manner, which implies a considerable amount of energy (Di Gregorio et al. 2014). Physicochemical interactions between the surface of the adsorbent and the mycotoxin, such as physical adsorption and ion exchange help form a stable complexation mycotoxin-adsorbent. Some adsorbents are composed mainly of polysaccharides, proteins, lipids, and various functional groups such as carboxyl, hydroxyl, phosphate, and amino, as well as hydrophobic adsorption sites such as aliphatic carbon chains and aromatic rings. These physicochemical phenomena are rapid and can be reversible (Ringot et al. 2007). This is how a stable mycotoxin–adsorbent complex is formed, even at variable pH throughout the entire gastrointestinal tract of the animal until it is eliminated in the feces. The stability of the complex is influenced by the physical properties of the adsorbent: total charge and charge distribution, the size of the pores and the accessible surface area, and the physicochemical properties of the toxins: polarity, solubility, molecular weight, and shape; in the case of ionized compounds, the charge distribution and dissociation constants are also important (Ringot et al. 2007; Vila-Donat et al. 2018). The most common criteria to consider for the evaluation of an adsorbent include the characterized by attenuated total reflectance-Fourier transform infrared spectroscopy (ATR-FTIR), energy-dispersive X-ray spectroscopy (EDS), zeta potential (ζ-potential), scanning electron microscopy (SEM), and point of zero charge (pHpzc) and some other techniques that involve little material and are easily repeatable, non-destructive, and reasonably simple (Vázquez-Durán et al. 2022). Additionally, the effectiveness of these adsorbents should be evaluated over a wide pH range (as it is expected to function throughout the entire gastrointestinal tract). Firstly, by performing in vitro assays to demonstrate the efficiency of mycotoxin-adsorbent bonding. Finally evaluated in vivo experiments that consider the sum of multiple variables between different species.

4.3 Inorganic Aflatoxin Adsorbents

Aluminosilicates are the most abundant group of rock-forming minerals. The molecular structure of aluminosilicates consists of a combination of sheets of tetrahedral silica and octahedral aluminum sheets, both structures linked by oxygen and hydroxyl groups (Ramos et al. 1996). This type of clay is highly sensitive to presenting isomorphic substitutions in the sheets, generating negative charges, therefore, smectites have a high negative charge and a high degree of expansion (Avantaggiato et al. 2007; Zabiulla et al. 2021). The group of aluminosilicates is divided into two: phyllosilicates (bentonites, montmorillonites, smectites, kaolinites, and illites) and tectosilicates (zeolites) (De Mil et al. 2015; Swaddle et al. 1994). Phyllosilicates adsorb on their surface or within their lamellar space and tectosilicates provide a long and specific bond and are suitable for the distinction of different molecules by size, shape, and charge selectivity. Phyllosilicates are characterized by a sheet-shaped crystalline structure with a common chemical formula $Si_2O_5{}^{2-}$. Tectosilicates include hydrated aluminosilicates with three-dimensional structures, consisting of SiO_4 and AlO_4 tetrahedra (Di Gregorio et al. 2014). Over the years, innumerable studies of inorganic adsorbents have been carried out to verify their effectiveness in adsorbing mycotoxins. However, many of these inorganic adsorbents have some disadvantages, they simultaneously indiscriminately adsorb micronutrients (Vitamins, minerals, and trace elements) from the diet and present the possibility of releasing toxic components (dioxins and heavy metals) (Elliott et al. 2020; Zavala-Franco et al. 2018)).

4.4 Organic Aflatoxin Adsorbents

In recent years, great interest has been shown in the study of organic mycotoxin adsorbents, since they have been characterized as safe, profitable, and effective. There are numerous organic adsorbents (plant extracts and fibers, yeast and bacterial cell wall extracts, enzymes, and organic modified materials) that have been promising in their mycotoxin adsorption efficiency in vitro and in vivo studies (Baglieri et al. 2013; Ghamsari et al. 2022; Ji et al. 2016).

The cell wall of yeast (*Saccharomyces cerevisiae*) has been used in the area of animal feeding as part of food supplementation since they improve the productive parameters of animals (Weaver et al. 2024). The cell wall of yeast has numerous binding mechanisms with mycotoxin such as hydrogen, ionic, and hydrophobic interactions. It has been suggested that the three-dimensional structure of the cell wall of yeast, which mainly consists of oligosaccharides, is capable of reacting with mycotoxin, by trapping it between its structures (Jouany et al. 2005).

Humic acids derived from vermicompost have demonstrated a high efficacy in adsorbing aflatoxins in in vitro and in vivo (Turkey poults) experiments. The in vitro experiment indicated that humic acids derived from vermicompost have a high potential in the adsorption of aflatoxins at different pH values (Maguey-Gonzalez et al. 2023a). Meanwhile, the in vivo study demonstrated that humic acids derived from vermicompost mitigated the adverse effects of aflatoxins in the turkey poults (Maguey-Gonzalez et al. 2023b, 2024).

5 Bio-Adsorbents

5.1 Plant Extract

Biosorption technology has been considered an optimal alternative, characterized by high efficiency, low costs, and the use of minimum inclusion levels (Luo et al. 2020). Additionally, it prevents the reduction of the nutritional value of contaminated diets and promotes the sustainable treatment of waste plants, which are generally considered disposed (Bočarov-Stančić et al. 2018; Greco et al. 2019). Bio-adsorbents have a multi-layer porous structure, with cavities and channels that provide a large volume per unit of adsorbent surface, which is favorable in the adsorption process. The average diameter of the pores is less than 1 μm, and could favor the diffusion and adsorption of mycotoxins. The biosorption mechanisms are simple or multiple ion exchanges, complex constructions, electrostatic interactions, and chelate constructions (Bočarov-Stančić et al. 2018). In addition, the plant components of plants also include phenolics, flavonoids, coumarins, chlorophyll, and its derivatives such as chlorophyllin. All of these have chemo-protective properties against carcinogenic compounds of mycotoxins. Many studies have observed the effect of chlorophyll and chlorophyllin in the presence of mycotoxins and the protective results of these compounds have been similar. Chlorophyll (plant pigment) and its water-soluble derivative, chlorophyllin (which is a semi-synthetic mixture of sodium and copper salts), have been widely studied as antioxidant, and anticancer agents, as well (Galvano et al. 2001; Jubert et al. 2009; Simonich et al. 2007).

Numerous plant extracts (*Silybum marianum*, *Solanum lycopersicum*, *Camellia sinensis* L., *Thymus vulgaris*, *Curcuma longa* L.) (Makhuvele et al. 2020) have been used to degrade mycotoxins, these types of strategies have demonstrated good effectiveness in the decontamination of aflatoxins; however, there is a necessity for standardizing the production, formulation, commercialization, and application of these plant extracts (Anjorin et al. 2013). Some studies (in vitro and in vivo) report high percentages of aflatoxin adsorption from the banana peel (28%), *Pyracantha* leaves (46%), aloe vera powder (69%), *Pyracantha koidzumii* leaves (86%), *Pyracantha koidzumii* leaves+berries (82%), *Pyracantha koidzumii* berries (46%), *Equisetum arvense* L. (37%), *Curly kale* (93.6%), *Lactuca sativa* L. (83.7%), and *Cuscuta corymbose* (97%) (Ramales-Valderrama et al. 2016; Zavala-Franco et al. 2018; Nava-Ramírez et al. 2021; Duran et al. 2021, 2023). Additionally, powdered alfalfa adsorbent was efficient for the adsorption of aflatoxins (in vitro), and improved the production parameters, reduced the histological lesions in turkey poults, therefore is an effective material to adsorb aflatoxins, and prevented its toxic effects (Nava-Ramírez et al. 2023; Nava-Ramírez et al. 2024).

5.2   Bacteria and Enzymes

The use of bacteria and their enzymes in the feed substrates with bacterial strains that reduce mycotoxin contamination through their metabolism (adsorption or degradation) is a novel method with several advantages compared to non-biological methods (Muhialdin et al. 2020; Solis-Cruz et al. 2019a, b). However, the main disadvantage of the use of microorganisms is time-consuming, only applicable to feed not for food, and use unknown or pathogenic strains (Piotrowska 2021). Researchers have experimented with various genera of bacteria such as *Bacillus*, *Lactobacillus*, *Propionibacterium*, *Rhodococcus*, and *Clostridium*. Although yeast such as *Candida*, *Rhodosporidium*, and *Saccharomyces*; Molds such as *Aspergillus*, *Saccharomyces*, or *Cladosporium* have the ability to degrade mycotoxins (Hernandez-Patlan et al. 2020; Piotrowska 2021; Solis-Cruz et al. 2019a, b).

The degradation capacity of enzymes has also been studied exhaustively; it has been shown that they can modify mycotoxins into less toxic derivative compounds (Alberts et al. 2019). However, like plant extracts, they are used in long-term techniques, and in some cases, their effectiveness has not been clearly demonstrated (Rushing and Selim 2019). Bacteria and enzymatic activities against mycotoxins are considered a very friendly control method compared with physical and chemical methods (Lyagin and Efremenko 2019). However, more studies are needed to elucidate mycotoxin detoxification mechanisms, dosage, time of microbial detoxification, and how to use these new microbial preparations to maximize the prevention and beneficial effects on adverse effects of mycotoxins (Loi et al. 2017).

6   Conclusions

Mycotoxins are toxic substances, with a worldwide distribution in nature, produced by various fungi species. Their occurrence in the food chain is inevitable and presents a serious problem on a global scale. Although there is not a definitive anti-mycotoxin strategy that eliminates these toxins, a variety of feed additives can significantly reduce their impact on poultry health. When combined with good nutrition, management, and biosecurity policies, producers guarantee the safety of poultry products for consumers. The present chapter reviews a comprehensive exploration of the detrimental impact of mycotoxins, as well as the current technologies used for controlling mycotoxins with emphasis on eco-friendly anti-mycotoxin products that reduce their environmental impact, promoting a responsible feed production.

References

Afsah-Hejri L, Hajeb P, Ehsani RJ (2020) Application of ozone for degradation of myco-toxins in food: A review. Compr Rev Food Sci Food Saf 19(4):1777–1808. https://doi.org/10.1111/1541-4337.12594

Agriopoulou S, Stamatelopoulou E, Varzakas T (2020) Advances in occurrence, importance, and mycotoxin control strategies: prevention and detoxification in foods. Foods 9(2):2. https://doi.org/10.3390/foods9020137

Aiko V, Mehta A (2015) Occurrence, detection and detoxification of mycotoxins. J Biosci 40(5):943–954. https://doi.org/10.1007/s12038-015-9569-6

Alberts J, Schatzmayr G, Moll W-D, Davids I, Rheeder J, Burger H-M, Shephard G, Gelderblom W (2019) Detoxification of the Fumonisin mycotoxins in maize: an enzymatic approach. Toxins 11(9):Article 9. https://doi.org/10.3390/toxins11090523

Aldars-García L, Berman M, Ortiz J, Ramos AJ, Marín S (2018) Probability models for growth and aflatoxin B1 production as affected by intraspecies variability in aspergillus flavus. Food Microbiol 72:166–175. https://doi.org/10.1016/j.fm.2017.11.015

Ali T (2020) Mycotoxins and mycotoxicosis. Pakistan J Sci 72(4):Article 4. https://doi.org/10.57041/pjs.v72i4.267

Anjorin TS, Salako EA, Makun HA, Anjorin TS, Salako EA, Makun HA (2013) Control of toxi-genic fungi and mycotoxins with phytochemicals: potentials and challenges. in mycotoxin and food safety in developing countries. IntechOpen. https://doi.org/10.5772/53477

Avantaggiato G, Havenaar R, Visconti A (2007) Assessment of the multi-mycotoxin-binding effi-cacy of a carbon/Aluminosilicate-based product in an in vitro gastrointestinal model. J Agric Food Chem 55(12):4810–4819. https://doi.org/10.1021/jf0702803

Awuchi CG, Ondari EN, Nwozo S, Odongo GA, Eseoghene IJ, Twinomuhwezi H, Ogbonna CU, Upadhyay AK, Adeleye AO, Okpala COR (2022) Mycotoxins' toxicological mechanisms involving humans, livestock and their associated health concerns: a review. Toxins 14(3):167. https://doi.org/10.3390/toxins14030167

Baglieri A, Reyneri A, Gennari M, Nègre M (2013) Organically modified clays as binders of fumonisins in feedstocks. J Environ Sci Health B 48(9):776–783. https://doi.org/10.1080/03601234.2013.780941

Bennett JW (1987) Mycotoxins, mycotoxicoses, mycotoxicology andMycopathologia. Mycopathologia 100(1):3–5. https://doi.org/10.1007/BF00769561

Bhat R, Rai RV, Karim, a. (2010) Mycotoxins in food and feed: present status and future concerns. Compr Rev Food Sci Food Saf 9(1):57–81. https://doi.org/10.1111/j.1541-4337.2009.00094.x

Bočarov-Stančić A, Lopičić ZR, Bodroža-Solarov MI, Stanković S, Janković S, Milojković JV, Krulj JA (2018) In vitro removing of mycotoxins by using different inorganic adsorbents and organic waste materials from Serbia. Food Feed Res 45(7):87–96. https://doi.org/10.5937/FFR1802087B

Boudergue C, Burel C, Dragacci S, Favrot M-C, Fremy J-M, Massimi C, Prigent P, Debongnie P, Pussemier L, Boudra H, Morgavi D, Oswald I, Perez A, Avantaggiato G (2009) Review of mycotoxin-detoxifying agents used as feed additives: mode of action, efficacy and feed/food safety. EFSA Supporting Publications 6(9):22E. https://doi.org/10.2903/sp.efsa.2009.EN-22

Bullerman LB, Bianchini A (2007) Stability of mycotoxins during food processing. Int J Food Microbiol 119(1):140–146. https://doi.org/10.1016/j.ijfoodmicro.2007.07.035

Calado T, Venâncio A, Abrunhosa L (2014) Irradiation for Mold and mycotoxin control: A review. Compr Rev Food Sci Food Saf 13(5):1049–1061. https://doi.org/10.1111/1541-4337.12095

Carvajal-Moreno M (2022) Mycotoxin challenges in maize production and possible control methods in the 21st century. J Cereal Sci 103:103293. https://doi.org/10.1016/j.jcs.2021.103293

Castro-Ríos K, Montoya-Estrada CN, Martínez-Miranda MM, Hurtado Cortés S, Taborda-Ocampo G (2021) Physicochemical treatments for the reduction of aflatoxins and aspergillus Niger in corn grains (Zea mays). J Sci Food Agric 101(9):3707–3713. https://doi.org/10.1002/jsfa.11001

Chen C, Frank K, Wang T, Wu F (2021) Global wheat trade and codex Alimentarius guidelines for deoxynivalenol: A mycotoxin common in wheat. Glob Food Sec 29:100538. https://doi.org/10.1016/j.gfs.2021.100538

Cleveland TE, Dowd PF, Desjardins AE, Bhatnagar D, Cotty PJ (2003) United States Department of agriculture—agricultural research service research on pre-harvest prevention of mycotoxins and mycotoxigenic fungi in US crops. Pest Manag Sci 59(6–7):629–642. https://doi.org/10.1002/ps.724

Čolović R, Puvača N, Cheli F, Avantaggiato G, Greco D, Đuragić O, Kos J, Pinotti L (2019) Decontamination of mycotoxin-contaminated feedstuffs and compound feed. Toxins 11(11):11. https://doi.org/10.3390/toxins11110617

Conte G, Fontanelli M, Galli F, Cotrozzi L, Pagni L, Pellegrini E (2020) Mycotoxins in feed and food and the role of ozone in their detoxification and degradation: an update. Toxins 12(8):8. https://doi.org/10.3390/toxins12080486

Coulombe RA (1993) Biological action of Mycotoxins1. J Dairy Sci 76(3):880–891. https://doi.org/10.3168/jds.S0022-0302(93)77414-7

Dänicke S (2002) Prevention and control of mycotoxins in the poultry production chain: A European view. Worlds Poult Sci J 58(4):451–474. https://doi.org/10.1079/WPS20020033

Daou R, Joubrane K, Maroun RG, Khabbaz LR, Ismail A, Khoury AE, Daou R, Joubrane K, Maroun RG, Khabbaz LR, Ismail A, Khoury AE (2021) Mycotoxins: factors influencing production and control strategies. AIMS Agric Food 6(1):416. https://doi.org/10.3934/agrfood.2021025

De Mil T, Devreese M, De Baere S, Van Ranst E, Eeckhout M, De Backer P, Croubels S (2015) Characterization of 27 mycotoxin binders and the relation with in vitro Zearalenone adsorption at a single concentration. Toxins 7(1). https://doi.org/10.3390/toxins7010021

Di Gregorio MC, de Neeff DV, Jager AV, Corassin CH, de Carão ÁCP, de Albuquerque R, de Azevedo AC, Oliveira CAF (2014) Mineral adsorbents for prevention of mycotoxins in animal feeds. Toxin Rev 33(3):125–135. https://doi.org/10.3109/15569543.2014.905604

Di Stefano V, Pitonzo R, Cicero N, D'Oca MC (2014) Mycotoxin contamination of animal feedingstuff: detoxification by gamma-irradiation and reduction of aflatoxins and ochratoxin A concentrations. Food Addit Contam Part A Chem Anal Control Expo Risk Assess 31(12):2034–2039. https://doi.org/10.1080/19440049.2014.968882

Diao E, Li X, Zhang Z, Ma W, Ji N, Dong H (2015) Ultraviolet irradiation detoxification of aflatoxins. Trends Food Sci Technol 42(1):64–69. https://doi.org/10.1016/j.tifs.2014.12.001

Dowd PF (2003) Insect management to facilitate Preharvest mycotoxin management. J Toxicol Toxin Rev 22(2–3):327–350. https://doi.org/10.1081/TXR-120024097

Duran A, Nava-Ramírez M, Hernandez-Patlan D, Solís-Cruz B, Hernández Gómez V, Tellez G, Albores A (2021) Potential of kale and lettuce residues as natural adsorbents of the carcinogen aflatoxin B1 in a dynamic gastrointestinal tract-simulated model. Toxins 13:771. https://doi.org/10.3390/toxins13110771

Duran A, Nava-Ramírez M, Martínez Escutia R, Figueroa J, Lòpez Coello C, Tellez G, Albores A (2023) Highly effcient adsorptive removal of the carcinogen aflatoxin B1 using the parasitic plant Cuscuta corymbosa Ruiz & Pavon. Environ Sci Pollut Res. https://doi.org/10.1007/s11356-023-30992-w

Elliott CT, Connolly L, Kolawole O (2020) Potential adverse effects on animal health and performance caused by the addition of mineral adsorbents to feeds to reduce mycotoxin exposure. Mycotoxin Res 36(1):115–126. https://doi.org/10.1007/s12550-019-00375-7

El-Sayed RA, Jebur AB, Kang W, El-Demerdash FM (2022) An overview on the major mycotoxins in food products: characteristics, toxicity, and analysis. J Future Foods 2(2):91–102. https://doi.org/10.1016/j.jfutfo.2022.03.002

Eskola M, Kos G, Elliott CT, Hajšlová J, Mayar S, Krska R (2020) Worldwide contamination of food-crops with mycotoxins: validity of the widely cited 'FAO estimate' of 25%. Crit Rev Food Sci Nutr 60(16):2773–2789. https://doi.org/10.1080/10408398.2019.1658570

Filazi A, Yurdakok-Dikmen B, Kuzukiran O, Sireli UT, Filazi A, Yurdakok-Dikmen B, Kuzukiran O, Sireli UT (2017) Mycotoxins in poultry. In: Poultry science. IntechOpen. https://doi.org/10.5772/66302

Galvano F, Piva A, Ritieni A, Galvano G (2001) Dietary strategies to counteract the effects of mycotoxins: a review. J Food Prot 64(1):120–131. https://doi.org/10.4315/0362-028X-64.1.120

Ghamsari FA, Ebrahimi MT, Varzaneh MB, Iranbakhsh A, Sepahi AA (2022) Effect of adding an organic binder on health of cows fed with mycotoxins contaminated diet. J Hellenic Vet Med Soc 73(4):Article 4. https://doi.org/10.12681/jhvms.28266

Goldblatt LA (1971) Control and removal of aflatoxin. J Am Oil Chem Soc 48(10):605–610. https://doi.org/10.1007/BF02544572

Gómez-Espinosa D, Cervantes-Aguilar FJ, Del Río-García JC, Villarreal-Barajas T, Vázquez-Durán A, Méndez-Albores A (2017) Ameliorative effects of neutral electrolyzed water on growth performance, biochemical constituents, and histopathological changes in Turkey Poults during Aflatoxicosis. Toxins 9(3):3. https://doi.org/10.3390/toxins9030104

Greco D, D'Ascanio V, Santovito E, Logrieco AF, Avantaggiato G (2019) Comparative efficacy of agricultural by-products in sequestering mycotoxins. J Sci Food Agric 99(4):1623–1634. https://doi.org/10.1002/jsfa.9343

Haque MA, Wang Y, Shen Z, Li X, Saleemi MK, He C (2020) Mycotoxin contamination and control strategy in human, domestic animal and poultry: A review. Microb Pathog 142:104095. https://doi.org/10.1016/j.micpath.2020.104095

Hernandez-Patlan D, Solis-Cruz B, Hargis M, B., & Tellez, G. (2020) The use of probiotics in poultry production for the control of bacterial infections and aflatoxins. In: Franco-Robles E, Ramírez-Emiliano J (eds) Prebiotics and probiotics—potential benefits in nutrition and health. IntechOpen. https://doi.org/10.5772/intechopen.88817

Hoerr FJ (2020) Mycotoxicoses. In: Diseases of poultry. John Wiley & Sons, Ltd., pp 1330–1348. https://doi.org/10.1002/9781119371199.ch31

Huwig A, Freimund S, Käppeli O, Dutler H (2001) Mycotoxin detoxication of animal feed by different adsorbents. Toxicol Lett 122(2):179–188. https://doi.org/10.1016/S0378-4274(01)00360-5

Jamil M, Khatoon A, Saleemi MK, Abidin ZU, Abbas RZ, Ul-Hassan Z, Bhatti SA, Irshad H, Imran M, Raza QS (2024) Use of phytochemicals to control the Mycotoxicosis in poultry. Worlds Poult Sci J 80(1):237–250. https://doi.org/10.1080/00439339.2023.2255575

Jand SK, Kaur P, Sharma NS (2005) Mycoses and mycotoxicosis in poultry: a review. Indian J Anim Sci 75(4):4. https://epubs.icar.org.in/index.php/IJAnS/article/view/9487

Janik E, Niemcewicz M, Ceremuga M, Stela M, Saluk-Bijak J, Siadkowski A, Bijak M (2020) Molecular aspects of mycotoxins—A serious problem for human health. Int J Mol Sci 21(21):8187. https://doi.org/10.3390/ijms21218187

Janssen EM, Mourits MCM, van der Fels-Klerx HJ, Lansink AGJMO (2019) Pre-harvest measures against fusarium spp. infection and related mycotoxins implemented by Dutch wheat farmers. Crop Prot 122:9–18. https://doi.org/10.1016/j.cropro.2019.04.005

Ji C, Fan Y, Zhao L (2016) Review on biological degradation of mycotoxins. Anim Nutr 2(3):127–133. https://doi.org/10.1016/j.aninu.2016.07.003

Jouany J-P, Yiannikouris A, Bertin G (2005) How yeast cell wall components can alleviate mycotoxicosis in animal production and improve the safety of edible animal products. J Anim Feed Sci 14(Suppl. 1):171–190. https://doi.org/10.22358/jafs/70361/2005

Jubert C, Mata J, Bench G, Dashwood R, Pereira C, Tracewell W, Turteltaub K, Williams D, Bailey G (2009) Effects of chlorophyll and Chlorophyllin on low-dose aflatoxin B1 pharmacokinetics in human volunteers. Cancer Prev Res 2(12):1015–1022. https://doi.org/10.1158/1940-6207. CAPR-09-0099

Kamle M, Mahato DK, Devi S, Lee KE, Kang SG, Kumar P (2019) Fumonisins: impact on agriculture, food, and human health and their management strategies. Toxins 11(6):328. https://doi. org/10.3390/toxins11060328

Karlovsky P, Suman M, Berthiller F, De Meester J, Eisenbrand G, Perrin I, Oswald IP, Speijers G, Chiodini A, Recker T, Dussort P (2016) Impact of food processing and detoxification treatments on mycotoxin contamination. Mycotoxin Res 32(4):179–205. https://doi.org/10.1007/ s12550-016-0257-7

Kaushik G (2015) Effect of processing on mycotoxin content in grains. Crit Rev Food Sci Nutr 55(12):1672–1683. https://doi.org/10.1080/10408398.2012.701254

Kolosova A, Stroka J (2011) Substances for reduction of the contamination of feed by mycotoxins: a review. World Mycotoxin J 4(3):225–256. https://doi.org/10.3920/WMJ2011.1288

Kumar A, Pathak H, Bhadauria S, Sudan J (2021) Aflatoxin contamination in food crops: causes, detection, and management: a review. Food Prod Process Nutr 3(1):17. https://doi.org/10.1186/ s43014-021-00064-y

Kumar P, Mahato DK, Kamle M, Mohanta TK, Kang SG (2017) Aflatoxins: a global concern for food safety, human health and their management. Front Microbiol 7. https://doi.org/10.3389/ fmicb.2016.02170

Lee J, Her J-Y, Lee K-G (2015) Reduction of aflatoxins (B1, B2, G1, and G2) in soybean-based model systems. Food Chem 189:45–51. https://doi.org/10.1016/j.foodchem.2015.02.013

Liu M, Zhao L, Gong G, Zhang L, Shi L, Dai J, Han Y, Wu Y, Khalil MM, Sun L (2022) Invited review: remediation strategies for mycotoxin control in feed. J Anim Sci Biotechnol 13(1):19. https://doi.org/10.1186/s40104-021-00661-4

Loi M, Fanelli F, Liuzzi VC, Logrieco AF, Mulè G (2017) Mycotoxin biotransformation by native and commercial enzymes: present and future perspectives. Toxins 9(4):Article 4. https://doi. org/10.3390/toxins9040111

Luo Y, Liu X, Li J (2018) Updating techniques on controlling mycotoxins—A review. Food Control 89:123–132. https://doi.org/10.1016/j.foodcont.2018.01.016

Luo Y, Liu X, Yuan L, Li J (2020) Complicated interactions between bio-adsorbents and mycotoxins during mycotoxin adsorption: current research and future prospects. Trends Food Sci Technol 96:127–134. https://doi.org/10.1016/j.tifs.2019.12.012

Lyagin I, Efremenko E (2019) Enzymes for detoxification of various mycotoxins: origins and mechanisms of catalytic action. Molecules 24(13):13. https://doi.org/10.3390/molecules24132362

Magan N, Aldred D (2007) Post-harvest control strategies: minimizing mycotoxins in the food chain. Int J Food Microbiol 119(1):131–139. https://doi.org/10.1016/j.ijfoodmicro.2007.07.034

Maguey-Gonzalez J, Nava-Ramírez M, Gomez-Rosales S, Ángeles M, Solís-Cruz B, Hernandez-Patlan D, Merino R, Hernandez-Velasco X, Figueroa J, Duran A, Hargis B, Tellez G, Albores A (2023a) Humic acids preparation, characterization, and their potential adsorption capacity for aflatoxin B1 in an in vitro poultry digestive model. Toxins 15:83. https://doi.org/10.3390/ toxins15020083

Maguey-Gonzalez J, Nava-Ramírez M, Gomez-Rosales S, Ángeles M, Solís-Cruz B, Hernandez-Patlan D, Merino R, Hernandez-Velasco X, Hernández-Ramírez J, Loeza I, Senas-Cuesta R, Latorre J, Duran A, Du X, Albores A, Hargis B, Tellez G (2023b) Corrigendum: evaluation of the efficacy of humic acids to counteract the toxic effects of aflatoxin B1 in Turkey poults. Front Vet Sci 10. https://doi.org/10.3389/fvets.2023.1346080

Maguey-Gonzalez J, Tellez G, Gomez-Rosales S, Nava-Ramírez M, Solís-Cruz B, Hernandez-Velasco X, Merino R, Latorre J, Hernandez-Patlan D, Albores A (2024) Assessment of the impact of humic acids on intestinal microbiota, gut integrity, ileum morphometry, and cellular immunity of Turkey Poults fed an aflatoxin B1 contaminated diet. Toxins 16. https://doi.org/10.3390/toxins16030122

Mahato DK, Lee KE, Kamle M, Devi S, Dewangan KN, Kumar P, Kang SG (2019) Aflatoxins in food and feed: an overview on prevalence, detection and control strategies. Front Microbiol 10. https://doi.org/10.3389/fmicb.2019.02266

Makhuvele R, Naidu K, Gbashi S, Thipe VC, Adebo OA, Njobeh PB (2020) The use of plant extracts and their phytochemicals for control of toxigenic fungi and mycotoxins. Heliyon 6(10):e05291. https://doi.org/10.1016/j.heliyon.2020.e05291

Mandal R, Mohammadi X, Wiktor A, Singh A, Pratap Singh A (2020) Applications of pulsed light decontamination technology in food processing: an overview. Appl Sci 10(10):10. https://doi.org/10.3390/app10103606

Marin S, Ramos AJ, Cano-Sancho G, Sanchis V (2013) Mycotoxins: occurrence, toxicology, and exposure assessment. Food Chem Toxicol 60:218–237. https://doi.org/10.1016/j.fct.2013.07.047

Mir SA, Dar BN, Shah MA, Sofi SA, Hamdani AM, Oliveira CAF, Hashemi Moosavi M, Mousavi Khaneghah A, Sant'Ana AS (2021) Application of new technologies in decontamination of mycotoxins in cereal grains: challenges, and perspectives. Food Chem Toxicol 148:111976. https://doi.org/10.1016/j.fct.2021.111976

Muhialdin BJ, Saari N, Meor Hussin AS (2020) Review on the biological detoxification of mycotoxins using lactic acid bacteria to enhance the sustainability of foods supply. Molecules 25(11):11. https://doi.org/10.3390/molecules25112655

Nava-Ramírez MDJ, Vázquez-Durán A, Figueroa-Cárdenas JDD, Hernández-Patlán D, Solís-Cruz B, Téllez-Isaías G, López-Coello C, Méndez-Albores A (2023) Removal of aflatoxin B1 using alfalfa leaves as an adsorbent material: a comparison between two in vitro experimental models. Toxins 15(10):604. https://doi.org/10.3390/toxins15100604

Nava-Ramírez M, Maguey-Gonzalez J, Gomez-Rosales S, Hernández-Ramírez, Latorre J, Lòpez Coello C, Hargis, Tellez G, Duran A, Albores A (2024) Efficacy of powdered alfalfa leaves to ameliorate the toxic effects of aflatoxin B 1 in Turkey poults. Mycotoxin Res 40:269. https://doi.org/10.1007/s12550-024-00527-4

Nava-Ramírez M, Salazar A, Sordo M, Lòpez Coello C, Tellez G, Albores A, Duran A (2021) Ability of low contents of biosorbents to bind the food carcinogen aflatoxin B1 in vitro. Food Chem 345:128863. https://doi.org/10.1016/j.foodchem.2020.128863

Nigam PS, Singh A (2014) Metabolic pathways | production of secondary metabolites—fungi. In: Batt CA, Tortorello ML (eds) Encyclopedia of food microbiology, 2nd edn. Academic Press, pp 570–578. https://doi.org/10.1016/B978-0-12-384730-0.00202-0

Ochieng PE, Scippo M-L, Kemboi DC, Croubels S, Okoth S, Kangethe EK, Doupovec B, Gathumbi JK, Lindahl JF, Antonissen G (2021) Mycotoxins in poultry feed and feed ingredients from sub-Saharan Africa and their impact on the production of broiler and layer chickens: A review. Toxins 13(9):633. https://doi.org/10.3390/toxins13090633

Okasha H, Song B, Song Z (2024) Hidden hazards revealed: mycotoxins and their masked forms in poultry. Toxins 16(3):3. https://doi.org/10.3390/toxins16030137

Omotayo OP, Omotayo AO, Mwanza M, Babalola OO (2019) Prevalence of mycotoxins and their consequences on human health. Toxicol Res 35(1):1–7. https://doi.org/10.5487/TR.2019.35.1.001

Peng W-X, Marchal JLM, van der Poel AFB (2018) Strategies to prevent and reduce mycotoxins for compound feed manufacturing. Anim Feed Sci Technol 237:129–153. https://doi.org/10.1016/j.anifeedsci.2018.01.017

Perrone G, Ferrara M, Medina A, Pascale M, Magan N (2020) Toxigenic fungi and mycotoxins in a climate change scenario: ecology, genomics, distribution, prediction and prevention of the risk. Microorganisms 8(10):10. https://doi.org/10.3390/microorganisms8101496

Piotrowska M (2021) Microbiological decontamination of mycotoxins: opportunities and limitations. Toxins 13(11):819. https://doi.org/10.3390/toxins13110819

Puvača N, Ljubojević Pelić D, Tufarelli V (2023) Mycotoxins adsorbents in food animal production. J Agron Technol Eng Manag 6:944. https://doi.org/10.55817/GYIC7602

Ramales-Valderrama RA, Vázquez-Durán A, Méndez-Albores A (2016) Biosorption of B-aflatoxins using biomasses obtained from Formosa firethorn [Pyracantha koidzumii (Hayata) Rehder]. Toxins 8(7):7. https://doi.org/10.3390/toxins8070218

Ramos A-J, Fink-Gremmels J, Hernández E (1996) Prevention of toxic effects of mycotoxins by means of nonnutritive adsorbent compounds. J Food Prot 59(6):631–641. https://doi.org/1 0.4315/0362-028X-59.6.631

Rawal S, Kim JE, Coulombe R (2010) Aflatoxin B1 in poultry: toxicology, metabolism and prevention. Res Vet Sci 89(3):325–331. https://doi.org/10.1016/j.rvsc.2010.04.011

Ringot D, Lerzy B, Chaplain K, Bonhoure J-P, Auclair E, Larondelle Y (2007) In vitro biosorption of ochratoxin A on the yeast industry by-products: comparison of isotherm models. Bioresour Technol 98(9):1812–1821. https://doi.org/10.1016/j.biortech.2006.06.015

Rushing BR, Selim MI (2019) Aflatoxin B1: a review on metabolism, toxicity, occurrence in food, occupational exposure, and detoxification methods. Food Chem Toxicol 124:81–100. https://doi.org/10.1016/j.fct.2018.11.047

Shanakhat H, Sorrentino A, Raiola A, Romano A, Masi P, Cavella S (2018) Current methods for mycotoxins analysis and innovative strategies for their reduction in cereals: an overview. J Sci Food Agric 98(11):4003–4013. https://doi.org/10.1002/jsfa.8933

Simonich MT, Egner PA, Roebuck BD, Orner GA, Jubert C, Pereira C, Groopman JD, Kensler TW, Dashwood RH, Williams DE, Bailey GS (2007) Natural chlorophyll inhibits aflatoxin B 1 -induced multi-organ carcinogenesis in the rat. Carcinogenesis 28(6):1294–1302. https://doi.org/10.1093/carcin/bgm027

Sipos P, Peles F, Brassó DL, Béri B, Pusztahelyi T, Pócsi I, Győri Z (2021) Physical and chemical methods for reduction in aflatoxin content of feed and food. Toxins 13(3):3. https://doi.org/10.3390/toxins13030204

Smith M-C, Madec S, Coton E, Hymery N (2016) Natural co-occurrence of mycotoxins in foods and feeds and their in vitro combined toxicological effects. Toxins 8(4):94. https://doi.org/10.3390/toxins8040094

Solis-Cruz B, Hernandez-Patlan D, Hargis BM, Tellez G (2019a) Control of Aflatoxicosis in poultry using probiotics and polymers. In: Berka Njobeh P, Stepman F (eds) Mycotoxins—impact and management strategies. IntechOpen. https://doi.org/10.5772/intechopen.76371

Solis-Cruz B, Hernandez-Patlan D, Petrone VM, Pontin KP, Latorre JD, Beyssac E, Hernandez-Velasco X, Merino-Guzman R, Arreguin MA, Hargis BM, Lopez-Arellano R, Tellez-Isaias G (2019b) Evaluation of a bacillus-based direct-fed microbial on aflatoxin B1 toxic effects, performance, immunologic status, and serum biochemical parameters in broiler chickens. Avian Dis 63(4):659. https://doi.org/10.1637/aviandiseases-D-19-00100

Stahr HM, Obioha WO (1982) Detoxification of aflatoxin contaminated corn by methanol extraction. Vet Hum Toxicol 24(1):16–17

Steyn PS (1995) Mycotoxins, general view, chemistry and structure. Toxicol Lett 82–83:843–851. https://doi.org/10.1016/0378-4274(95)03525-7

Swaddle TW, Salerno J, Tregloan PA (1994) Aqueous aluminates, silicates, and aluminosilicates. Chem Soc Rev 23(5):319–325. https://doi.org/10.1039/CS9942300319

Vázquez-Durán A, Nava-Ramírez M, Tellez G, Albores A (2022) Removal of aflatoxins using agrowaste-based materials and current characterization techniques used for biosorption assessment. Front Vet Sci 9:897302. https://doi.org/10.3389/fvets.2022.897302

Vila-Donat P, Marín S, Sanchis V, Ramos AJ (2018) A review of the mycotoxin adsorbing agents, with an emphasis on their multi-binding capacity, for animal feed decontamination. Food Chem Toxicol 114:246–259. https://doi.org/10.1016/j.fct.2018.02.044

Villarreal-Barajas T, Vázquez-Durán A, Méndez-Albores A (2022) Effectiveness of electrolyzed oxidizing water on fungi and mycotoxins in food. Food Control 131:108454. https://doi.org/10.1016/j.foodcont.2021.108454

Vörösházi J, Neogrády Z, Mátis G, Mackei M (2024) Pathological consequences, metabolism and toxic effects of trichothecene T-2 toxin in poultry. Poult Sci 103(3):103471. https://doi.org/10.1016/j.psj.2024.103471

Wan J, Chen B, Rao J (2020) Occurrence and preventive strategies to control mycotoxins in cereal-based food. Compr Rev Food Sci Food Saf 19(3):928–953. https://doi.org/10.1111/1541-4337.12546

Wang Y, Shang J, Cai M, Liu Y, Yang K (2023) Detoxification of mycotoxins in agricultural products by non-thermal physical technologies: A review of the past five years. Crit Rev Food Sci Nutr 63(33):11668–11678. https://doi.org/10.1080/10408398.2022.2095554

Weaver AC, Weaver DM, Adams N, Yiannikouris A (2024) Meta-analysis of the effects of yeast Cell Wall extract supple-mentation during mycotoxin challenges on the performance of laying hens. Toxins 16(4):4. https://doi.org/10.3390/toxins16040171

Xiong K, Li X, Guo S, Li L, Liu H (2014) The antifungal mechanism of electrolyzed oxidizing water against aspergillus flavus. Food Sci Biotechnol 23(2):661–669. https://doi.org/10.1007/s10068-014-0090-8

Zabiulla I, Malathi V, Swamy HVLN, Naik J, Pineda L, Han Y (2021) The efficacy of a Smectite-based mycotoxin binder in reducing aflatoxin B1 toxicity on performance, health and histopathology of broiler chickens. Toxins 13(12):12. https://doi.org/10.3390/toxins13120856

Zain ME (2011) Impact of mycotoxins on humans and animals. J Saudi Chem Soc 15(2):129–144. https://doi.org/10.1016/j.jscs.2010.06.006

Zavala-Franco A, Hernández-Patlán D, Solís-Cruz B, López-Arellano R, Tellez-Isaias G, Vázquez-Durán A, Méndez-Albores A (2018) Assessing the aflatoxin B1 adsorption capacity between biosorbents using an in vitro multicompartmental model simulating the dynamic conditions in the gastrointestinal tract of poultry. Toxins 10(11):11. https://doi.org/10.3390/toxins10110484

Strategies to Control Coccidiosis and Parasitic Diseases in Biofarms

13

Lauren Laverty, Juan D. Latorre, Jesus A. Maguey-Gonzalez, Inkar Castellanos-Huerta, Awad A. Shehata, Wolfgang Eisenreich, Guillermo Tellez-Isaias, and Billy M. Hargis

Contents

L. Laverty (✉) · J. D. Latorre · J. A. Maguey-Gonzalez · I. Castellanos-Huerta ·
G. Tellez-Isaias · B. M. Hargis
Department of Poultry Science, University of Arkansas Division of Agriculture, Fayetteville, AR, USA
e-mail: lmlavert@uark.edu

A. A. Shehata · W. Eisenreich
Structural Membrane Biochemistry, Bavarian NMR Center, Technical University of Munich (TUM), Garching, Bayern, Germany

Abstract

The use of antibiotics in commercial poultry is being phased out due in part to changing consumer preferences and increased concern for antibiotic resistance. Consumer preference is a major market driver and the use of antibiotics growth promoters over time has led to antimicrobial resistance. This is also true in the case of anticoccidial drugs in *Eimeria* infections, the causative agent of coccidiosis. Coccidiosis is a widespread and expensive parasitic disease in poultry that results in significant economic losses yearly from performance losses. Because the use of antibiotics is being discouraged, poultry producers are seeking out new methods to mitigate disease from *Eimeria* in addition to focusing on existing methods such as vaccination. Prevention of coccidiosis is key in minimizing losses and methods that align with organic standards include vaccination, careful flock management, biosecurity, and promoting strong gut health using prebiotics, probiotics, and phytogenic substances. Promoting gut health, which can bolster the overall health of a flock is a high interest and promising avenue for coccidiosis prophylaxis. Phytogenic substances, while shown to aid in gut health, also have merit in reducing coccidiosis severity. Additionally, there are challenges with studying coccidiosis as *Eimeria* oocysts require a living host to reproduce and amplify. For this reason, in vitro storage is widely used to preserve oocysts for research. To understand how to counter the challenges of coccidiosis in poultry, it is important to examine all phases of *Eimeria* research, including methods for in vitro research, processing, and storage. The use of avian intestinal enteroids is an interesting method for studying *Eimeria* host interactions, and autofluorescent microscopy is a promising tool for assessing oocyst integrity in vitro. Current methods for suspending and storing *Eimeria* oocysts involve aqueous potassium dichromate (PDC), which is a dangerous chemical for lab personnel and has risks associated with its use. Chlorhexidine (CHX) salts, which are gentle enough to be utilized commonly in dental products, demonstrated strong antimicrobial properties and were well tolerated by *Eimeria* oocysts in suspension. The purpose of this chapter is to address the growing concern of coccidiosis in response to the shift toward organic practices, to discuss potential strategies that can be used to control coccidiosis and to explore new methods for in vitro research and storage of *Eimeria* oocysts.

Keywords
Eimeria · Coccidiosis · Organic · Chlorhexidine · Potassium dichromate · Poultry

1 Introduction

Coccidiosis is a parasitic disease caused by protozoa of the *Eimeria* spp. that results in billions of dollars in losses annually in the poultry industry. The poultry industry, which in 2018 was valued at $46 billion USD, is estimated to lose $3 billion USD yearly from *Eimeria* infections (Jordan et al. 2020). The majority of costs are associated with subclinical disease that results in reduced performance, but the cost of prevention and anticoccidials are also significant (Jordan et al. 2020). Coccidiosis occurs when *Eimeria* oocysts are ingested by the host and infect the gastrointestinal tract (Mcdougald and Fitz-Coy 2008). Coccidia oocysts are ubiquitous in poultry, and it is exceedingly difficult to raise poultry without exposure to oocysts during their lifetime (Mcdougald and Fitz-Coy 2008). Subclinical *Eimeria* infections result in weight reduction from decreased feed intake and reduced feed conversion (López-Osorio et al. 2020). In severe cases, intestinal lesions, mucosal sloughing, and bloody enteritis from epithelial cell lysing can occur, putting a flock at risk of increased mortality and secondary infections (López-Osorio et al. 2020). In chickens, *E. acervuline*, *E. maxima*, and *E. tenella* are prevalent species, with *E. tenella* being among the most pathogenic (Chapman 2014). In turkeys, *E. adenoeides*, *E. gallopavonis* and *E. meleagrimitis* are prevalent (Gadde et al. 2020). The location in the gastrointestinal tract infected is dependent on the species of *Eimeria*; for this reason, infection of the host by one species will not provide immunological protection for another species (Hazef 2008). Because it is advantageous for a parasite to keep its host alive to allow it to reproduce effectively, *Eimeria* demonstrate a self-limiting life cycle (Hazef 2008). Non-domesticated birds in the wild will also likely encounter oocysts during their lifetime. However, the dose that they encounter will be markedly lower than the dose of oocysts present in commercial poultry houses. Oocysts can be introduced into poultry houses by personnel as well as vertebrate or invertebrate vectors (Belli et al. 2006). Introduction to a small number of oocysts over a period of time allows for immunity to develop gradually in wild birds. There are multiple factors that work in tandem that affect the severity of *Eimeria* outbreaks (Fig. 13.1). Commercial poultry, in contrast, are reared in high-density housing and are more likely to ingest a high dose of oocysts that are shed into the feces by the infected flock. For immunologically naïve poultry, this is more likely to cause severe infection. Coccidiosis prevention is essential to control infection, methods

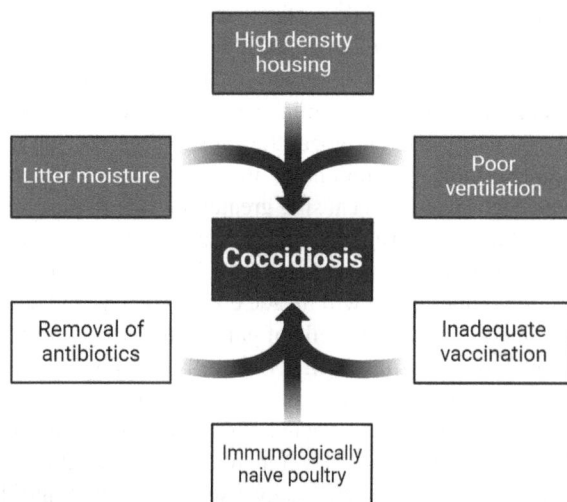

Fig. 13.1 Factors influencing coccidiosis severity. Created with BioRender.com

include biosecurity, careful flock management, vaccination, and coccidiostat feed additives (Mcdougald and Fitz-Coy 2008; Lan et al. 2017).

Oocyst walls are durable, making *Eimeria* resistant to many commonly used disinfectants and allowing them to persist in the environment (Dubey et al. 2020). Even when introduced to hypochlorite solution at 5–6% concentration, which damages the outermost veil, *Eimeria* may remain viable and have the potential to infect (Dubey et al. 2020). This resistance to environmental factors makes oocyst numbers difficult to control in commercial operations. *Eimeria* has both exogenous and endogenous stages of reproduction (Allen and Fetterer 2002). The endogenous phase involves asexual and sexual phases, which cause damage to the gastrointestinal tract by lysing epithelial cells in the host. Sporulation is the reproductive phase of *Eimeria* which occurs outside of the host and in the litter of commercial poultry houses. Because this phase is exogenous, careful management can help to reduce the number of oocysts present. Favorable temperature, moisture, and ample oxygen are all required for oocysts to sporulate and become infective. The ideal temperature for sporulation is between 25 °C and 30 °C (Hazef 2008). Desiccation, inadequate levels of oxygen and temperatures above 40 °C have been demonstrated to be fatal to oocysts (Dubey et al. 2020).

2   *Eimeria* Reproduction

The endogenous phase of the *Eimeria* life cycle begins when a mature, sporulated, infective oocyst is ingested (Chapman 2014). Chemical and mechanical activities of the gizzard and stomach will cause the release of sporozoites contained within sporocysts (López-Osorio et al. 2020). Sporozoites are motile infective agents of *Eimeria* and will migrate to the site of infection with aid from peristalsis (Dubey et al. 2020). Merogony, the asexual phase of reproduction, occurs when these sporozoites penetrate, infect, and proliferate within the epithelial cells. After penetration into epithelial cells, sporozoites will develop into trophozoites (Marugan-Hernandez et al. 2020). Trophozoites will develop into mature meronts that will release a large number of motile merozoites into the lumen after rupturing (Deplazes et al. 2016; Marugan-Hernandez et al. 2020). Merozoites will continue to invade new cells and repeat the cycle before starting the sexual phase of reproduction. Multiple generations of merogony will occur during the *Eimeria* life cycle with each generation causing greater damage to the gastrointestinal tract of the host.

Following merogony, the sexual phase known as gamogony or gametogony occurs (López-Osorio et al. 2020). The majority of merozoites from the final generation of merogony will invade epithelial cells and develop into mononuclear macrogametes, the female cells of gamogony (Madden and Vetterling 1977). A smaller number of merozoites will invade cells and develop into polynucleated male microgamonts, which will rupture and release flagellated microgametes to fertilize macrogametes and form zygotes (López-Osorio et al. 2020; Madden and Vetterling 1977). A zygote will develop into a sporont after forming an oocyst wall and becomes an unsporulated oocyst when it is shed into the environment through the feces of the host (Deplazes et al. 2016). The life cycle described is depicted in Fig. 13.2.

Sporulation, also known as sporogony, is the exogenous phase of the *Eimeria* life cycle. Unsporulated oocysts are non-infective and require oxygen, moisture, and warm temperatures to sporulate and mature (Li and Ooi 2008). During poultry production, sporulation will occur in the litter of barns and will take approximately 72 h. For research purposes, sporulation can also be achieved in controlled conditions in a laboratory.

In the case of *E. acervuline* and *E. maxima*, proliferation and cell lysing will occur in the villi where cells are naturally sloughed after a short time. With *E. tenella*, however, infection occurs in the ceca within crypt cells, causing more substantial damage (Mesa-Pineda et al. 2021). Damage to enough epithelial cells will cause reduced nutrient absorption in the host and in turn reduce weight gain. As mentioned previously, *Eimeria* demonstrates a self-limiting life cycle (Hazef 2008). Interestingly, as the dose of *Eimeria* oocysts increases, reproductive potential will eventually plateau in an occurrence known as the crowding effect (Horton-Smith 1947; Brackett and Bliznick 1952; Williams 1973; Johnston et al. 2001). The reproductive potential of an *Eimeria* oocyst ordinarily would be consistent until a threshold of ingested oocysts is reached (Williams 2001). After this threshold is reached, increasing the dose of oocysts will no longer increase oocyst output and may even decrease it (Williams 2001). It is theorized that as the dose of oocysts increases, the incidence of cell sloughing also increases which reduces the availability of host

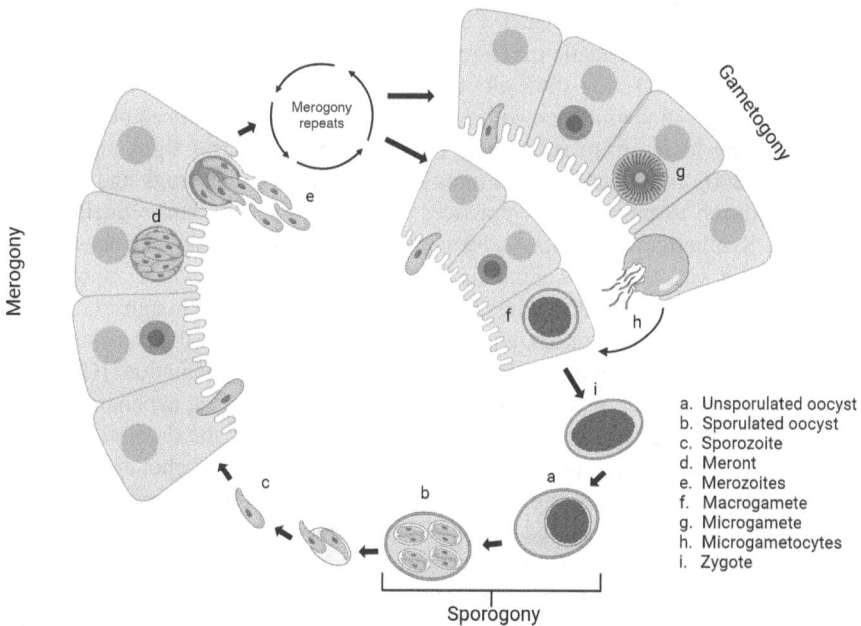

a. Unsporulated oocyst
b. Sporulated oocyst
c. Sporozoite
d. Meront
e. Merozoites
f. Macrogamete
g. Microgamete
h. Microgametocytes
i. Zygote

Fig. 13.2 *Eimeria* life cycle. Created with BioRender.com

epithelial cells to invade, thus reducing the reproductive potential of the oocysts (Tyzzer et al. 1932; Johnston et al. 2001). Fecundity of *Eimeria* depends on the dose of oocysts, species, and strain of *Eimeria* but also depends on several host factors, including age, breed, sex, diet, and even size (Johnston et al. 2001; Williams 2001). The crowding effect is most likely to occur in poultry that has not been previously exposed to oocysts as they are immunologically naïve. In commercial operations, poultry flocks will gain natural immunity after about 3 cycles of *Eimeria* through the host or 5–7 weeks (Chapman 1999).

3 Regulations of Antimicrobials on Organic Farms

Organic agriculture is an immensely fast-growing industry that has gained considerable momentum in recent years. In the United States, organic food sales, including animal products, were predicted to have increased from $28.4 billion to $35 billion USD between 2012 and 2014 (Greene 2013; Salim et al. 2018; Abd El-Hack et al. 2022). In 2021, "organic retail sales" were estimated to be upward of $52 billion USD (USDA-ERS 2024). Organic is a term used in farming when food is produced in a way that meets USDA "organic" standards. At its core, organic farming encompasses food production that integrates "cultural, biological and mechanical practices that foster cycling of resources, promote ecological balance and conserve biodiversity" (USDA-AMS 2024).

The use of Ionophores in the United States is one method for controlling coccidiosis that is affected by the shift toward antibiotic-free poultry-rearing practices (Cervantes and McDougald 2023). Current methods for controlling coccidiosis in poultry focus on prophylactic measures. As it is nearly impossible to raise poultry without them encountering some amount of *Eimeria* oocysts in their lifetime, prevention is vital. A flock infected with *Eimeria* will likely have reduced performance and economic losses from subclinical infection before clinical signs are evident (Chapman 2003). Preventative measures include anticoccidial drugs and vaccination in addition to management practices such as biosecurity and reducing litter moisture. Synthetic compounds and ionophores, also called polyether antibiotics, are two categories of anticoccidials used in poultry production. It is important to note that not all anticoccidial drugs are considered antibiotics, though ionophores are. Synthetic drugs/chemicals are produced through chemical synthesis while polyether antibiotics/ionophores are produced through fermentation (Chapman 2014). The mechanism of action of synthetic compounds varies, depending on the compound or drug used. Mechanisms of actions that involve the disruption of cellular function include inhibition of the folic acid pathway, mitochondrial respiration, and thiamine uptake (Noack et al. 2020). Ionophores are another type of anticoccidial drug, which is used in livestock. Ionophores function by disrupting ion gradients across the cellular membrane of oocysts (Noack et al. 2020). One interesting advantage of ionophores is that they allow for a small amount of subclinical *Eimeria* cycling which encourages host immunity to develop, especially in breeders or layers that are longer lived than broilers (Noack et al. 2020). When an ionophore and synthetic compound are used in tandem on the same flock to increase immunity against *Eimeria* infections, this is called a shuttle program (Chapman and Jeffers 2014). A bioshuttle strategy is accomplished when a flock is vaccinated with a drug-susceptible and then fed an anticoccidial supplement after vaccinal oocysts have replicated sufficiently (Attree et al. 2021). Drug rotation for anticoccidials is routinely used, however, resistance to these drugs has emerged in *Eimeria* from extensive and prolonged use (Chapman 1997; Chapman 1999). Additionally, concerns about chemical residue in livestock have discouraged the development of new anticoccidials (Blake and Tomley 2014).

4 Coccidiosis Control Strategies

4.1 Vaccination

Vaccination is another prevention method but comes with some difficulties. It is difficult to uniformly dose flocks and *Eimeria* antigenic diversity is an additional challenge (Long and Millard 1979; Chapman et al. 2002).

Previously, vaccination for coccidiosis was more prevalent in poultry that is longer lived, such as breeders and layers than in broilers (Shapiro 2001; Chapman et al. 2002). This was due to concerns for chick growth cost of vaccines (Williams 2002). Methods for vaccination included administration through drinking water, and was eventually replaced by spraying vaccine on feed, which resulted in a more uniform exposure to oocysts (Chapman et al. 2002). Spraying vaccine into the eye of birds shortly after hatch was common before the introduction of in ovo vaccination for Marek's (Chapman et al. 2002). Originally, spraying vaccine in the eye was performed at the same time as Marek's vaccine was administered, so this method diminished after the shift to in ovo Marek's vaccination (Chapman et al. 2002). Spray cabinets involve aerosolized vaccines that are introduced to chicks in a partially enclosed space at the hatchery (Chapman et al. 2002). When sprayed with the vaccine, chicks will preen themselves and other chicks and ingest oocysts during the process (Bafundo and Jeffers 1990; Williams 2002). Oocysts containing edible gels, which are colored to interest the bird, are another method of vaccination more commonly in breeders (Chapman et al. 2002). A study conducted by Chapman and Cherry (1997) found that the administration of a vaccine by spraying the feed was more effective at inducing some infection and therefore subsequent immunity than spray cabinets or gels. Advances with in ovo or embryo immunizations have been made in recent years where the *Eimeria* vaccine is delivered to the amniotic cavity of 18-day chicken embryos (Williams 2005; Ahmad et al. 2024). The first generation of vaccines that became available were manufactured using a formula of live wild-type oocysts dosed in controlled amounts (Blake and Tomley 2014). Recombinant vaccines for *Eimeria* have been explored for decades as the development of a vaccine that can be administered without using live parasites would be highly advantageous (Shivaramaiah et al. 2014). Recombinant vaccines have not yet been widely implemented commercially, but recent efforts show promise for future developments (Allen and Fetterer 2002; Shivaramaiah et al. 2014). The reproductive stages and virulence of live wild-type oocysts are unaltered, but live vaccines using precocious oocysts also exist (Williams 2002). Precocious vaccines are made by selecting *Eimeria* oocysts that have fewer endogenous reproductive stages or have fewer reproductive numbers (Shirley and Bedrník 1997). Because these selected oocysts have lower fecundity and fewer reproductive cycles, they cause less damage within the host during infection and allow immunity to develop (Allen and Fetterer 2002). Vaccine manufacturing is also challenging as it depends on developing precocious *Eimeria* oocysts, which is costly (Shivaramaiah et al. 2014).

4.2   Litter Management

As all commercial poultry will most likely encounter oocysts in their lifetime, rearing poultry free from coccidia is unrealistic in most operations. With mindful house management and biosecurity, the severity of *Eimeria* infections can be reduced, though it is usually not adequate for total control or prevention (Blake and Tomley 2014). Although additional methods such as vaccination and feed additives are used to mitigate coccidiosis, the importance of proper barn management is vital in reducing the challenge of *Eimeria* in a flock. After completion of the endogenous phases of the life cycle, immature oocysts are shed through the feces and reside in poultry litter. Although *Eimeria* has a self-limiting life cycle that exists to avoid killing its host (Hazef 2008), high stocking density concentrates the number of oocysts shed in litter to levels that can overwhelm a flock. Careful litter management and proper ventilation can reduce moisture levels and mitigate *Eimeria* sporulation and cycling, which in turn reduces the number of oocysts a flock is exposed to in its environment. During litter composting, high temperatures, elevated ammonia levels, and decreased oxygen are destructive to oocysts (Mcdougald and Fitz-Coy 2008).

4.3   Gut Health

Production losses due to *Eimeria* are concerning, and equally concerning is the emergence of enteric diseases in flocks left vulnerable to coccidiosis. The most concerning disease associated with coccidiosis is necrotic enteritis, caused by *Clostridium perfringens* (Skinner et al. 2010; Adhikari et al. 2020). *C. perfringens* is commonly found in the environment in soil, water, and poultry feed and also exists naturally in the gastrointestinal tract of poultry (Llanco et al. 2017; Fasina and Lillehoj 2019; Adhikari et al. 2020). When *Eimeria* disrupts and ruptures enterocytes, opportunistic *C. perfringens* will proliferate and cause necrotic enteritis (Adhikari et al. 2020). Disruption of tissue leading to serum leaking into the lumen and accelerated mucus production in turn provides additional nutrients for *C. perfringens* proliferation, worsening the disease (Forder et al. 2012; Adhikari et al. 2020). Necrotic enteritis is an example of how multiple factors can compound to worsen flock health as a result of stress from existing and secondary infections.

With the removal of previous, conventional control methods, necrotic enteritis is an additional complication from coccidiosis that the poultry industry must adapt to. However, regardless of pathogenic bacteria, when gut health is subpar in a flock it can lead to reduced performance and increased stress (Hargis et al. 2021).

4.4   Phenols

Phenols or phenolic disinfectants are a class of disinfectants that are used primarily to inhibit the exogenous sporulation phase of the *Eimeria* life cycle (Rajendran and Nabila Fatima 2023). As these disinfectants reduce sporulation rate, they have the potential to be utilized in commercial operations to reduce the number of infective oocysts and reduce the need for antibiotics as feed additives (Rajendran and Nabila Fatima 2023). Phenols as prospective coccidiocidals began with studies assessing the effect of ammonia on *Eimeria* oocysts and interest in using chemicals safer for handling grew (Williams 1997). As Williams (1997) indicates, this process of disrupting the sporulation process with disinfectants is nonreversible. Phenols, therefore, show promise as a method of reducing an *Eimeria* dose within a flock when used for disinfection.

4.5   Prebiotics and Probiotics

One method of reducing the presence of pathogenic bacteria in the gastrointestinal tract is to promote a healthy gut microbiome (Getachew 2016) such as through the addition of living microorganisms to feed known as probiotics (Al-Khalaifah 2018). The addition of probiotics to poultry diets is a recent practice that has been demonstrated to improve gut health and nutrition of poultry by encouraging growth of beneficial microorganisms in the GI tract (Getachew 2016; Hargis et al. 2021; Abd El-Hack et al. 2022). Probiotics have multiple different mechanisms of action to improve gut health, one of which is suppression of pathogenic bacteria through competitive exclusion (Mountzouris et al. 2009). Probiotics when added to feed were demonstrated to reduce clinical signs and bolster host immunity in the case of broiler chickens challenged with *E. maxima* (Lee et al. 2010). This was apparent from increased splenocyte production and changes in serum nitric oxide levels using *Bacillus*-based probiotics (Lee et al. 2010). Furthermore, broilers challenged with *E. tenella* showed increased CD3+, CD4+, and CD8+ T-lymphocyte counts when supplemented with *Saccharomyces cerevisiae*, which also improved body weight gain and feed intake during a 42-day trial (Gao et al. 2009).

In the case of prebiotics, these are typically carbohydrate-based feed additives that are non-digestible and assist the growth of beneficial bacteria in the gut, which also induces competitive exclusion of pathogenic bacteria (Al-Khalaifah 2018). In a study conducted by Angwech et al. (2019), prebiotics introduced in ovo to chicken embryos at day 12 of incubation appeared to decrease the severity of *Eimeria* infection as intestinal lesions and oocyst output were reduced.

4.6   Phytogenic Substances

Another emerging category of research which aims to improve poultry gut health is that of phytogenic feed additives. Phytogenics generally consists of plant derivatives such as essential oils, spices, herbs, oleoresins, and flavonoids and other compounds that are intended to improve the gut health of poultry (Alçiçek et al. 2004; Mountzouris et al. 2011; Abd El-Hack et al. 2022) which in turn is related to increased performance (Getachew 2016; Hargis et al. 2021; Abd El-Hack et al. 2022).

4.7   Essential Oils

Essential oils are more prevalent and although they are not widely used, data suggests that their inclusion positively influences nutrient digestibility, feed conversion ratio, and body weight gain (Mountzouris et al. 2011). However, results from different studies vary in their findings as many factors, such as age and gut ecology, seem to influence the efficacy of phytogenic additives (Mountzouris et al. 2011). With regard to *Eimeria*, phytogenic additives have been demonstrated to protect broilers challenged with *E. tenella* (Allen et al. 1997; Youn and Noh 2001; Giannenas et al. 2003). Allen et al. (1997) reported that a supplement of dried *Artemisia annua* leaves reduced the severity of lesions for *E. tenella* but had no significant effect on lesions from *E. maxima* or *E. acervulina*. Youn and Noh (2001) reported that broilers supplemented with various herbal extracts exhibited lower oocyst output, improved survivability, and reduced instances of bloody diarrhea and intestinal lesions. Bozkurt et al. (2016) more recently reported reduced oocyst output from broilers supplemented with oregano essential oil, though was not as effective as monensin, a conventional anticoccidial. Although treatment success varies with coccidiosis studies, the antimicrobial and immunity-boosting properties of phytogenics make them a promising option for *Eimeria* control (Adhikari et al. 2020).

4.8   Antioxidants

One natural response to *Eimeria* infection is the production of free radicals and oxidative species during host cellular immune response (Allen et al. 1997; Vladimirov 2004; Masood et al. 2013). While this helps to defend against parasitic infections, in excess it also causes harmful conditions for host cells, leading to cytotoxicity and higher incidence of tissue damage (Masood et al. 2013). Antioxidants as an additive to poultry feed have been used to protect cells and reduce damage from oxidative stress caused by coccidiosis and other parasitic infections (Allen et al. 1998; Masood et al. 2013).

5   Challenges with *Eimeria* In Vitro Research and Future Approaches

Eimeria are obligate intracellular parasites and require a host in order to reproduce. Currently, there is no reliable or sustainable method to amplify *Eimeria* for research without passing oocysts through a living host. As such, in vitro storage is vital for conducting research on *Eimeria* oocysts. Potassium dichromate (PDC) is a chemical commonly used to suspend oocysts for in vitro sporulation, storage, and research. Although well tolerated by oocysts, PDC is hazardous to handle which could be a potential drawback to uses in research. To better understand how to combat coccidiosis, safe and efficient research in a laboratory setting is important.

6   Avian Intestinal Enteroids

Stem cells when isolated from the gastrointestinal tract and allowed to culture have been demonstrated to form gut models that are remarkably similar to living hosts (Ranganathan et al. 2020). These gut models, known as enteroids, have been achieved using cells derived from humans (Spence et al. 2011) and more recently from chickens (Nash et al. 2021). Nash et al. (2021) have demonstrated that enteroids cultured using cells derived from the ceca of chickens can act as a model for *E. tenella* interactions. Such models show promise for future research in understanding and visualizing how *Eimeria* behaves within hosts.

7   Autofluorescence in *Eimeria*

One challenge with in vitro studies involving *Eimeria* is determining whether oocysts are viable or non-viable, as morphologically they are difficult to distinguish. This creates problems with dosing poultry for *Eimeria* challenges and verifying the health of oocyst stocks in storage. Beer et al. (2018) previously showed that *Eimeria* oocysts when excited by UV light emit autofluorescence visible to the eye during photomicroscopy that can be used as an indication of non-viability. Sporocysts of non-viable oocysts when excited will have a strong autofluorescent light, whereas autofluorescence in the sporocysts of viable oocysts will be faint or absent (Fig. 13.3) (Augustine 1980; Beer et al. 2018; Laverty et al. 2023). Autofluorescent microscopy used in this way presents a potential method for determining oocyst viability and therefore infectivity, which has a multitude of uses, including making informed decisions on doses for in vivo studies.

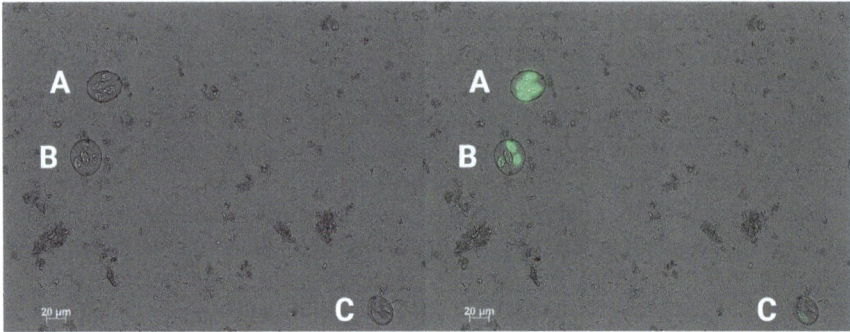

Fig. 13.3 200x magnification conventional DIC photomicrograph (left) and green autofluorescent photomicrograph (right) with sporulated *Eimeria maxima* oocysts. Autofluorescent oocyst (**a**), partially autofluorescent oocyst (**b**), and non-autofluorescent oocyst (**c**). Laverty et al. 2023 (Created with BioRender.com)

8   Potassium Dichromate

PDC is a powerful oxidizing agent that enters cells through sulfate channels, where it causes damage to DNA via reduction (Alexander and Aaseth 1995; Salnikow and Zhitkovich 2008; Salama et al. 2022). PDC is a chemical that is used to preserve *Eimeria* species oocysts for future use, including for the maturation process known as sporulation. It is advantageous as it demonstrates strong antimicrobial properties and does not adversely affect *Eimeria* oocyst sporulation or viability (Gong et al. 2021). *Eimeria* oocysts are tolerant to storage in PDC but require oxygen to survive while in suspension, especially during sporulation when oxygen is rapidly consumed. When stored in an aqueous solution appropriately and allowed oxygen, *Eimeria* oocysts will survive for a long period of time (Williams et al. 2010). Although infectivity will decline the longer *Eimeria* oocysts are stored and eventually cease, oocysts can remain morphologically intact and if allowed oxygen will survive in storage until energy in the form of amylopectin depletes (Vetterling and Doran 1969; Ruff et al. 1981; Cha et al. 2018). Infectivity has been shown to decline after 6–8 months in storage at 4 °C (Ruff et al. 1981; Cha et al. 2018), but DNA has been preserved for years (Williams et al. 2010). The purpose of *Eimeria* oocyst suspensions using aqueous PDC is to reduce the number of oxygen-consuming microorganisms. The depletion of oxygen-consuming microorganisms in turn reduces the competition for oxygen in suspensions, allowing oocysts to respire and survive. The disadvantage of PDC arises from its risk to human health as it is carcinogenic and can potentially damage the kidneys, liver, and brain due to oxidative injury (Dayan and Paine 2001; Salama et al. 2022). Research exploring the potential to utilize chlorhexidine (CHX) salts in the place of PDC has demonstrated how CHX salts can reduce bacterial recovery while still preserving the integrity of oocysts.

9   Chlorhexidine as an Alternative to PDC

CHX is insoluble in water and is combined with a water-soluble salt when made into a solution; gluconate, digluconate, and diacetate are salts mentioned here. CHX salts will be used to refer to a water-soluble solution containing CHX, whereas CHX will be used to refer to the CHX compound. Notably, CHX salts are used as an active ingredient in dental products such as toothpastes, mouthwashes, and varnishes, and as an antiseptic agent in medicine. CHX salts are advantageous as they are effective without causing harm to the patient, long lasting, and have a broad spectrum of activity (Bescos et al. 2020). For products intended for direct use on oral surfaces, the concentration of CHX is between 0.10% and 0.20% (Poppolo Deus and Ouanounou 2022). At higher concentrations, between 2.50% and 5.40%, contact with skin or oral surfaces is more likely to cause contact dermatitis (Lim and Kam 2008; Cheung et al. 2012). The effectiveness of CHX salts for killing bacteria is dependent on concentration, exposure time, and the species of bacteria. Generally, CHX salts are estimated to be bacteriostatic at low concentrations of <0.05% and bactericidal at high concentrations of >0.05% (Lim and Kam 2008). It has been reported that ingestion at low concentrations was well tolerated but caused some adverse effects when ingested by patients accidentally at 0.05% (Lim and Kam 2008). When investigating CHX in animal models of infection, CHX salts were shown to have no systemic absorption (Barrett-Bee et al. 1994).

CHX is a cationic biguanide compound noted for its broad range of activity against bacteria, fungi, and viruses to an extent (Barrett-Bee et al. 1994). The mechanism of action of CHX was first investigated in 1994. Initial investigations revealed that when bacteria are introduced to a bactericidal concentration of CHX salts, CHX causes the cytoplasmic components of the cell to leak (Barrett-Bee et al. 1994). As mentioned previously, the concentration at which CHX is bacteriostatic and bactericidal is dependent on the species of bacteria but is generally considered to be bactericidal at >0.05%, and bacteriostatic at <0.05% (Lim and Kam 2008). At bacteriostatic concentrations, CHX will form covalent bonds to sulfate and phosphate groups, which are negatively charged, on the bacterial cell wall (Solderer et al. 2019). Interestingly, the bacteriostatic phase of CHX action is reversible (Solderer et al. 2019). Increasing the concentration of CHX will cause progressively greater and irreversible damage as a result of inner cell membrane penetration (Mathur et al. 2011; Thangavelu et al. 2020). The binding on the bacterial cell surface causes strong absorption of proteins and subsequent penetration of the cell wall (Łukomska-Szymańska et al. 2017; Poppolo Deus and Ouanounou 2022). The penetration of the cell wall causes inner cell membrane integrity to deteriorate and permeability to increase, resulting in the leakage of cytoplasmic components (Fig. 13.4) (Thangavelu et al. 2020; Poppolo Deus and Ouanounou 2022). When CHX is allowed to freely enter the cytoplasm through the damaged membrane, it will irreversibly precipitate with ATP and nucleic acids present in the cell (Lim and Kam 2008). CHX also has notable substantivity when it binds to negatively charged surfaces on tissue, causing a long-lasting "pin cushion" effect (Lim and Kam 2008; Thangavelu et al. 2020). This can occur when cationic CHX binds to surfaces in the mouth, making it a

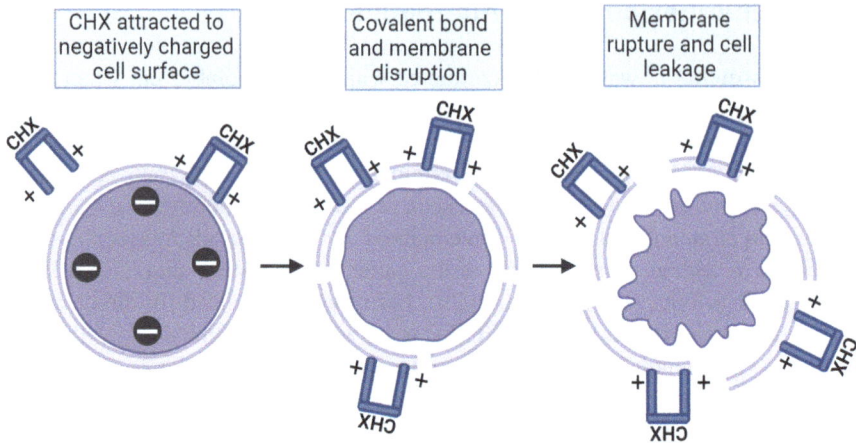

Fig. 13.4 Chlorhexidine interaction with cellular membrane. Created with BioRender.com

useful agent in the dental industry. The fact that CHX salts demonstrate strong antimicrobial effects but are gentle enough to be used in dental products as high as 0.20% prompted an investigation into how such solutions would behave with EM oocysts. In experiments conducted by Laverty et al. (2023), CHX salts have demonstrated great antibacterial properties when introduced to EM oocyst-containing poultry fecal slurries. These findings support the idea that CHX salts are capable of reducing the recovery of bacteria present while retaining oocyst integrity.

10   CHX in Research In Vitro and In Vivo Results

At present, CHX salts in agriculture are used in some commercial detergents, but other potential uses should be evaluated. In exploring chemical alternatives to PDC, the antimicrobial properties of CHX salts have been demonstrated to be very effective when tested on poultry fecal slurries and EM oocyst suspensions (Laverty et al. 2023). CHX gluconate and CHX digluconate were both tested and compared to PDC in terms of reducing bacterial recovery in a poultry fecal slurry. In further studies, their effect on EM oocyst integrity, infectivity, and sporulation rate was evaluated and also found to be similar to PDC. In particular, a commercially available solution of CHX gluconate (VETone brand, 2% solution), effectively preserved the infectivity of EM oocysts stored in a suspension containing 0.20% CHX gluconate. When reintroduced to chickens as an oral challenge, EM oocysts stored for 5 months at 4 °C in 0.20% CHX gluconate caused a total oocyst output statistically comparable to EM oocysts that were stored under the same conditions in 2.50% PDC. The average total oocyst output was 44.62 million and 48.92 million CHX gluconate and PDC respectively (Laverty et al. 2023). Under the same storage conditions, reinfection with EM oocysts stored in 0.20% CHX digluconate (Sigma, C9394, 20% aqueous solution) caused significantly lower oocyst output when compared to PDC and

CHX gluconate and the oocyst output was not statistically different from EM oocysts stored in saline for the same amount of time. The average total oocyst output was 29.51 million and 28.86 million for CHX digluconate and saline, respectively (Laverty et al. 2023).

The purpose of the study conducted by Laverty et al. (2023) was to test methods of preserving oocysts that could potentially be as effective as PDC. Although CHX gluconate and digluconate have notable antimicrobial properties in oocyst suspensions, their effect on the EM oocysts was relatively benign as the oocysts tolerated in vitro storage in the CHX salts well. For this reason, CHX gluconate could potentially be used to process, sporulate, and preserve *Eimeria* oocysts, but likely is not a viable method for destroying oocysts. However, CHX diacetate (Sigma C6143, salt hydrate) at 0.14% when tested was described to have both marked antibacterial effects and was also deleterious to oocysts (Laverty et al. 2023). CHX gluconate-containing detergent was more effective at reducing bacterial recovery than CHX gluconate or CHX digluconate alone solubilized in saline. The CHX gluconate-containing detergent contains additional ingredients, including citric acid. Presumably, citric acid is being added as a chelating agent. Therefore, the presence of additional components may increase the effectiveness of CHX gluconate. It is also possible that the presence of oocysts may have an effect on bacterial reduction in CHX salt suspensions and solutions.

The different effects on EM oocysts between CHX gluconate, CHX digluconate, and CHX diacetate suggest that to some extent, the salt that CHX is combined with may affect how CHX behaves, as do other components present. The difference between CHX salt solutions and CHX diacetate's ability to degrade oocysts could be explored as a potential tool for controlling coccidiosis infections. It is been demonstrated that CHX salts when combined with an alcohol in solution increase the antimicrobial effects (Lim and Kam 2008). This synergy could be utilized to develop a method to kill oocysts and possibly used in footbaths. More research is needed to understand the effect of CHX diacetate on oocysts and how its effects can be utilized, but its use should not be disregarded.

11 Conclusion

Consumer preferences and industry practices are continually shifting toward organic and will continue to do so for the foreseeable future. With labels like "raised without antibiotics" and "no antibiotics ever," exploring new alternatives for disease prevention and treatment is essential. Because of the ubiquitous nature of *Eimeria*, exploring every new or existing tool available gives the industry the best chance of managing outbreaks. Vaccination is a known technique to encourage host immune development against *Eimeria*. Sanitation and litter management are important steps in reducing the dose of infectious oocysts in the environment. Prebiotics, probiotics, and phytogenics have demonstrated valuable health-boosting properties to poultry, and possibly reduce the severity of coccidiosis. More research is needed to understand the full extent of how these new techniques can aid poultry production in the future as their use could benefit flocks while aligning with organic practices.

References

Abd El-Hack ME, El-Saadony MT, Salem HM, El-Tahan AM, Soliman MM, Youssef GBA, Taha AE, Soliman SM, Ahmed AE, El-kott AF, Al Syaad KM, Swelum AA (2022) Alternatives to antibiotics for organic poultry production: types, modes of action and impacts on bird's health and production. Poult Sci 101(4):101696. https://doi.org/10.1016/j.psj.2022.101696

Adhikari P, Kiess A, Adhikari R, Jha R (2020) An approach to alternative strategies to control avian coccidiosis and necrotic enteritis. J Appl Poult Res 29(2):515–534. https://doi.org/10.1016/j.japr.2019.11.005

Ahmad R, Yu Y-H, Hua K-F, Chen W-J, Zaborski D, Dybus A, Hsiao FS-H, Cheng Y-H (2024) Management and control of coccidiosis in poultry—a review. Anim Biosci 37(1):1–15. https://doi.org/10.5713/ab.23.0189

Alçiçek A, Bozkurt M, Cabuk M (2004) The effect of a mixture of herbal essential oils, an organic acid or a probiotic on broiler performance. South African J Anim Sci 34

Al-Khalaifah HS (2018) Benefits of probiotics and/or prebiotics for antibiotic-reduced poultry. Poult Sci 97(11):3807–3815. https://doi.org/10.3382/ps/pey160

Allen P, Lydon J, Danforth H (1997) Effects of components of Artemisia annua on coccidia infections in chickens. Poult Sci 76(8):1156–1163. https://doi.org/10.1093/ps/76.8.1156

Allen P, Danforth H, Augustine P (1998) Diet modulation of avian coccidiosis. Int J Parasitol 28:1131–1140. https://doi.org/10.1016/S0020-7519(98)00029-0

Allen PC, Fetterer RH (2002) Recent advances in biology and Immunobiology of Eimeria species and in diagnosis and control of infection with these coccidian parasites of poultry. Clin Microbiol Rev 15(1):58–65. https://doi.org/10.1128/CMR.15.1.58-65.2002

Alexander J, Aaseth J (1995) Uptake of chromate in human red blood cells and isolated rat liver cells: the role of the anion carrier. Analyst 120(3):931–933. https://doi.org/10.1039/AN9952000931

Angwech H, Tavaniello S, Ongwech A, Kaaya AN, Maiorano G (2019) Efficacy of in Ovo delivered prebiotics on growth performance, meat quality and gut health of Kuroiler chickens in the face of a natural coccidiosis challenge. Animals 9(11):11. https://doi.org/10.3390/ani9110876

Attree E, Sanchez-Arsuaga G, Jones M, Xia D, Marugan-Hernandez V, Blake D, Tomley F (2021) Controlling the causative agents of coccidiosis in domestic chickens; an eye on the past and considerations for the future. CABI Agric Biosci 2(1):37. https://doi.org/10.1186/s43170-021-00056-5

Augustine PC (1980) Effects of storage time and temperature on amylopectin levels and oocyst production of Eimeria meleagrimitis oocysts. Parasitology 81(Pt 3):519–524. https://doi.org/10.1017/s0031182000061904

Bafundo KW, Jeffers TK (1990) Anticoccidial method. United States Patent 4(935):007

Barrett-Bee K, Newboult L, Edwards S (1994) The membrane destabilising action of the antibacterial agent chlorhexidine. FEMS Microbiol Lett 119(1–2):1–2. https://doi.org/10.1111/j.1574-6968.1994.tb06896.x

Beer LC, Bielke LR, Barta JR, Faulkner OB, Latorre JD, Briggs WN, Wilson KM, Baxter MFA, Tellez G, Hargis BM (2018) Evaluation of autofluorescent Eimeria maxima oocysts as a potential indicator of non-viability when enumerating oocysts. Poult Sci 97(8):2684–2689. https://doi.org/10.3382/ps/pey124

Belli SI, Smith NC, Ferguson DJP (2006) The coccidian oocyst: a tough nut to crack! Trends Parasitol 22(9):416–423. https://doi.org/10.1016/j.pt.2006.07.004

Bescos R, Ashworth A, Cutler C, Brookes ZL, Belfield L, Rodiles A, Casas-Agustench P, Farnham G, Liddle L, Burleigh M, White D, Easton C, Hickson M (2020) Effects of chlorhexidine mouthwash on the oral microbiome. Sci Rep 10(1):1. https://doi.org/10.1038/s41598-020-61912-4

Brackett S, Bliznick A (1952) The reproductive potential of five species of coccidia of the chicken as demonstrated by oocyst production. J Parasitol 38(2):133–139

Blake DP, Tomley FM (2014) Securing poultry production from the ever-present Eimeria challenge. Trends Parasitol 30(1):12–19. https://doi.org/10.1016/j.pt.2013.10.003

Bozkurt M, Ege G, Aysul N, Akşit H, Tüzün AE, Küçükyılmaz K, Borum AE, Uygun M, Akşit D, Aypak S, Şimşek E, Seyrek K, Koçer B, Bintaş E, Orojpour A (2016) Effect of anticoccidial monensin with oregano essential oil on broilers experimentally challenged with mixed Eimeria spp. Poult Sci 95(8):1858–1868. https://doi.org/10.3382/ps/pew077

Cervantes HM, McDougald LR (2023) Raising broiler chickens without ionophore anticoccidials. J Appl Poult Res 32(2):100347. https://doi.org/10.1016/j.japr.2023.100347

Cha JO, Zhao J, Yang MS, Kim WI, Cho HS, Lim CW, Kim B (2018) Oocyst-shedding patterns of three Eimeria species in chickens and shedding pattern variation depending on the storage period of Eimeria tenella oocysts. J Parasitol 104(1):18–22. https://doi.org/10.1645/16-132

Chapman HD (1997) Biochemical, genetic and applied aspects of drug resistance in Eimeria parasites of the fowl. Avian Pathol 26(2):221–244. https://doi.org/10.1080/03079459708419208

Chapman HD (1999) The development of immunity to Eimeria species in broilers given anticoccidial drugs. Avian Pathol 28(2):155–162. https://doi.org/10.1080/03079459994885

Chapman HD (2003) Origins of coccidiosis research in the fowl—the first fifty years. Avian Dis 47(1):1–20

Chapman HD (2014) Milestones in avian coccidiosis research: a review. Poult Sci 93(3):501–511. https://doi.org/10.3382/ps.2013-03634

Chapman HD, Cherry TE (1997) Eyespray vaccination: infectivity and development of immunity to Eimeria acervulina and Eimeria tenella1. J Appl Poult Res 6(3):274–278. https://doi.org/10.1093/japr/6.3.274

Chapman HD, Cherry TE, Danforth HD, Richards G, Shirley MW, Williams RB (2002) Sustainable coccidiosis control in poultry production: the role of live vaccines. Int J Parasitol 32(5):617–629. https://doi.org/10.1016/S0020-7519(01)00362-9

Chapman HD, Jeffers TK (2014) Vaccination of chickens against coccidiosis ameliorates drug resistance in commercial poultry production. Int J Parasitol Drugs Drug Resist 4(3):214–217. https://doi.org/10.1016/j.ijpddr.2014.10.002

Cheung H-Y, Wong MM-K, Cheung S-H, Liang LY, Lam Y-W, Chiu S-K (2012) Differential actions of chlorhexidine on the Cell Wall of Bacillus subtilis and Escherichia coli. PLoS One 7(5):5. https://doi.org/10.1371/journal.pone.0036659

Dayan AD, Paine AJ (2001) Mechanisms of chromium toxicity, carcinogenicity and allergenicity: review of the literature from 1985 to 2000. Hum Exp Toxicol 20(9):439–451. https://doi.org/10.1191/096032701682693062

Deplazes P, Eckert J, Mathis A, von Samson-Himmelstjerna G, Zahner H (2016) Parasitology in veterinary medicine. In: Parasitology in veterinary medicine. Wageningen Academic Press, Wageningen. https://doi.org/10.3920/978-90-8686-274-0

Dubey JP, Lindsay DS, Jenkins MC, Bauer C (2020) Biology of intestinal Coccidia. In: Coccidiosis in livestock, poultry, companion animals, and humans. CRC Press, pp 1–36

Fasina YO, Lillehoj HS (2019) Characterization of intestinal immune response to Clostridium perfringens infection in broiler chickens. Poult Sci 98(1):188–198. https://doi.org/10.3382/ps/pey390

Forder REA, Nattrass GS, Geier MS, Hughes RJ, Hynd PI (2012) Quantitative analyses of genes associated with mucin synthesis of broiler chickens with induced necrotic enteritis. Poult Sci 91(6):1335–1341. https://doi.org/10.3382/ps.2011-02062

Getachew T (2016) A review on effects of probiotic supplementation in poultry performance and cholesterol levels of egg and meat. J World's Poul Res 6(1):31–36

Gadde UD, Rathinam T, Finklin MN, Chapman HD (2020) Pathology caused by three species of Eimeria that infect the Turkey with a description of a scoring system for intestinal lesions. Avian Pathol 49(1):80–86. https://doi.org/10.1080/03079457.2019.1669767

Gao J, Zhang HJ, Wu SG, Yu SH, Yoon I, Moore D, Gao YP, Yan HJ, Qi GH (2009) Effect of Saccharomyces cerevisiae fermentation product on immune functions of broilers challenged with Eimeria tenella. Poult Sci 88(10):2141–2151. https://doi.org/10.3382/ps.2009-00151

Giannenas I, Florou-Paneri P, Papazahariadou M, Christaki E, Botsoglou NA, Spais AB (2003) Effect of dietary supplementation with oregano essential oil on performance of broilers after experimental infection with *eimeria tenella*. Arch Anim Nutr 57(2):99–106. https://doi.org/10.1080/0003942031000107299

Gong Z, Wei H, Chang F, Yin H, Cai J (2021) Sporulation rate and viability of Eimeria tenella oocysts stored in potassium sorbate solution. Parasitol Res 120(6):2297–2301. https://doi.org/10.1007/s00436-020-06792-3

Greene C (2013) Growth patterns in the U.S. organic industry. USDA Economic Research Service. http://www.ers.usda.gov/amberwaves/2013-october/growth-patterns-in-the-usorganic-industry.aspx#.VPdRlvnF98E. Accessed Feb 2024

Hargis BM, Tellez GL, Latorre JD, Wolfenden R (2021) Compositions, probiotic formulations and methods to promote digestion and improve nutrition in poultry. United States Patent 10,959:447 B2

Hazef MH (2008) Poultry coccidiosis: Prevention and control approaches—European Poultry Science. https://www.european-poultry-science.com/poultry-coccidiosis-prevention-and-control-approaches,QUlEPTQyMTg3ODEmTUlEPTE2MTAxNA.html.   Accessed   30 July 2023

Horton-Smith C (1947) Coccidiosis, some factors influencing its epidemiology. Vet Rec 59(47):645

Johnston WT, Shirley MW, Smith AL, Gravenor MB (2001) Modelling host cell availability and the crowding effect in Eimeria infections. Int J Parasitol 31(10):1070–1081. https://doi.org/10.1016/s0020-7519(01)00234-x

Jordan B, Albanese G, Tensa L (2020) Coccidiosis in chickens (*Gallus gallus*). In: Coccidiosis in livestock, poultry, companion animals, and humans. CRC Press, pp 169–173

Lan L-H, Sun B-B, Zuo B-X-Z, Chen X-Q, Du A-F (2017) Prevalence and drug resistance of avian Eimeria species in broiler chicken farms of Zhejiang province, China. Poul Sci 96(7):2104–2109. https://doi.org/10.3382/ps/pew499

Laverty L, Beer LC, Martin K, Hernandez-Velasco X, Juarez-Estrada MA, Arango-Cardona M, Forga AJ, Coles ME, Vuong CN, Latorre JD, Señas-Cuesta R, Loeza I, Gray LS, Barta JR, Hargis BM, Tellez-Isaias G, Graham BD (2023) In vitro and in vivo evaluation of chlorhexidine salts as potential alternatives to potassium dichromate for Eimeria maxima M6 oocyst preservation. Front Vet Sci 10:1226298. https://doi.org/10.3389/fvets.2023.1226298

Lee K-W, Lillehoj HS, Jang SI, Li G, Lee S-H, Lillehoj EP, Siragusa GR (2010) Effect of bacillus-based direct-fed microbials on Eimeria maxima infection in broiler chickens. Comp Immunol Microbiol Infect Dis 33(6):e105–e110. https://doi.org/10.1016/j.cimid.2010.06.001

Lim K-S, Kam PCA (2008) Chlorhexidine—pharmacology and clinical applications. Anaesth Intensive Care 36(4):4. https://doi.org/10.1177/0310057X0803600404

Li M-H, Ooi H-K (2008) Effect of chromium compounds on sporulation of *Eimeria piriformis* oocysts. Exp Anim 57(1):79–83. https://doi.org/10.1538/expanim.57.79

Llanco LA, Nakano V, de Moraes CTP, Piazza RMF, Avila-Campos MJ (2017) Adhesion and invasion of Clostridium perfringens type a into epithelial cells. Braz J Microbiol 48(4):764. https://doi.org/10.1016/j.bjm.2017.06.002

Long PL, Millard BJ (1979) Immunological differences in Eimeria maxima: effect of a mixed immunizing inoculum on heterologous challenge. Parasitology 79(3):451–457

López-Osorio S, Chaparro-Gutiérrez JJ, Gómez-Osorio LM (2020) Overview of poultry Eimeria life cycle and host-parasite interactions. Front Vet Sci 7. https://www.frontiersin.org/articles/10.3389/fvets.2020.00384

Łukomska-Szymańska M, Sokołowski J, Łapińska B (2017) Chlorhexidine—mechanism of action and its application to dentistry. J Stomatol 70(4):4. https://doi.org/10.5604/01.3001.0010.5698

Madden PA, Vetterling JM (1977) Scanning electron microscopy of Eimeria tenella microgametogenesis and fertilization. J Parasitol 63(4):607–610

Marugan-Hernandez V, Jeremiah G, Aguiar-Martins K, Burrell A, Vaughan S, Xia D, Randle N, Tomley F (2020) The growth of Eimeria tenella: characterization and application of quantitative methods to assess Sporozoite invasion and endogenous development in cell culture. Front Cell Infect Microbiol 10:579833. https://doi.org/10.3389/fcimb.2020.579833

Masood S, Abbas RZ, Iqbal Z, Mansoor MK, Sindhu Z-D, Anjum M, Khan JA (2013) Role of natural antioxidants for the control of coccidiosis in poultry. Pak Vet J 8

Mathur S, Mathur T, Srivastava R, Khatri R (2011) Chlorhexidine: the gold standard in chemical plaque control. Nat J Physiol, Pharm Pharmacol 1

Mcdougald L, Fitz-Coy S (2008) Coccidiosis. Dis Poul:1068–1085

Mesa-Pineda C, Navarro-Ruíz JL, López-Osorio S, Chaparro-Gutiérrez JJ, Gómez-Osorio LM (2021) Chicken coccidiosis: from the parasite lifecycle to control of the disease. Front Vet Sci 8:787653. https://doi.org/10.3389/fvets.2021.787653

Mountzouris KC, Balaskas C, Xanthakos I, Tzivinikou A, Fegeros K (2009) Effects of a multispecies probiotic on biomarkers of competitive exclusion efficacy in broilers challenged with salmonella enteritidis. Br Poult Sci 50(4):467–478. https://doi.org/10.1080/00071660903110935

Mountzouris KC, Paraskevas V, Tsirtsikos P, Palamidi I, Steiner T, Schatzmayr G, Fegeros K (2011) Assessment of a phytogenic feed additive effect on broiler growth performance, nutrient digestibility and caecal microflora composition. Anim Feed Sci Technol 168(3):223–231. https://doi.org/10.1016/j.anifeedsci.2011.03.020

Nash TJ, Morris KM, Mabbott NA, Vervelde L (2021) Inside-out chicken enteroids with leukocyte component as a model to study host–pathogen interactions. Commun Biol 4(1):Article 1. https://doi.org/10.1038/s42003-021-01901-z

Noack S, Chapman HD, Seizer PM (2020) Anticoccidial drugs of livestock and poultry industries. In: Coccidiosis in livestock, poultry, companion animals, and humans. CRC Press, pp 66–77

Poppolo Deus F, Ouanounou A (2022) Chlorhexidine in dentistry: pharmacology, uses, and adverse effects. Int Dental J 72(3):3. https://doi.org/10.1016/j.identj.2022.01.005

Rajendran RM, Nabila Fatima S (2023) Effect of phenolic disinfectant on sporulation inhibition of Eimeria tenella for prevention of coccidiosis. J Appl Biol Biotechnol 11(Issue):228–232. https://doi.org/10.7324/JABB.2023.135101

Ranganathan S, Smith EM, Foulke-Abel JD, Barry EM (2020) Research in a time of enteroids and organoids: how the human gut model has transformed the study of enteric bacterial pathogens. Gut Microbes 12(1):1795389. https://doi.org/10.1080/19490976.2020.1795389

Ruff MD, Doran DJ, Wilkins GC (1981) Effect of aging on survival and pathogenicity of Eimeria acervulina and Eimeria tenella. Avian Dis 25(3):595–599. https://doi.org/10.2307/1589989

Salama AAA, Mostafa RE, Elgohary R (2022) Effect of L-carnitine on potassium dichromate-induced nephrotoxicity in rats: modulation of PI3K/AKT signaling pathway. Res Pharm Sci 17(2):153–163. https://doi.org/10.4103/1735-5362.335174

Salim HMD, Huque KS, Kamaruddin KM, Haque Beg A (2018) Global restriction of using antibiotic growth promoters and alternative strategies in poultry production. Sci Prog 101(1):52–75. https://doi.org/10.3184/003685018X15173975498947

Salnikow K, Zhitkovich A (2008) Genetic and epigenetic mechanisms in metal carcinogenesis and Cocarcinogenesis: nickel, arsenic, and chromium. Chem Res Toxicol 21(1):28–44. https://doi.org/10.1021/tx700198a

Shapiro D (2001) Coccidiosis control in replacement pullets. Int Hatchery Pract 15:13–17

Shirley MW, Bedrník P (1997) Live attenuated vaccines against avian coccidiosis: success with precocious and egg-adapted lines of Eimeria. Parasitol Today (Personal Ed) 13(12):481–484. https://doi.org/10.1016/s0169-4758(97)01153-8

Shivaramaiah C, Barta JR, Hernandez-Velasco X, Téllez G, Hargis BM (2014) Coccidiosis: recent advancements in the immunobiology of Eimeria species, preventive measures, and the importance of vaccination as a control tool against these apicomplexan parasites. Vet Med: Res Rep 5:23–34. https://doi.org/10.2147/VMRR.S57839

Skinner JT, Bauer S, Young V, Pauling G, Wilson J (2010) An economic analysis of the impact of subclinical (mild) necrotic enteritis in broiler chickens. Avian Dis 54(4):1237–1240. https://doi.org/10.1637/9399-052110-Reg.1

Solderer A, Kaufmann M, Hofer D, Wiedemeier D, Attin T, Schmidlin PR (2019) Efficacy of chlorhexidine rinses after periodontal or implant surgery: a systematic review. Clin Oral Investig 23(1):Article 1. https://doi.org/10.1007/s00784-018-2761-y

Spence JR, Mayhew CN, Rankin SA, Kuhar MF, Vallance JE, Tolle K, Hoskins EE, Kalinichenko VV, Wells SI, Zorn AM, Shroyer NF, Wells JM (2011) Directed differentiation of human pluripotent stem cells into intestinal tissue in vitro. Nature 470(7332):105–109. https://doi.org/10.1038/nature09691

Thangavelu A, Kaspar SS, Kathirvelu RP, Srinivasan B, Srinivasan S, Sundram R (2020) Chlorhexidine: an elixir for periodontics. J Pharm Bioallied Sci 12(Suppl 1):Suppl 1. https://doi.org/10.4103/jpbs.JPBS_162_20

Tyzzer EE, Theiler H, Jones EE (1932) Coccidiosis in gallinaceous birds: II. A comparative study of species of Eimeria of the chicken. Am J Epidemiol 15(2):319–393. https://doi.org/10.1093/oxfordjournals.aje.a117823

USDA-AMS (2024) Organic market review. USDA-ERS, 2014. Organic market review. https://www.ers.usda.gov/topics/natural-resources-environment/organic-agriculture/.   Accessed Feb 2024

USDA-ERS (2024) Organic market review. https://www.ers.usda.gov/topics/natural-resources-environment/organic-agriculture/. Accessed Feb 2024

Vetterling JM, Doran DJ (1969) Storage polysaccharide in Coccidial Sporozites after Excystation and penetration of cells. J Protozool 16(4):772–775. https://doi.org/10.1111/j.1550-7408.1969.tb02341.x

Vladimirov Y (2004) Reactive oxygen and nitrogen species: diagnostic, preventive and therapeutic values. Biokhimiya/Biochemistry 69:1–3. https://doi.org/10.1023/B:BIRY.0000016343.21774.c4

Williams RB (1973) The effect of Eimeria acervulina on the reproductive potentials of four other species of chicken coccidia during concurrent infections. Br Vet J 129(3):xxix–xxxi. https://doi.org/10.1016/s0007-1935(17)36498-9

Williams R (1997) Laboratory tests of phenolic disinfectants as oocysticides against the chicken coccidium Eimeria tenella. Vet Rec 141:447–448. https://doi.org/10.1136/vr.141.17.447

Williams RB (2001) Quantification of the crowding effect during infections with the seven Eimeria species of the domesticated fowl: its importance for experimental designs and the production of oocyst stocks. Int J Parasitol 31(10):1056–1069. https://doi.org/10.1016/S0020-7519(01)00235-1

Williams RB (2002) Anticoccidial vaccines for broiler chickens: pathways to success. Avian Pathol 31(4):317–353. https://doi.org/10.1080/03079450220148988

Williams RB (2005) Intercurrent coccidiosis and necrotic enteritis of chickens: rational, integrated disease management by maintenance of gut integrity. Avian Pathol 34:159–180. https://doi.org/10.1080/03079450500112195

Williams RB, Thebo P, Marshall RN, Marshall JA (2010) Coccidian oöcysts as type-specimens: Long-term storage in aqueous potassium dichromate solution preserves DNA. Syst Parasitol 76:69. https://doi.org/10.1007/s11230-010-9234-2

Youn HJ, Noh JW (2001) Screening of the anticoccidial effects of herb extracts against Eimeria tenella. Vet Parasitol 96(4):257–263. https://doi.org/10.1016/S0304-4017(01)00385-5

An Overview on Nanotechnology and Its Poultry Applications: Opportunities and Challenges

14

Abdulaziz M. Alanazi, Nehal Eid, Ismail Dergaa, Shereen Basiouni, Awad A. Shehata, and Hesham R. El-Seedi

Contents

A. M. Alanazi · H. R. El-Seedi (✉)
Department of Chemistry, Faculty of Science, Islamic University of Madinah, Madinah, Saudi Arabia
e-mail: hesham.el-seedi@fkog.uu.se

N. Eid
Department of Chemistry, Faculty of Science, Menoufia University, Shebin El-Kom, Egypt

I. Dergaa
Primary Health Care Corporation (PHCC), Doha, Qatar

S. Basiouni
Structural Membrane Biochemistry, Bavarian NMR Center, Technical University of Munich (TUM), Garching, Germany

A. A. Shehata
Structural Membrane Biochemistry, Bavarian NMR Center, Technical University of Munich (TUM), Garching, Bayern, Germany

© The Author(s), under exclusive license to Springer Nature Switzerland AG 2024
A. A. Shehata et al. (eds.), *Alternatives to Antibiotics against Pathogens in Poultry*, https://doi.org/10.1007/978-3-031-70480-2_14

Abstract

Nanotechnology is a very promising science with great potential and wide applications in different fields such as catalysts, biosensors, pharmaceutics, biomedicine, drug delivery, food and chemical industry, electronics, human health, and poultry. Nanoparticles possess unique physical and chemical properties and bioavailability that can be enhanced by increasing the surface area of respective minerals. Additionally, nanoparticles have various magnetic, electrical, mechanical, and optical properties distinguishing them from the bulk materials, and this can be owned to the control over the molecular scale. Recently, the application of nanoparticles in poultry nutrition become highly interesting. Various nanoparticles such as zinc, copper, silver, selenium, iron, chromium, and manganese in the diets of broilers, layers, turkeys, and quails showed promising results regarding the biological function and biochemical reaction, as they help to improve the digestive efficiency, immunity, and performance of livestock and poultry. There are various results in the applied studies on comparative enhancement of production performance, reproductive properties, and other biological functions that can be explained by the superior properties of NPs, such as sizes, shapes, and direct interactions on basal diets. The supplementation of zinc oxide NPs showed a higher laying performance, egg quality, and nutrient digestibility. AgNPs can be considered antimicrobial agents because of their effect against viruses and bacteria and boosting the immune system, making them potential health-promoting operators for poultry. As well, FeNPs have displayed great abilities in weight gain, hemato biochemical profile, egg production, fertility, and reduced feed conversion ratio in the poultry industry. Collectively, the different nanoparticles have an efficient impact on poultry health. However, more research is still recommended to explore the different implications of NPs in this field and to underline the unexpected side effects.

Keywords

Nanotechnology · Synthesis · Mode of actions

1   Introduction

Historically, the nanotechnology science is backed by Reynman's famous lecture in 1959, where the Feynman (φnman) (1 Feynman $[\varphi] \equiv 10^{-9}$ meter $=10^{-3}$ Micron $[\mu]$ $=10$ Angstroms $[\mathring{A}]$) was suggested primarily instead of the nanoscale matter. Nanotechnology is an art of pharmaceutical science that deals with the building of materials, devices, and systems at atomic and molecular levels in the range 1–100 nanometers (one billionth of a meter). Nanoscale materials have at least one dimension and are characterized by various unique, fundamental, and building block abilities. They have magnetic, electrical, mechanical, and optical properties, distinguishing them from the bulk materials. These advantages are owned by the processing of them with efficient chemical and physical techniques that control the molecular scale (Mansoori 2005; Mansoori and Soelaiman 2005; Roco et al. 2000). Metal nanoparticles have various applications in different fields Fig. 14.1. They exhibited broad spectral applications in many fields such as industry, catalysts, biosensors, pharmaceutics, biomedicine, drug delivery, food industry, electronics, and chemical industry. Green synthesis of nanoparticles (NPs) is eco-friendly, less expensive, and free of chemical contaminants for medical and biological applications; for example, silver NPs (AgNPs) are known for their versatile applications in medical industries (Park 2007; Rashidi and Khosravi-Darani 2011). Nowadays, it is, widely used in various fields such as nutrition, therapy, targeted drug delivery, preparations of vaccines, and various purification processes in textile industries (Mekuye and Abera 2023).

Nanotechnology is a very promising and interesting strategy with a great scope of applications in the poultry industry sector and has potential in nutrition, therapy, targeted drug delivery, preparations of vaccines, and various purification processes in textile industries (Abd El-Ghany et al. 2021). They have high absorption and bioavailability potential with more efficient apportionment to the tissue than their bulk particles. Various nanomaterials are available in different properties such as sizes, shapes, surface modifications, charges, and natures that make them beneficial in drug delivery and can be utilized in the diagnosis and prevention of diseases (Ahmad et al. 2022). The importance of mineral supplementation in poultry diets is much recommended due to its essential role in many biological functions and biochemical reactions in the body, as the presence of minerals in diets is poor, and they are needed in large amounts (Park et al. 2004; Suttle 2010). Additionally, they are not only characterized by high bioavailability, but they decrease the extraction of

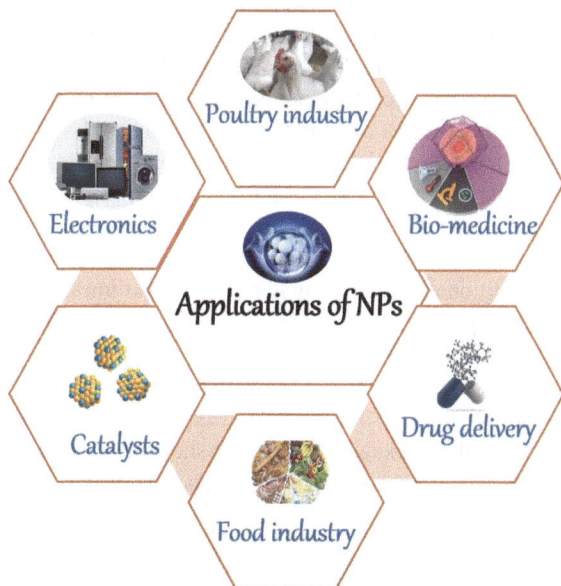

Fig. 14.1 The different applications of nanomaterials

minerals in the environment, which makes them eco-friendly. Various mineral nanoparticles, mainly copper, iron, zinc, titanium, selenium, and silver, have been used for animals and mainly for poultry nutrition (Hassan et al. 2020). The different studies demonstrated that nanoparticles improved digestive efficiency, immunity, and performance in livestock and poultry (Marappan Gopi Marappan and Govindasamy 2017; Scrinis and Lyons 2007).

2 Types of Nanoparticles

Nanomaterials must have at least one dimension at the nanoscale. Practically, they have been classified according to the dimension into four classes, which are zero, one, two, and three-dimensional nanomaterials, as seen in Fig. 14.2 (Lalitha et al. 2019). Quantum dots, fullerenes, and nanoparticles belong to the zero-dimensional nanomaterials.

Nanotubes, nanofibers, Nanorods, nanowires, and nano horns are among the two dimensions, and nanosheets, nanofilms, and nanolayers are one-dimensional, whereas bulk powders, dispersions of nanoparticles, and arrays of nanowires are three dimensional in the nanoscale (lalitha et al. 2019; Joudeh and Linke 2022).

Additionally, nanomaterials can be classified into three classes according to their composition, which include organic, inorganic, and carbon-based nanomaterials (Ibrahim Khan and Khan 2019). Organic nanoparticles consist of any organic substance like protein, carbohydrate, and lipid, and the most common examples are dendrimers, liposomes, micelles, and protein complexes (Pan and Zhong 2016; Joudeh and Linke 2022). The carbon-based nanoparticles mainly consist of carbon atoms such as fullerenes, carbon black NPs, and carbon quantum dots (Ibrahim Khan and Khan 2019), (Long et al. 2013), and on the other hand, the inorganic nanoparticles are represented by the nanoparticles that are not made of carbon or organic compounds. Metal (monometallic, bimetallic, and polymetallic), ceramic, and semiconductor NPs can be created in different morphologies such as spheres, cylinders, sheets, and nanotubes (Yuan et al. 2019; Nascimento et al. 2018). Inorganic nanoparticles have many applications in food products, for instance, titanium dioxide, which acts as a feed colorant and can be used as an ultraviolet barrier in packaging. Different minerals like silver, magnesium, and calcium are being used as antimicrobial agents, as water purifiers, and in food storage.

Additionally, organic nanoparticles also improve their nutritional value as they can encapsulate nutrients and transport them through the bloodstream, which are now referred to as nanocapsules. Due to the increased bioavailability, nanocapsules are used to deliver the nutrients without altering the taste and appearance. These encapsulated nanomaterials are incorporated into food as liposomes and biosensors useful in the packaging system, as shelf-life extenders, identification markers, and antimicrobial agents in the stored food (Ahmad et al. 2022; Mageswari et al. 2016).

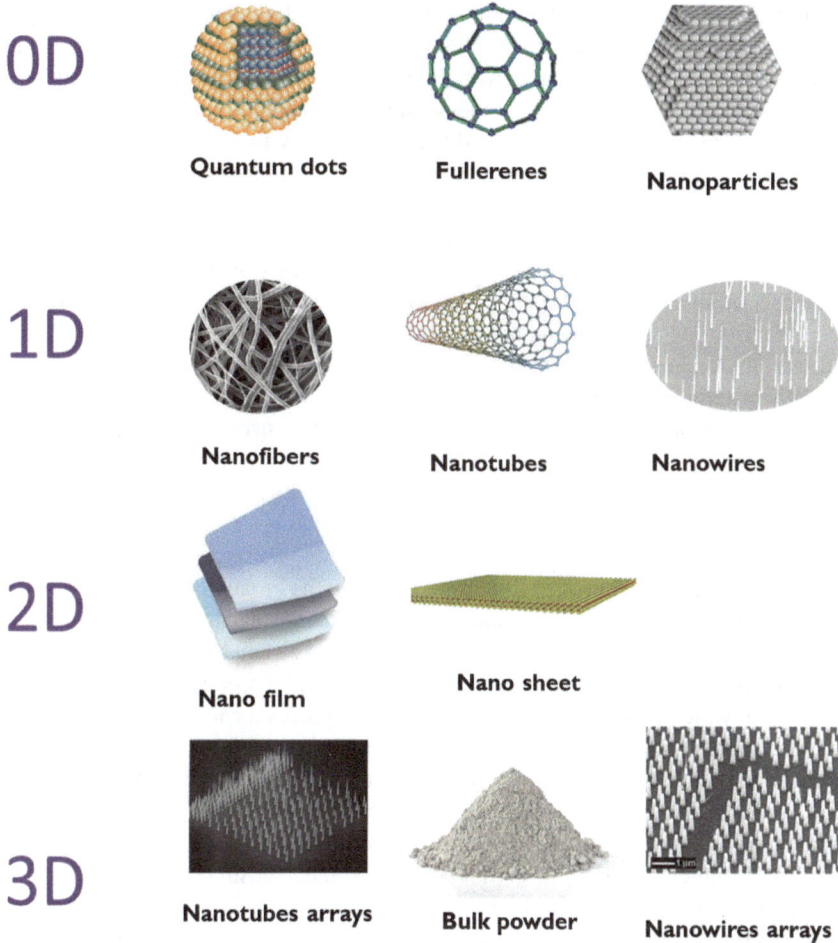

Fig. 14.2 The classification of nanomaterials according to their dimensionality and various examples

3   Synthesis and Characterization of Nanoparticles

The nanoparticles can be generated through various ways, including physical, chemical, and biological techniques (Rajput 2015). "Top-down" and "bottom-up" are the two main strategies for the creation of NPs. The chemical and biological methods are based on the self-assembly of atoms to new nuclei, then grow into nano-sized particles. The physical approach mainly depends on is "top-down" strategy, which can be applied in different ways like ultra-sonication, microwave (MW) irradiation, and electrochemical methods. The physical and chemical approaches of NPs creation involve the utilization of harmful reducing and capping agents that are expensive, toxic, cause harmful effects on the environment, and also of low profile in clinical and biomedical applications (Pantidos and Horsfall 2014), for example, Ag, Au, PbS, and Cd NPs are metal NPs synthesized physically (Yadav et al. 2012). Cyclodextrin, citrate, polyvinyl pyrrolidone, polyvinyl alcohol, and quaternary ammonium salts are different examples of stabilizing agents that help in the prevention of the agglomeration of NPs and provide stability through the synthesis process. Recently, the green synthesis of NPs has become a very interesting tool to overcome the disadvantages of traditional methods (Gour and Jain 2019).

The concept of producing green nanoparticles was first developed by Raveendran et al., who used β-D-glucose as a reducing as well as a stabilizing agent for the synthesis of silver nanoparticles. Green synthesis of NPs is very attractive due to their contribution as a reducing and capping agent without the need for other harmful, and toxic chemicals, as well as its cheap costs, simple, economical, and eco-friendly criteria. Silver, gold, palladium, iron, and zinc oxide are various examples of metallic nanoparticles that have been synthesized by using green chemistry. The green chemistry in NPs synthesis mainly depends on the exploitation of various natural products like plant extracts, polysaccharides, algae, fungi, bacteria, yeasts, and marine (Ijaz et al. 2020; Yosri et al. 2021; Guan et al. 2022). Natural resources have remarkable properties owed to the richness of bioactive compounds like flavonoids, phenolics, terpenoids, and steroids, allowing the biogenic reduction of metallic and nonmetallic NPs. These capping and reducing agents can act as growth terminators for engineering the growth, morphology, and size and decreasing agglomeration process of NPs. Additionally, they can be produced via a bottom-up strategy in which an extract of natural products replaces a chemical agent to cause the reduction of precursors (Zhang et al. 2020). Green synthesis of NPs has exhibited a remarkable material strength, solubility, conductivity, optical properties, thermal behavior, and catalytic activity and have a higher ratio of surface-to-volume and proportion of atoms at the surface, which determine their predominant properties compared to their pluck particles. The suggested mechanism of green synthesis of nanoparticles is shown in Fig. 14.3.

The study of nanoparticle characterization is very paramount to understanding and controlling their synthesis and applications. In this line, there are various techniques such as transmission and scanning electron microscopy (TEM and SEM), atomic force microscopy (AFM), dynamic light scattering (DLS), X-ray photoelectron spectroscopy (XPS), powder X-ray diffractometric (XRD), Fourier transform infrared spectroscopy (FTIR), and UV–Vis spectroscopy that are adopted for in-depth characterization. Particle size, shape, crystallinity, fractal dimensions, pore size, and surface area, in addition to orientation, intercalation, and dispersion, are different properties that can also be detected by these techniques (Hodoroaba et al. 2019; Ijaz et al. 2020).

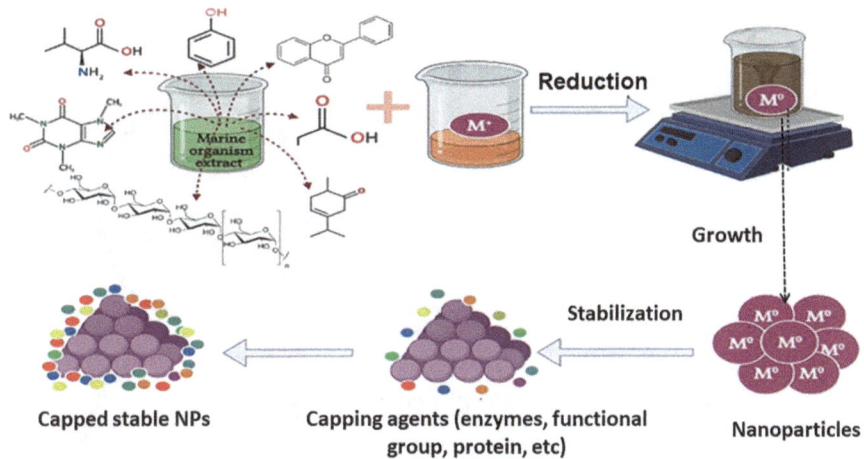

Fig. 14.3 Schematic diagram of the proposed mechanism of the green synthesis of metallic NPs via marine organisms (Yosri et al. 2021)

4 Mode of Action of Nanoparticles

Many studies have been carried out to investigate the actual mode of action of nanoparticles, which have contradictory results as far as the growth improvement is concerned Fig. 14.4. As these are small-sized particles, they are efficiently absorbed through the gastric intestinal tract and blood; then, they can exert immense biological effects on certain tissues. These mineral nanoparticles result in increased carcass yield, growth performance, egg-laying ability, less toxicity, and improved distribution and bioavailability (Sadr et al. 2023). They also improve the amount of respective trace minerals in poultry meat and eggs, which helps in providing new chances for the development of enriched products and functional foods for humans. Moreover, certain nanoparticles like AgNPs and ZnNPs also have a role in controlling the risk of foodborne pathogen *Campylobacter*, along with an increase in the population of beneficial gut microflora. Also, there are suggested that nanoparticle feeding has also enhanced the immunity, digestibility, and growth performance in broilers (Duffy et al. 2018). They have the potential ability to increase the surface area of the compounds that help in the processing of many biological reactions in the body, and their small size causes them to efficiently penetrate the tissues through the tiny capillaries and cross organs with epithelial tissue (e.g., liver). This can be related to the small size of nanoparticles that makes their passage very fast through the walls of the gastrointestinal tract, creating many efficient effects in various body systems that give the opportunity for researchers to deal with nanomaterials by studying many various fields such as production, reproduction, disease control, dealing with biological materials such as the study of DNA and cellular molecules (Sheikhalipour et al. 2022).

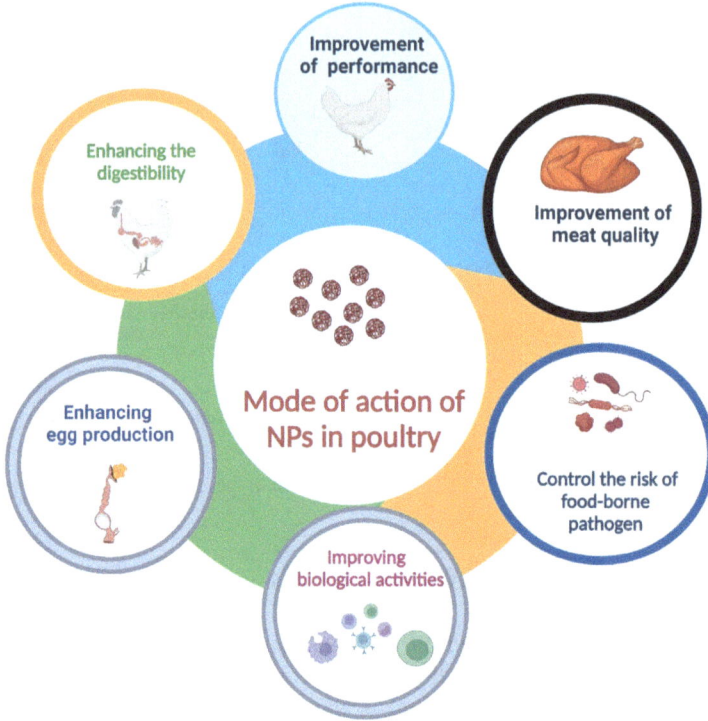

Fig. 14.4 The suggested mode of action of NPs in poultry

5 Applications of Nanoparticles in Poultry

In the poultry field, NPs have many various applications in the diagnosis of many diseases by using different techniques, vaccine preparations and immuno-stimulation, disinfection, production enhancement, detection of food adulteration, and antimicrobial activities (antiviral, antibacterial, antiparasitic, antifungal), anti-mycotoxins, hepatoprotective, and antioxidant effect (Abd El-Ghany et al. 2021). Moreover, these applications involve surface biocides, protective clothing, air and water filters, packaging, biosensors, and rapid detection methods for contaminants (King et al. 2018). The Ca, Zn, Cu, Cr, Fe, Mn, and Se NPs are various examples of metallic nanoparticles that have been employed in poultry nutrition (Patra and Lalhriatpuii 2020). Additionally, the antimicrobial activity of different metallic nanoparticles such as iron, zinc, silver, and selenium have been demonstrated and have been used to help in inhibiting the growth of mycotoxigenic molds and preven-tion of the production of their respective mycotoxins. Also, they help in the improve-ment of growth performance (Mohammadi et al. 2015). It appears the antifungal effects damage the cell wall of the microbial cells and the adhesion of the cell con-tents, causing cell death. The combination therapy of metal nanoparticles, chemical antibiotics, herbs, and ozone fumigation has been highly recommended to decrease microbial resistance to traditional antibiotics (Hassan et al. 2020). In conclusion, applications of nano-trace minerals are promising in the poultry industry.

5.1   Applications of Nano-Zinc in Poultry

Zinc NPs are one of the common metallic nanoparticles that have an essential role in the improvement of poultry health and performance. Zinc is considered a co-factor for about 300 metalloenzymes and has a fundamental function in the metabolism process, normal growth, skeletal development, and reproduction in birds (Pandav and Puranik 2015; Hidayat et al. 2020). Zinc and selenium caused the amelioration of toxic effects of aflatoxins in rabbits and rats and showed hepatoprotective effects through scavenging of free radicals and elevation of antioxidant activity in the liver against adverse effects of mycotoxins. The supplementation of zinc oxide NPs improves the laying performance, egg quality, nutrient digestibility, and zinc retention have been approved at 20, 40, or 60 mg/kg. Also, hen day egg production, egg mass, shell thickness, and eggshell percentage were significantly enhanced. The serum biochemical analyses revealed that serum cholesterol, glutamic oxaloacetic transaminase (GOT), glutamic pyruvic transaminase (GPT), urea, and somewhat creatinine linearly decreased with increasing Nano-ZnO levels in the diets (Fawaz et al. 2019).

5.2   Applications of Nano-Iron in Poultry

Iron has many important physiological functions, such as the synthesis of hemoglobin, which takes part in the catalytic activity, peroxidases, and hydroxylases process. FeNPs at different doses of (0.3, 0.75, 1.5, 4 mg/kg,) supported the weight gain, hemato biochemical profile, egg production, fertility, and reduced feed conversion ratio in poultry (Nikonov et al. 2011; Miroshnikov et al. 2017). A mixture of xylanase and FeONPs was tested on poultry growth. They demonstrated high potential to be used in poultry feed for large-scale meat production without any side effects (Rehman et al. 2020). Additionally, an increase in blood parameters such as hemoglobin, total protein, and white blood cells was observed after the injection of chickens by FeNPs and CuNPs, as well as a gain in weight by 9–12% in comparison to the control group (Yausheva 2021). It was also found that iron oxide chitosan nanocomposite exhibited antiviral activity that could significantly reduce the infectious bursal disease virus (IBDV) load in the bursa and decrease pathological changes in lymphoid organs (Nasef et al. 2021). In rabbits, the nanocapsulated ivermectin treatment had the lowest ivermectin residues in edible tissues, with the shortest withdrawal duration (14 days) below the maximum residue limits (Abdelmoteleb et al. 2022).

5.3   Applications of Nano-Chromium in Poultry

Chromium is one of the essential elements that have a significant health value, promoting the biological activities and health of animals and the human body, although it risks toxicity in high amounts. It is also essential for various metabolic processes. Many studies have shown the potential healthy effect of CrNPs on birds. Cr NPs at 0.2, 0.4, 0.8, and 1.2 mg/kg helped increase the feed intake ratio (Berenjian et al. 2018), while 0.50 and 3.0 mg/kg of Cr NPs with a size of 81 nm assisted in enhancing the egg quality and increasing minerals concentration of Cr in the liver, yolk, and eggshell in layer chickens (Sirirat et al. 2013). Also, 0.50 mg/kg Cr NPs with a size of 30–60 nm demonstrated feed efficiency, carcass yields, and lean muscles. In addition, it decreased the fat and cholesterol contents in the abdominal and thigh muscles in heat-stressed broiler chicks (Malathi 2015).

5.4   Applications of Nano-Sliver in Poultry

AgNPs have a very interesting role in poultry applications as they can act as carriers for nutrients such as amino acids, vitamins, and trace elements to certain functions. As well, AgNPs also showed antimicrobial potential to combat viruses and bacteria and boost the immune system, acting as health-promoting operators for poultry (Bhanja et al. 2019). Also, their supplementation in the drinking water (DW) of broiler chickens during the fattening period at concentrations up to 2 ppm showed an improvement in growth performance, body weight gain, and food conversion ratio (Fouda et al. 2021). Supplementation of AgNPs (5, 15, and 25 ppm) to quail in water increased the lactic acid bacteria, however, no effect on other gut microflora (Sawosz et al. 2007). Moreover, the insertion of nano-Ag hydrocolloid at a concentration of 5 ppm in the diet of poultry enhanced the leukocytes' phagocytosis, metabolic activity, and oxidative stress (lipid peroxidation products) while significantly reducing serum antioxidant enzyme activity and hemoglobin content (Ognik et al. 2016).

6   Future Prospective

Nanotechnology is a science that can be used in livestock and poultry diets to improve the bioavailability of nutrients, production performance, growth improvement, immunity response, and healthy state of the body. Various metallic nanoparticles have been used in many different applications but much research is needed to investigate the toxic effects of NPs as well as the effective doses in relation to their usefulness and efficiency. There needs to be more research due to the virtually unlimited possibilities for nanotechnology element preparations and the utilization of green synthesis methods of nanoparticles, which can increase their effect on poultry applications and determination of the efficient mechanism of action of various nanoparticles.

References

Abd El-Ghany WA, Shaalan M, Salem HM (2021) Nanoparticles applications in poultry production: An updated review. Worlds Poult Sci J 77(4):1001–1025

Abdelmoteleb AMM, Elmasry DMA, Amro FH, Mahmoud RAA (2022) Comparative effect of dose escalation of nanocapsulated ivermectin against mange in rabbits. German J Vet Res 2(4):8–15. https://doi.org/10.51585/gjvr.2022.4.0043

Ahmad I, Mashwani Z-U-R, Raja NI, Kazmi A, Wahab A, Ali A, Younas Z, Yaqoob S, Rahimi M (2022) Comprehensive approaches of nanoparticles for growth performance and health benefits in poultry: An update on the current scenario. BioMed Res Int 2022

Berenjian A, Sharifi SD, Mohammadi-Sangcheshmeh A, Ghazanfari S (2018) Effect of chromium nanoparticles on physiological stress induced by exogenous dexamethasone in Japanese quails. Biol Trace Elem Res 184(2):474–481

Bhanja SK, Mehra M, Goel A (2019) Application of silver nanoparticles in poultry production. Ind J Poult Sci 54(3):185–192

Duffy LL, Osmond-McLeod MJ, Judy J, King T (2018) Investigation into the antibacterial activity of silver, zinc oxide and copper oxide nanoparticles against poultry-relevant isolates of salmonella and campylobacter. Food Control 92:293–300

Fawaz MA, Abdel-Wareth AAA, Hassan HA, Südekum KH (2019) Applications of nanoparticles of zinc oxide on productive performance of laying hens. SVU-Int J Agric Sci 1(1):34–45

Fouda MMG, Dosoky WM, Radwan NS, Abdelsalam NR, Taha AE, Khafaga AF (2021) Oral administration of silver nanoparticles–adorned starch as a growth promotor in poultry: Immunological and histopathological study. Int J Biol Macromol 187:830–839

Gour A, Jain NK (2019) Advances in green synthesis of nanoparticles. Artif Cells Nanomed Biotechnol 47(1):844–851

Guan Z, Ying S, Ofoegbu P, Clubb P, Rico C, He F, Hong J (2022) Green synthesis of nanoparticles: Current developments and limitations. Environ Technol Innov 26:102336

Hassan AA, Sayed-Elahl RM, Oraby NH, El-Hamaky AMA (2020) Chapter 11–metal nanoparticles for management of mycotoxigenic fungi and mycotoxicosis diseases of animals and poultry. In: Rai M, Abd-Elsalam KA (eds) Nanomycotoxicology. Academic Press, pp 251–269

Hidayat C, Sumiati AJ, Wina E (2020) Effect of zinc addition on the immune response and production performance of broilers: Ameta-analysis. Asian Australas J Anim Sci 33(3):465–479

Hodoroaba V-D, Unger W, Shard A (2019) Characterization of nanoparticles: measurement processes for nanoparticles. Elsevier

Ibrahim Khan KS, Khan I (2019) Nanoparticles: properties, applications and toxicities. Arab J Chem 12(7):908–931

Ijaz I, Gilani E, Nazir A, Bukhari A (2020) Detail review on chemical, physical and green synthesis, classification, characterizations and applications of nanoparticles. Green Chem Lett Rev 13(3):223–245

Joudeh N, Linke D (2022) Nanoparticle classification, physicochemical properties, characterization, and applications: A comprehensive review for biologists. J Nanobiotechnol 20(1):262

King T, Osmond-McLeod MJ, Duffy LL (2018) Nanotechnology in the food sector and potential applications for the poultry industry. Trends Food Sci Technol 72:62–73

Lalitha AK, Viswanath IV, Bhagavathula D, Boddeti G, Venu R, Murthy YLN (2019) Review on nanomaterials: Synthesis and applications. Materials Today: Proceedings 18:2182–2190

Long CM, Nascarella MA, Valberg PA (2013) Carbon black vs. black carbon and other airborne materials containing elemental carbon: Physical and chemical distinctions. Environ Pollut 181:271–286

Mageswari A, Srinivasan R, Subramanian P, Ramesh N, Gothandam KM (2016) Nanomaterials: Classification, biological synthesis and characterization. Nanosci Food Agric 3:31–71

Malathi V (2015) Performance of dual purpose chicken supplemented with chromium yeast and Nano chromium

Mansoori GA (2005) Principles of nanotechnology: Molecular-based study of condensed matter in small systems

Mansoori GA, Soelaiman TAF (2005) Nanotechnology—an introduction for the standards community. J ASTM Int 2:1–21

Marappan Gopi MG, Beulah PBP, Kumar RD, Muthuvel SMS, Govindasamy PGP (2017) Role of nanoparticles in animal and poultry nutrition: Modes of action and applications in formulating feed additives and food processing. Int J Pharmacol 13(7):724–731

Mekuye B, Abera B (2023) Nanomaterials: an overview of synthesis, classification, characterization, and applications. Nano Select 4(8):486–501

Miroshnikov SA, Donnik IM, Yausheva EV, Kosyan DB, Sizova EA (2017) Research of opportunities for using iron nanoparticles and amino acids in poultry nutrition. Geomate J 13(40):124–131

Mohammadi V, Ghazanfari S, Mohammadi-Sangcheshmeh A, Nazaran MH (2015) Comparative effects of zinc-nano complexes, zinc-sulphate and zinc-methionine on performance in broiler chickens. Br Poult Sci 56(4):486–493

Nascimento MA, Cruz JC, Rodrigues GD, de Oliveira AF, Lopes RP (2018) Synthesis of polymetallic nanoparticles from spent lithium-ion batteries and application in the removal of reactive blue 4 dye. J Clean Prod 202:264–272

Nasef S, Ayoub M, Selim K, Elmasry D (2021) Trial control of infectious bursal disease virus isolated from broiler chicken using iron oxide chitosan nanocomposite. German J Vet Res 3(2):17–27. https://doi.org/10.21203/rs.3.rs-1077607/v1

Nikonov IN, Folmanis YG, Folmanis GE, Kovalenko LV, Laptev GY, Egorov IA, Fisinin VI, Tananaev IG (2011) Iron nanoparticles as a food additive for poultry. Dokl Biol Sci 440:328–331

Ognik K, Cholewińska E, Czech A, Kozłowski K, Wlazło Ł, Nowakowicz-Dębek B, Szlązak R, Tutaj K (2016) Effect of silver nanoparticles on the immune, redox, and lipid status of chicken blood. Czech J Anim Sci 61(10):450–461. https://doi.org/10.17221/80/2015-CJAS

Pan K, Zhong Q (2016) Organic nanoparticles in foods: Fabrication, characterization, and utilization. Annu Rev Food Sci Technol 7:245–266

Pandav P, Puranik P (2015) Trials on metal enriched Spirulina platensis supplementation on poultry growth. Glob J Bio-Science Technol 4:128–134

Pantidos N, Horsfall LE (2014) Biological synthesis of metallic nanoparticles by bacteria, fungi and plants. J Nanomed Nanotechnol 5(5):1

Park K (2007) Nanotechnology: What it can do for drug delivery. J Control Release 120(1–2):1–3

Park SY, Birkhold SG, Kubena LF, Nisbet DJ, Ricke SC (2004) Review on the role of dietary zinc in poultry nutrition, immunity, and reproduction. Biol Trace Elem Res 101(2):147–163

Patra A, Lalhriatpuii M (2020) Progress and prospect of essential mineral nanoparticles in poultry nutrition and feeding—a review. Biol Trace Elem Res 197(1):233–253

Rajput N (2015) Methods of preparation of nanoparticles-a review. Int J Adv Engg Technol 7(6):1806

Rashidi L, Khosravi-Darani K (2011) The applications of nanotechnology in food industry. Crit Rev Food Sci Nutr 51(8):723–730

Rehman H, Akram M, Kiyani MM, Yaseen T, Ghani A, Saggu JI, Shah SSH, Khalid ZM, Bokhari SAI (2020) Effect of Endoxylanase and iron oxide nanoparticles on performance and histopathological features in broilers. Biol Trace Elem Res 193(2):524–535

Roco MC, Williams RS, Alivisatos P (2000) Nanotechnology research directions: IWGN workshop report: vision for nanotechnology in the next decade. Springer Science & Business Media

Sadr S, Lotfalizadeh N, Ghafouri SA, Delrobaei M, Komeili N, Hajjafari A (2023) Nanotechnology innovations for increasing the productivity of poultry and the prospective of nanobiosensors. Vet Med Sci 9(5):2118–2131

Sawosz E, Binek M, Grodzik M, Zielińska M, Sysa P, Szmidt M, Niemiec T, Chwalibog A (2007) Influence of hydrocolloidal silver nanoparticles on gastrointestinal microflora and morphology of enterocytes of quails. Arch Anim Nutr 61(6):444–451

Scrinis G, Lyons K (2007) The emerging nano-corporate paradigm: nanotechnology and the transformation of nature, food and Agri-food systems. Int J Sociol Agric Food 15(2):22–44

Sheikhalipour A, Taghizadeh A, Hosseinkhani A, Palangi V (2022) Application of nanotechnology in animal nutrition. NanoEra 2(1):10–13

Sirirat N, Lu J-J, Hung A, Lien T-F (2013) Effect of different levels of nanoparticles chromium picolinate supplementation on performance, egg quality, mineral retention, and tissues minerals accumulation in layer chickens. J Agric Sci 5:150

Suttle N (2010) Mineral nutrition of livestock: Fourth edition. Mineral Nutr Livestock, Fourth Edition, pp 1–547

Yadav TP, Yadav RM, Singh DP (2012) Mechanical milling: a top down approach for the synthesis of nanomaterials and nanocomposites. Nanosci Nanotechnol 2(3):22–48

Yausheva EV (2021) Increasing efficiency in the poultry meat production when using iron and copper nanoparticles in nutrition. IOP Conference Series: Earth Environ Sci 624(1):012046

Yosri N, Khalifa SAM, Guo Z, Xu B, Zou X, El-Seedi HR (2021) Marine organisms: Pioneer natural sources of polysaccharides/proteins for green synthesis of nanoparticles and their potential applications. Int J Biol Macromol 193:1767–1798

Yuan X, Zhang X, Sun L, Wei Y, Wei X (2019) Cellular toxicity and immunological effects of carbon-based nanomaterials. Part Fibre Toxicol 16:1–27

Zhang D, Ma X-L, Gu Y, Huang H, Zhang G-W (2020) Green synthesis of metallic nanoparticles and their potential applications to treat cancer. Front Chemistry 8:799